HORIZONS IN WORLD PHYSICS SERIES

HORIZONS IN WORLD PHYSICS, VOLUME 268

Horizons in World Physics Series

Volume 268 **Horizons in World Physics**, Edited by Michael Everett and Louis Pedroza
Volume 267 **Superconductivity and Superconducting Wires,** Edited by Dominic Matteri and Leone Futino
Volume 266 **Applied Physics in the 21st Century,** Edited by Raymond P. Valencia
Volume 265 **The Physics of Quarks: New Research,** Edited by Nicolas L. Watson and Theo M. Grant
Volume 264 **Lev Davidovich Landau and his Impact on Contemporary Theoretical Physics** Edited by Ammar Sakaji and Ignazio Licata
Volume 263 **New Trends in Quantum Coherence and Nonlinear Optics** Edited by Rafael Drampyan
Volume 262 **Astrophysics and Condensed Matter** Edited Thomas G. Hardwell
Volume 261 **Superconductivity Research Advances** Edited by James E. Nolan
Volume 260 **Instabilities of Relativistic Electron Beam in Plasma,** Edited by Valery B. Krasovitskii
Volume 259 **Self-Focusing of Relativistic Electron Bunches in Plasma,** Edited by Valery B. Krasovitskii
Volume 258 **New Topics in Theoretical Physics,** Edited by Henk F. Arnoldus and Thomas George
Volume 257 **Boson's Ferromagnetism and Crystal Growth Research,** Edited by Emerson D. Seifer
Volume 256 **Horizons in World Physics,** Edited by Victor H. Marselle
Volume 255 **Frontiers in General Relativity and Quantum Cosmology Research,** Edited by Victor H. Marselle
Volume 254 **Optics and Electro-Optics Research,** Edited by Albert V. Berzilla
Volume 253 **Trends in General Relativity and Quantum Cosmology,** Edited by Albert Reimer
Volume 252 **Chemical Physics Research Trends,** Edited by Benjamin V. Arnold
Volume 251 **Quantum Dots: Research Developments,** Edited by Peter A. Ling
Volume 250 **Quantum Gravity Research Trends,** Edited by Albert Reimer
Volume 249 **General Relativity Research Trends,** Edited by Albert Reimer
Volume 248 **Spacetime Physics Research Trends,** Edited by Albert Reimer
Volume 247 **New Developments in Quantum Cosmology Research,** Edited by Albert Reimer
Volume 246 **Quantum Cosmology Research Trends,** Edited by Albert Reimer
Volume 245 **Horizons in World Physics,** Edited by Tori V. Lynch
Volume 244 **Horizons in World Physics,** Edited by Albert Reimer
Volume 243 **Theoretical Physics 2002, Part 2,** Edited by Henk F. Arnoldus and Thomas F. George
Volume 242 **Models and Methods of High-T_c Superconductivity: Some Frontal Aspects, Volume 2,** Edited by J. K. Srivastava and S. M. Rao
Volume 241 **Models and Methods of High-T_c Superconductivity: Some Frontal Aspects, Volume 1,** Edited by J. K. Srivastava and S. M. Rao
Volume 240 **Horizons in World Physics,** Edited by Albert Reimer
Volume 239 **Theoretical Physics 2002, Part 1,** Edited by Henk F. Arnoldus and Thomas F. George
Volume 238 **Theoretical Physics 2001,** Edited by Henk F. Arnoldus and Thomas F. George
Volume 237 **Generalized functions in mathematical physics: Main ideas and concepts,** Edited by A.S. Demidov
Volume 236 **Edge Excitations of Low-Dimensional Charged Systems,** Edited by Oleg Kirichek
Volume 235 **On the Structure of the Physical Vacuum and a New Interaction in Nature (Theory, Experiment, Applications),** Edited by Yu. A. Baurov
Volume 234 **Electron Scattering on Complex Atoms (Ions),** Edited by V. Lengyel, O. Zatsarinny and E. Remeta
Volume 233 **An Introduction to Hot Laser Plasma Physics,** Edited by A.A. Andreev, A.A. Mak and N.A. Solovyev
Volume 232 **Electron-Atom Scattering: An Introduction,** Maurizio Dapor
Volume 231 **Theory of Structure Transformations in Non-equilibrium Condensed Matter**, Edited by Alexander L. Olemskoi
Volume 230 **Topics in Cosmic-Ray Astrophysics**, Edited by M. A. DuVernois
Volume 229 **Interactions of Ions in Condensed Matter**, Edited by A. Galdikas and L. Pranevièius
Volume 228 **Optical Vortices**, Edited by M. Vasnetsov and K. Staliunas
Volume 227 **Frontiers in Field Theory**, Edited by Quantum Gravity and Strings, R.K. Kaul, J. Maharana, S. Mukhi and S. K. Rama
Volume 226 **Quantum Groups, Noncommutative Geometry and Fundamental Physical Interactions**, Edited by D. Kastler, M. Russo and T. Schücker
Volume 225 **Aspects of Gravitational Interactions**, Edited by S.K. Srivastava and K.P. Sinha
Volume 224 **Dynamics of Transition Metals and Alloys**, Edited by S. Prakash
Volume 223 **Fields and Transients in Superhigh Pulse Current Devices** Edited by G.A. Shneerson
Volume 222 **Gaussian Beams and Optical Resonators,** Edited by A. N. Oraevsky

Volume 221	**Laser Cathode-Ray Tubes,** Edited by Yu. M. Popov
Volume 220	**Quantum Electrodynamics with Unstable Vacuum,** Edited by V.L. Ginzburg
Volume 219	**The Muon Method in Science,** Edited by V.P. Smilga and Yu.M. Belousov
Volume 218	**Radiative and Collisional Characteristics of Ions in Hot Plasmas,** Edited by I.I. Sobel'man
Volume 217	**Nonlinear and Quantum Optical Phenomena in Nonequilibrium Media** , Edited by V.A. Shcheglov
Volume 216	**Semiconductors and Insulators: Optical and Spectroscopic ...,** Edited by Yu.I. Koptev
Volume 215	**Relativistic Theory of Gravity,** Edited by A.A. Logunov
Volume 214	**Research in Quantum Field Theory**, Edited by V.I. Man'ko
Volume 213	**Nonlinear Theory of Strong Electromagnetic Wave-plasma...,** Edited by O.N. Krokhin
Volume 212	**Dynamics of Elementary Atomic-Molecular...,** Edited by O. S. Scheglov
Volume 211	**Optics and Lasers,** Edited by G.G. Petrash
Volume 210	**Theory and Technology of High Temperature Superconductors,** Edited by Yu.I. Koptev
Volume 209	**Theory of Interaction of Multilevel Systems...,** Edited by V.I. Man'ko
Volume 208	**Physical Properties of High Temperature Superconductors,** Edited by Yu.I. Koptev
Volume 207	**Generalized Mayer Series in Classical Statistical Mechanics**, Edited by Igor J. Ivanchik
Volume 206	**Mathematical Physics, Applied Mathematics and Informatics,** Edited by Yu.I. Koptev
Volume 205	**Squeezed and Correlated States of Quantum Systems**, Edited by M.A. Markov
Volume 204	**Sakharov Memorial Lectures in Physics**, Edited by L.V. Keldysh and V.Ya. Fainberg
Volume 203	**Fiber Optics: Research and Development,** Edited by Ye.M. Dianov
Volume 202	**Gas Dynamics**, Edited by Yu.I. Koptev
Volume 201	**Generation of Nonlinear Waves and Quasistationary ...**, Edited by L.M. Kovrizhnykh
Volume 200	**Laser Thermonuclear Targets and Superdurable Microballoons**, Edited by A.I. Isakov
Volume 199	**Spatiotemporal Characteristics of Laser Emission,** Edited by M.V. Pyatakhin and A.F. Suchkov
Volume 198	**Theory of Nonstationary Quantum Oscillators**, Edited by M.A. Markov
Volume 197	**Reliability Problems of Semiconductor Lasers**, P.G. Eliseev
Volume 196	**Pulsars,** Edited by A.D. Kuzmin
Volume 195	**Physical Characteristics and Critical Temperature of High Temperature Superconductors**, Edited by M.M. Sushchinskiy
Volume 194	**Biophysical Approach to Complex Biological Phenomena,** Edited by E. Volkov
Volume 193	**Research on Chemical Lasers** Edited by A.N. Orayevskiy
Volume 192	**Atomic and Ionic Spectra and Elementary Processes in Plasma** , Edited by I.I. Sobelman
Volume 191	**Polarization of the Vacuum and a Quantum Relativistic Gas ...**, A.Ye. Shabad
Volume 190, (Supp. Vol.)	**Superconductors with A15 Lattice and Bridge Contacts Based ...**, Edited by M. M. Sushchinskiy
Volume 190	**Metal-Optics and Superconductivity**, Edited by A.I. Golovashkin
Volume 189	**Clusters of Galaxies and Extragalactic Radio Sources**, Edited by A.D. Kuz'min
Volume 188	**Integral Systems, Solid State Physics and Theory of Phase Transitions. Pt. 2. Symm. and Alg. Structures in Physics**, Edited by V.V. Dodonov and V.I. Man'ko
Volume 187	**Quantum Field Theory, Quantum Mechanics and Quantum Optics: Pt. 1. Symm. and Alg. Structures in Physics,** Edited by V.V. Dodonov and V.I. Man'ko
Volume 186	**Photoproduction of Pions on Nucleons and Nuclei**, Edited by A.A. Komar
Volume 185	**Injection Lasers in Optical Communication and Information Processing Systems**, Edited by Yu. M. Popov
Volume 184	**Electron Processes in MIS-Structure Memories**, Edited by A.F. Plotnikov
Volume 183	**Invariants and the Evolution of Nonstationary Quantum Systems**, Edited by M.A. Markov
Volume 182	**Luminescence of Wideband Semiconductors**, Edited by M.D. Galanin
Volume 181	**Pulsed Metal and Metal Halogenide Vapor Lasers**, Edited by G.G. Petrash
Volume 180	**Inelastic Light Scattering in Crystals**, Edited by M.M. Sushchinskiy
Volume 179, (Supp. Vol. 2)	**Electron-Excited Molecules in Nonequilibrium Plasma**, Edited by N.N. Sobolev
Volume 179, (Supp. Vol. 1)	**Interaction of Ultrashort Pulses with Matter** , Edited by M.D. Galanin
Volume 179	**X-Ray Plasma Spectroscopy and the Properties of Multiply-Charged Ions**, Edited by I.I. Sobel'man
Volume 178	**The Delfin Laser Thermonuclear Installation: Operational Complex ...**, Edited by G.V. Sklizkov
Volume 177	**Stoichiometry in Crystal Compounds and Its Influence on Their Physical Properties**, Edited by Yu. M. Popov
Volume 176, (Supp. Vol)	**Quantization, Gravitation and Group Methods in Physics** Edited by A.A. Komar

Volume 176	**Classical and Quantum Effects in Electrodynamics** Edited by A.A. Komar
Volume 175, (Supp. Vol.)	**Luminescence and Anisotropy of Zinc Sulfide Crystals** Edited by M.D. Galanin
Volume 175	**Luminescence Centers of Rare Earth Ions in Crystal Phosphors**, Edited by M.D. Galanin
Volume 174, (Supp. Vol. 2)	**Electron Liquid Theory of Normal Metals** Edited by V.P. Silin
Volume 174, (Supp. Vol. 1)	**Nonequilibrium Superconductivity**, Edited by V.L. Ginzburg
Volume 174	**Thermodynamics and Electrodynamics of Superconductors,** Edited by V.L. Ginzburg
Volume 173	**Quantum Mechanics and Statistical Methods** , Edited by M.M. Sushchinskiy
Volume 172	**Phase Conjugation of Laser Emission** , Edited by N.G. Basov
Volume 171	**Research on Laser Theory** , Edited by A.N. Orayevskiy
Volume 170	**The Theory of Target Compression by Longwave Laser Emission** , Edited by G.V. Sklizkov
Volume 169	**The Physical Effects in the Gravitational Field Black Holes**, Edited by M.A. Markov
Volume 168	**Issues in Intense-Field Quantum Electrodynamics**, Edited by V.L. Ginzburg
Volume 167	**Group Theory, Gravitation and Elementary Particle Physics**, Edited by A.A. Komar
Volume 166	**The Nonlinear Optics of Semiconductor Lasers**, Edited by N.G. Basov
Volume 165	**Solitons and Instantons, Operator Quantization**, Edited by V.L. Ginzburg

HORIZONS IN WORLD PHYSICS SERIES

HORIZONS IN WORLD PHYSICS, VOLUME 268

MICHAEL EVERETT
AND
LOUIS PEDROZA
EDITORS

Nova Science Publishers, Inc.
New York

For permission to use material from this book please contact us:
Telephone 631-231-7269; Fax 631-231-8175
Web Site: http://www.novapublishers.com

LIBRARY OF CONGRESS CATALOGING-IN-PUBLICATION DATA
Available upon request.

ISBN 978-1-60692-861-5

Published by Nova Science Publishers, Inc. ✦ New York

CONTENTS

PREFACE

This book presents information on the general theory of critical sound propagation, which takes into account important nonasymptotic effects. Unsolved questions and future prospects in this field are outlined. The electronic and magnetic structure of pure and doped manganese clusters from density functional theory are also discussed using generalized gradient approximation for the exchange-correlation energy. Furthermore, a critical review of the applicability of the resonant ultrasound spectroscopy for determination of all independent elastic coefficients of anisotropic solids is presented. Grating and hybrid integration techniques and several types of hybrid integrated lasers are explained as well, with a variety of silica waveguide gratings on a Si-substrate. Finally, systems of free particles in a quantum theory bases on a Galois field are explored in detail. Aspects of the thermodynamics of black holes are also reviewed; the quantum entanglement between the degrees of freedom of a scalar field, traced inside the horizon, is particularly taken into account as the origin of black hole entropy.

The critical dynamics of sound is a very interesting field in which we can test modern concepts of the phase transition theory such as the universality of critical exponents, scaling or the crossover to another universality class etc. It is the aim of Chapter 1 to present a general theory of critical sound propagation, which takes also into account some important nonasymptotic effects. In metallic magnets the critical anomalies in the sound attenuation coefficient are of different types than in magnetic insulators. The difference in the critical exponents used to be explained by the occurrence of different kinds of magnetoelastic coupling in the two classes of magnets mentioned. We will show in this chapter that one should assume coexistence of both types of coupling in all magnets. A very important role is played by the ratio of the spin-lattice relaxation time to the characteristic time of spin fluctuations. It is a crucial parameter determining whether the sound attenuation coefficient reveals a strong or a weak singularity in a given material.

After a short introduction the fundamental concepts of the phase transition theory such as critical exponents, the scaling and universality hypothesis etc are reviewed in Section 2 of this chapter. Section 3 presents the idea of critical slowing down, dynamic scaling as well as the presentation of the basic dynamic universality classes. In Section 4, the model describing the static behavior of acoustic degrees of freedom is investigated. The expressions for the adiabatic and the isothermal sound velocity are also derived. The phenomenological theory of critical sound propagation is presented in very intuitive way in Section 5, while Section 6 contains a detailed description of the dynamic model based on the coupled nonlinear

Langevin equations of motion. Three basic regimes characterized by different critical exponents and scaling functions are distinguished in the sound attenuation coefficient. Crossover effects from the insulator-type regime to the metallic-type regime and to the high-frequency regime are demonstrated on the example of the ultrasonic data for MnF_2. The concept of the effective sound attenuation exponent is introduced using the data reported for FeF_2 and $RbMnF_3$. The frequency dependent longitudinal sound velocity and its relation to the static quantities are discussed. Finally, the unsolved questions and future prospects in this field are outlined.

In Chapter 2, we report electronic and magnetic structure of pure and (As-) doped manganese clusters from density functional theory using generalized gradient approximation for the exchange-correlation energy. Ferromagnetic to ferrimagnetic transition takes place at $n = 5$ for pure manganese clusters, Mn_n, and remarkable lowering of magnetic moment is found for Mn_{13} and Mn_{19} due to their closed icosahedral growth pattern and results show excellent agreement with experiment. On the other hand, for As-doped manganese clusters, Mn_nAs, ferromagnetic coupling is found only in Mn_2As and Mn_4As and inclusion of a single As stabilizes manganese clusters. Exchange coupling in the Mn_nAs clusters are anomalous and behave quite differently from the Ruderman-Kittel-Kasuya-Yosida like predictions. Finally, possible relevance of the observed magnetic behaviour is discussed in the context of Mn-doped GaAs semiconductor ferromagnetism.

Chapter 3 brings a critical review of the applicability of the resonant ultrasound spectroscopy (RUS) for determination of all independent elastic coefficients of anisotropic solids. Such applicability limits are sought which follow from the properties of the examined materials, i.e. from the strength and class of the anisotropy, etc.. After introducing the general theoretical background of RUS, particular limiting factors are illustrated on experimental results, namely on the investigation of extremely strongly anisotropic single crystals, of weakly anisotropic polycrystals (where neither the class nor the orientation of the anisotropy are known) and of single crystals with strong temperature-dependent magneto-elastic attenuation. In all these cases, a sensitivity analysis is carried out to show which elastic coefficients (their combinations) can be accurately determined form RUS measurements and which cannot, whereto the complementarity of the RUS and pulse-echo methods is shown and utilized. The general findings of both the theoretical introduction and the experimental part are summarized in a concluding section, which tries to formulate the most essential open questions of the RUS method.

As discussed in Chapter 4, grating technology has been well matured, and therefore successfully utilized to select the oscillation wavelength and polarization properties of conventional laser diodes. This is because the lasing characteristics of the laser can be easily controlled by simply writing a specific grating directly in a waveguide cavity. Grating and hybrid integration techniques are presented. Several types of hybrid integrated external cavity lasers (ECLs) demonstrated with a variety of silica waveguide gratings on a Si-substrate are discussed. They include highly polarized ECLs, a dual-wavelength ECL and ECLs with high sidemode suppression ratio.

Systems of free particles in a quantum theory based on a Galois field (GFQT) are discussed in detail. In Chapter 5 infinities cannot exist, the cosmological constant problem does not arise and one irreducible representation of the symmetry algebra necessarily describes a particle and its antiparticle simultaneously. As a consequence, well known results of the standard theory (spin-statistics theorem; a particle and its antiparticle have the same masses and spins but opposite charges etc.) can be proved without involving local covariant equations. The spin-statistics theorem is simply a requirement that quantum theory should be based on complex numbers. Some new features of GFQT are as follows: i) elementary particles cannot be neutral; ii) the Dirac vacuum energy problem has a natural solution and the vacuum energy (which in the standard theory is infinite and negative) equals zero as it should be. In the AdS version of the theory there exists a dilemma that either the notion of particles and antiparticles is absolute and then only particles with a half-integer spin can be elementary or the notion is valid only when energies are not asymptotically large and then supersymmetry is possible.

Chapter 6 reviews aspects of the thermodynamics of black holes and in particular take into account the fact that the quantum entanglement between the degrees of freedom of a scalar field, traced inside the event horizon, can be the origin of black hole entropy. The main reason behind such a plausibility is that the well-known Bekenstein-Hawking entropy-area proportionality—the so-called 'area law' of black hole physics—holds for entanglement entropy as well, provided the scalar field is in its ground state, or in other minimum uncertainty states, such as a generic coherent state or squeezed state. However, when the field is either in an excited state or in a state which is a superposition of ground and excited states, a power-law correction to the area law is shown to exist. Such a correction term falls off with increasing area, so that eventually the area law is recovered for large enough horizon area. On ascertaining the location of the microscopic degrees of freedom that lead to the entanglement entropy of black holes, it is found that although the degrees of freedom close to the horizon contribute most to the total entropy, the contributions from those that are far from the horizon are more significant for excited/superposed states than for the ground state. Thus, the deviations from the area law for excited/superposed states may, in a way, be attributed to the far-away degrees of freedom. Finally, taking the scalar field (which is traced over) to be massive, we explore the changes on the area law due to the mass. Although most of our computations are done in flat space-time with a hypothetical spherical region, considered to be the analogue of the horizon, we show that our results hold as well in curved space-times representing static asymptotically flat spherical black holes with single horizon.

Evaluation of variation of a Green's function in a gauge field theory with a gauge parameter θ involves field transformations that are (close to) singular. Recently, we had demonstrated [hep-th/0106264] some unusual results that follow from this fact for an interpolating gauge interpolating between the Feynman and the Coulomb gauge (formulated by Doust). We carry out further studies of this model. We study properties of simple loop integrals involved in an interpolating gauge. We find several unusual features not normally noticed in covariant Quantum field theories. We find that the proof of continuation of a Green's function from the Feynman gauge to the Coulomb gauge *via* such a gauge in a gauge-invariant manner seems obstructed by the lack of differentiability of the path-integral with respect to θ (at least at discrete values for a specific Green's function considered) and/or by additional contributions to the WTidentities. We show this by the consideration of simple loop diagrams for a simple scattering process. The lack of differentiability, alternately, produces a large change in the path-integral for a small enough change in θ near some values. We find several applications of these observations in a gauge field theory. We show that the usual procedure followed in the derivation of the WT-identity that leads to the evaluation of a gauge variation of a Green's function involves steps that are not always valid in the context of such interpolating gauges. We further find new results related to the need for keeping the ϵ-term in the in the derivation of the WT-identity and and a nontrivial contribution to gauge variation from it. We also demonstrate how arguments using Wick rotation cannot rid us of these problems. Chapter 7 brings out the pitfalls in the use of interpolating gauges in a clearer focus.

Quartz crystal microbalances (QCMs) are commonly well-known as high-resolution mass-sensitive transducers. They are known to be, in fact, a versatile category of physical, biological or chemical sensors sensitive to the mass of molecular analytes. They are composed of piezoelectric crystal plus at least one layer of organic coating to both improve and tune chemical sensitivity. In literature bio-inspired sensor systems, mimicking natural olfaction, have been engineered housing selected arrays of chemical sensors based on quartz crystal microbalances, in suitable measuring chambers. Potentials and limitations of such a system have been deeply studied. In Chapter 8, interesting advances to quartz performances have been introduced by designing and developing a proper integrated micro-heater on a quartz side. The temperature influence has been investigated in the resulting fundamental frequency of the crystal plate as well as in ad-desorption mechanisms occurring between some different chemical layers above the plate surfaces and selected volatile organic compounds. Three different suitable micro-heater designs have been projected to characterize the functioning, the stability, the desorption time and the sensing performances in general (sensitivity, selectivity, drift, noise, etc.) of selected chemical layers covering the quartzes. The micro-heaters designs have been supported by simulations performed by Finite Element Method (FEM) software in order to analyze the temperature distribution on the whole quartz plate and get a thermal ramp. An electronic circuit interface has been designed to control the developed micro-heater, setting the desired working temperature. Moreover, when environmental temperature and humidity are constant, if the flow is changing, the temperature controller regulates the working temperature either increasing or decreasing the heater supplied power to set the desired temperature. Based on this information, it's possible to estimate the flow velocity value. This is an useful parameter in order to calibrate the sensors, overall if they have to be employed as elements of an "electronic nose" chamber where the flow velocity adopted could vary. The potentialities of such a thermo-ruled artificial olfactory system have been analyzed, too, to support some peculiar sensing strategies.

In: Horizons in World Physics, Volume 268
Editors: M. Everett and L. Pedroza, pp. 1-63

ISBN 978-1-60692-861-5

Chapter 1

CRITICAL SOUND PROPAGATION IN MAGNETS

Andrzej Pawlak
Department of Physics, A. Mickiewicz University
Poznań, Poland

Abstract

The critical dynamics of sound is a very interesting field in which we can test modern concepts of the phase transition theory such as the universality of critical exponents, scaling or the crossover to another universality class etc. It is the aim of the study to present a general theory of critical sound propagation, which takes also into account some important nonasymptotic effects. In metallic magnets the critical anomalies in the sound attenuation coefficient are of different types than in magnetic insulators. The difference in the critical exponents used to be explained by the occurrence of different kinds of magnetoelastic coupling in the two classes of magnets mentioned. We will show in this chapter that one should assume coexistence of both types of coupling in all magnets. A very important role is played by the ratio of the spin-lattice relaxation time to the characteristic time of spin fluctuations. It is a crucial parameter determining whether the sound attenuation coefficient reveals a strong or a weak singularity in a given material.

After a short introduction the fundamental concepts of the phase transition theory such as critical exponents, the scaling and universality hypothesis etc are reviewed in Section 2 of this chapter. Section 3 presents the idea of critical slowing down, dynamic scaling as well as the presentation of the basic dynamic universality classes. In Section 4, the model describing the static behavior of acoustic degrees of freedom is investigated. The expressions for the adiabatic and the isothermal sound velocity are also derived. The phenomenological theory of critical sound propagation is presented in very intuitive way in Section 5, while Section 6 contains a detailed description of the dynamic model based on the coupled nonlinear Langevin equations of motion. Three basic regimes characterized by different critical exponents and scaling functions are distinguished in the sound attenuation coefficient. Crossover effects from the insulator-type regime to the metallic-type regime and to the high-frequency regime are demonstrated on the example of the ultrasonic data for MnF_2. The concept of the effective sound attenuation exponent is introduced using the data reported for FeF_2 and $RbMnF_3$. The frequency dependent longitudinal sound velocity and its relation to the static quantities are discussed. Finally, the unsolved questions and future prospects in this field are outlined.

1. Introduction

The sound attenuation coefficient and the sound velocity show anomalous behavior near the critical point of the magnetic systems. The singular behavior of these quantities is connected with very strong fluctuations of the magnetic order parameter near the critical temperature. These fluctuations give rise to a characteristic attenuation peak whose position is correlated with that of the minimum in the sound velocity. In Fig. 1. we show the temperature dependence of the longitudinal ultrasonic attenuation and changes in the sound velocity in Gd (Moran and Luthi [1]).

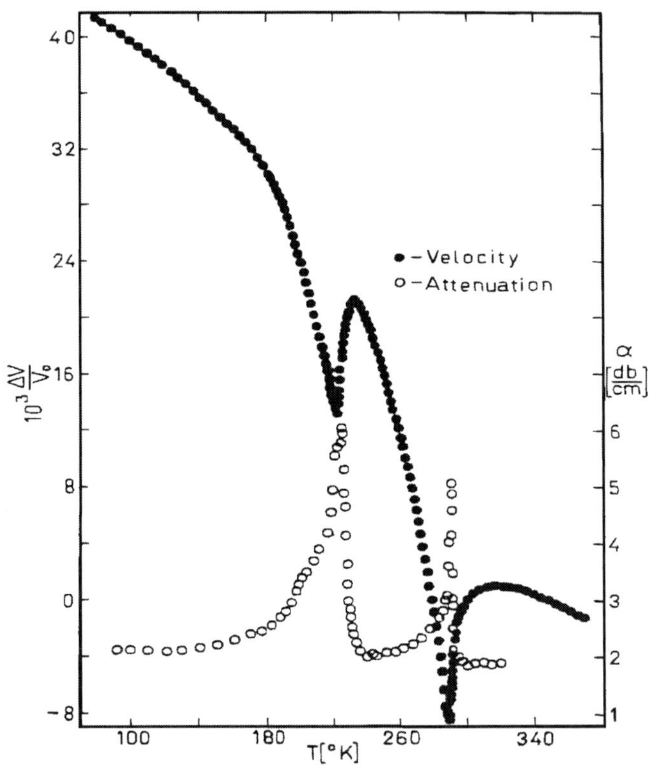

Figure 1. Temperature dependence of the ultrasonic attenuation and changes in the sound velocity for the longitudinal waves along c-axis with f=50 MHz (Moran and Luthi [1]).

The problem of strongly interacting fluctuations cannot be reduced to the problem of ideal gas even in the lowest order approximation. The general method of treating such issues has been shown by Wilson [2, 3] to be the renormalization group theory. Using this method we can find not only the critical exponents and the scaling functions but we can also study the nonasymptotic effects as the crossover from one universality class to another (crossover phenomena). Later it was possible to generalize the renormalization group formalism to the dynamic phenomena [4] such as transport coefficients and the relaxation rates. The studies of the critical dynamics of sound is a very important field where we can test the modern concepts of phase transition theory such as scaling, universality of the critical exponents or the crossover to another universality class. Moreover, the measurements of the sound attenuation coefficient and the sound velocity permit determination of the phase diagram or the symmetry of the coupling between the order parameter and the elastic degrees of freedom. It is the class of magnetic materials which is especially important from this point of view (although it is still not fully recognized in many details), being a prototype for many other systems. In magnets we meet in general three types of magneto-elastic coupling (in this paper they will also be called the spin-phonon couplings) [5] but usually only one called the volume magnetostriction dominates. In spite of this ostensible simplification we observe there a whole variety of possible behaviors which sometimes cannot be explained. This fact is connected with the coexistence in magnets of many different spin interactions of different symmetry and range and with very rich and complicated dynamics in some systems [6, 7]. These factors can be manifested over different temperature ranges, which sometimes makes it impossible to describe the system's dynamics with the aid of one set of critical exponents. In the magnets being electric insulators the acoustic singularities encountered are different than in those showing metallic properties [6, 7]. In insulators we usually observe a weak singularity characterized by a small sound attenuation exponent. Sometimes the singularity is even not observed in a given experimental frequency range. In magnetic metals the singularity is much more noticeable and the critical exponent is much higher (usually higher than one). It was initially explained by the fact that in insulators the spin exchange interactions are of short range nature and in this case the spin-phonon Hamiltonian which arises mostly via the strain modulation of the exchange interaction [6] is proportional to the exchange Hamiltonian[1]. This mechanism was proposed by Kawasaki [8] who noticed also that the energy fluctuations should decay only by the spin lattice relaxation. In this case we say that the sound wave couples to the energy fluctuations contrary to the metallic magnetic systems in which the long range exchange interactions generate a more general spin-phonon interaction which is linear in the sound mode and bilinear in the order parameter (spin) fluctuations. The different couplings should lead to different sound attenuation exponents. However, it was a simplified point of view as it was later shown [9, 10] that the energy fluctuations couple to the same bilinear combination of the order parameter fluctuations as for the magnetic metals. The general theory [9] which takes into account both types of magnetoelastic couplings as well as the proper coupling of energy to the order parameter fluctuations shows that both singularities: typical of the metallic as well as insulating systems appear in the acoustic self energy with the same effective coupling constant and the parameter which distinguishes the two types of behavior is the ratio of the spin-lattice relaxation time to the characteristic time of spin fluctuations[2]. For insulators this ratio is very high as the spin-lattice relaxation times are much longer than for metals. The long spin-lattice relaxation time favors the weak singularity. If these times are comparable, the strong singularity dominates. We will show also the existence of another high-frequency regime which is expected for some materials. The nonasymptotic effects showing the crossover from insulator-type regime to the metallic-type regime and to the high-frequency regime will be demonstrated on the example of the ultrasonic data for MnF_2. We will also show the usefulness of the concept of the effective sound attenuation exponent which is intro-

[1]However, it is true only when we can neglect the next nearest neighbor exchange coupling and only in the case of propagation along some symmetry directions. In general the sound mode couples only to the part of the spin energy density [6].

[2]It will be shown explicitly in Section 6 of this chapter.

duced using the experimental data for FeF_2 and $RbMnF_3$. Finally a summary of the sound attenuation exponents in magnetic metals and insulators will be given as well as an outlook for the future progress in this field will be outlined.

2. The Fundamental Concepts of the Phase Transition Theory

There is a huge variety of physical[3] systems which undergo phase transitions. The most interesting class of phase transformations seems to be that of the continuous phase transitions which show no latent heat but at which many physical quantities diverge to infinity or tend to zero when approaching the critical temperature T_c. The behavior of the specific heat of a ferromagnet near the critical temperature is shown in Fig. 2. The free energy in

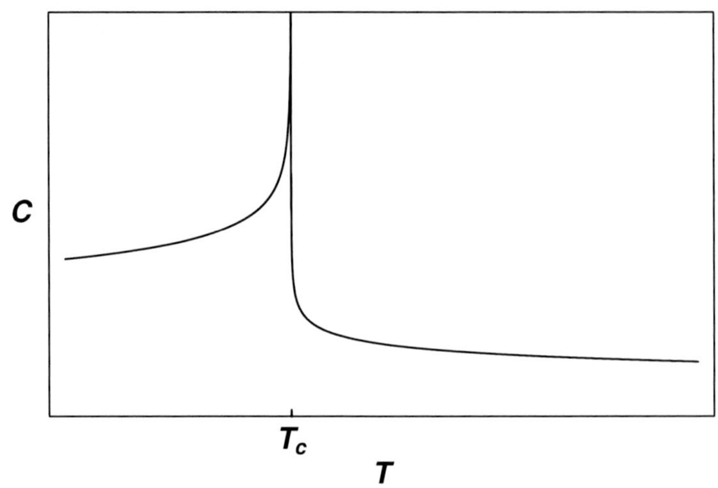

Figure 2. Specific heat C vs. temperature T in a ferromagnet.

such systems is a nonanalytical function of its arguments which is a manifestation of very strong fluctuations of a quantity called the order parameter. Usually, we define the order parameter as the quantity which is space and time dependent. It will be denoted by $S(\boldsymbol{x}, t)$ for anisotropic ferromagnets, and sometimes we will refer to it as the spin. The prototype of the continuous phase transition is that from the paramagnetic phase (disordered spins) to the ferromagnetic phase with nonzero average magnetization. In this case, the order parameter is the local magnetic moment whose average (magnetization) tends to zero when approaching the Curie temperature as shown in Fig. 3. For antiferromagnets the order parameter is given by the staggered local magnetization; in the case of gas-liquid transition it is proportional to the deviation of the mass density from its critical value, and for the superconducting transition it is a wave function of the Cooper pairs [11]. The order parameter can have more than one components as for example for isotropic ferromagnets in which it is a vector with three components. We say that in this case the order parameter

[3]In general the phase transitions can be found in economic, biological, social and many other systems. For example the collective motion of large groups of biological organisms like flocks of birds or fish schools (self-driven organisms) can develop a kinetic phase transition from an ordered to chaotic motion [12, 13].

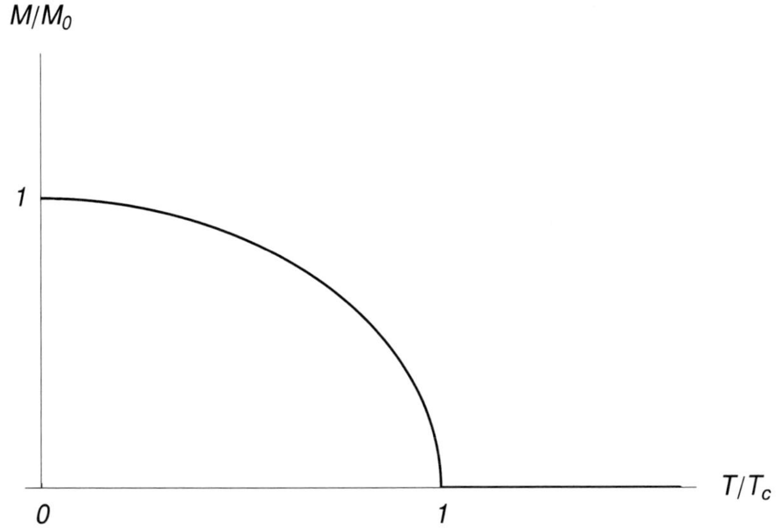

Figure 3. Magnetization as a function of temperature.

dimension is three: $n = 3$. If there is an anisotropy in the system such that the magnetization (staggered magnetization) vector is forced to lie within a given plane we deal with the XY ferromagnet (antiferromagnet) for which $n = 2$. For a magnet with only one easy axis $n = 1$ and we talk about the scalar order parameter. The order parameter can have much more components and a nature more complicated than a vector as for example in liquid He^3 [11]. In the theory of phase transitions and critical phenomena the key problem is the identification of the order parameter since the same system of atoms may exhibit in different temperature ranges the liquid-gas transition, many structural and/or liquid crystals transitions, paramagnet-ferromagnet transition etc. The physical intuition plays here a very important role indicating the most important features of a given phase transformation.

2.1. Critical Exponents

The rate at which physical quantities diverge to infinity or converge to zero when approaching a critical point is described by critical exponents. If the distance from the critical point is measured by the reduced temperature

$$t = \frac{T - T_c}{T_c}, \tag{1}$$

than the critical exponent, describing the quantity $F(t)$ is defined by:

$$x_F = - \lim_{t \to 0^+} \frac{\ln F(t)}{\ln t}. \tag{2}$$

We say that for $t \to 0^+$ the function $F(t)$ diverges (with a positive exponent x_F) as t^{-x_F}. We can also define the low-temperature exponent

$$x'_F = - \lim_{t \to 0^-} \frac{\ln F(t)}{\ln t}, \tag{3}$$

which corresponds to the ordered (low-temperature phase), or other critical exponents describing the power-law behavior with respect to the other thermodynamic quantities, distance or the wave vector etc. For some quantities like the average of the order parameter the corresponding exponent is defined with the minus sign in Eq. 3. We define the basic static critical exponents on the example of the Ising type ferromagnet ($n = 1$). In such a simple system the critical behavior of all thermodynamic quantities is controlled by only two parameters: the reduced temperature t and the magnetic field h. Let us consider the following quantities:

1. The specific heat C_h under constant magnetic field. Near T_c it is described by the relations:

$$C_h \approx A^+ t^{-\alpha} + B, \qquad t > 0,\ h = 0, \tag{4}$$

$$C_h \approx A^- |t|^{-\alpha'} + B, \qquad t < 0,\ h = 0. \tag{5}$$

In the case of two dimensional Ising model $\alpha = 0$ and the specific heat diverges logarithmically

$$C_h \approx -A^{\pm} \ln |t|. \tag{6}$$

The coefficients A^+ i A^- are called the critical amplitudes and α i α' are known as the specific heat critical exponents.

2. Susceptibility χ (the derivative of the magnetization with respect to the magnetic field). We observe the following power-law behavior:

$$\chi \approx C^+ t^{-\gamma}, \qquad t > 0,\ h = 0, \tag{7}$$

$$\chi \approx C^- |t|^{-\gamma'}, \qquad t < 0,\ h = 0. \tag{8}$$

For the systems with vector order parameter ($n \geq 2$) below T_c the susceptibility is infinite in agreement with the famous Goldstone theorem [14] which says that for the system with broken continuous symmetry $n - 1$ transversal modes appear whose frequencies tend to zero as the wave vector goes to zero. These massless modes imply that the transversal susceptibility diverges for a vanishing external field [15].

3. The order parameter

$$M \approx B'(-t)^{\beta}, \qquad t < 0,\ h = 0. \tag{9}$$

Another interesting critical exponent is the one connected with approaching the critical point for $T = T_c$ with $h \to 0$. Then the order parameter is described by the following scaling law:

$$M \approx B^c\, h^{1/\delta}, \qquad t = 0. \tag{10}$$

4. Two-point correlation function

$$C(x) = \langle S(\boldsymbol{x}) S(\boldsymbol{0}) \rangle - \langle S(\boldsymbol{0}) \rangle^2, \tag{11}$$

where $\langle ... \rangle$ denotes an average and $S(\boldsymbol{x})$ is a local value of the order parameter at point $\boldsymbol{x}$. At the critical point ($T = T_c$) it is characterized by the power-law behavior at large distances:

$$C(x) \propto x^{-d+2-\eta}, \qquad t = 0,\ h = 0, \tag{12}$$

where d is the space dimension and η is an anomalous critical exponent which measures the deviation from the classical Ornstein-Zernike behavior where $\eta = 0$.

In the neighborhood of the critical point (but not exactly at it) the correlation function decays exponentially

$$C(x) \propto \exp(-x/\xi), \tag{13}$$

where ξ denotes a correlation length of the system which diverges when approaching the critical temperature:

$$\xi \approx \xi_0^+ t^{-\nu}, \qquad t > 0,\ h = 0, \tag{14}$$

$$\xi \approx \xi_0^- \left(-t\right)^{-\nu'}, \qquad t < 0,\ h = 0. \tag{15}$$

2.2. Scaling Hypothesis

Already at very early stage of development of the phase transition theory, it was realized that the critical exponents are not fully independent of each another and fulfill a number of relations called the „scaling laws" [16]. These relations can be derived from the scaling hypothesis which says that near the critical point the correlation length ξ is the only characteristic length scale in terms of which all other quantities with dimensions of length are to be measured. In general a system has usually many intrinsic length scales, as for example the length of a system or the mean distance between nearest lattice points in a crystal. We say that the system near a critical point shows a *scale invariance*. Using the scaling hypothesis one can derive the above mentioned scaling laws which are in very good agreement with experiment. A mathematical manifestation of the scaling hypothesis is that the singular parts of the thermodynamic potentials or the correlation function etc. are generalized homogeneous[4] functions of their arguments [16–20]. For example the free energy of the magnetic system $F_{\text{sing}}(T, h)$ obeys the relation:

$$F_{\text{sing}}(\lambda^{x_t} t, \lambda^{x_h} h) = \lambda F_{\text{sing}}(t, h), \tag{16}$$

where λ is a rescaling factor (any real number) and x_t and x_h are the characteristic exponents of the phase transition. Choosing $\lambda = t^{-1/x_t}$ we obtain $F_{\text{sing}}(t, h) = t^{1/x_t}\phi(h/t^{x_h/x_t})$ where ϕ is a scaling function. Also a derivative of one homogeneous function is another homogeneous function. Thus differentiating expression (16) with respect to the reduced temperature or magnetic field and comparing it with the corresponding definitions of critical exponents we can express the critical exponents α, β, γ and δ by only two independent ones x_t and x_h. Analogous considerations applied to the correlation function [18] shows that also the exponents η and ν can be obtained from the two mentioned independent ones. A consequence of the scaling hypothesis is also the equality of low-temperature and high-temperature exponents: $\alpha = \alpha'$, $\gamma = \gamma'$ and $\nu = \nu'$. Eliminating x_t and x_h from the relations between the critical exponents one can obtain a number of exponent identities called the scaling laws [16]:

$$\alpha + 2\beta + \gamma = 2, \qquad \text{Rushbrooke's law}, \tag{17}$$

[4]In general, a function $f(y_1, y_2, ...)$ is homogeneous if $f(b^{x_1}y_1, b^{x_2}y_2, \cdots) = b^{x_f} f(y_1, y_2, \cdots)$ for any b. By a proper choice of the rescaling factor b one of the arguments of f can be removed, leading to a scaling forms used in this subsection. An important consequence of the scaling ideas is that the critical system has an additional dilatation symmetry.

$$\alpha + \beta(\delta + 1) = 2, \qquad \text{Griffiths' law,} \tag{18}$$

$$\gamma = (2 - \eta)\nu, \qquad \text{Fisher's law,} \tag{19}$$

$$\alpha = 2 - d\nu, \qquad \text{Josephson's law.} \tag{20}$$

The Josephson's identity is the only one which involves the space dimension. Such identities are known as *hyperscaling relations*. They are true only for $d < d_c$ where d_c is the upper critical dimension ($d_c = 4$ for models with n-vector order-parameter) above which the mean-field critical exponents are exact:

$$\alpha = 0, \qquad \gamma = 1, \qquad \nu = \frac{1}{2}, \qquad \eta = 0, \qquad \beta = \frac{1}{2}, \qquad \delta = 3. \tag{21}$$

The scaling laws were confirmed in many experiments, whereas the theoretical explanation was given by the renormalization group theory [3, 16, 21]. Moreover, this theory provided us also with the efficient tools for calculating the critical exponents and the scaling functions. Within this formalism one can also calculate the corrections to the asymptotic ($t \to 0$) power laws and assess their magnitude [22, 23]. As will be shown in the next section it is possible to generalize the scaling hypothesis onto the dynamic phenomena.

2.3. Universality Hypothesis

The main goal of the theory of phase transitions is to permit the calculations of the scaling exponents and the scaling functions. According to the universality hypothesis, diverse physical systems that share the same essential symmetry properties will exhibit the same physical behavior close to their critical points and the values of their critical exponents do not depend on the thermodynamic parameters, the strength of interactions, atomic structure of the system and other microscopic details of the interactions. For example a uniaxial ferromagnet is characterized by the same set of critical exponents as the liquid-gas phase transition and the planar ferromagnet's exponents are the same as for the liquid helium near the transition to superfluid phase. Very close to the critical point the most of the detailed information about the interactions in the system becomes irrelevant and even highly idealized models (and much simpler than the real system), which possess the important symmetries of the real system, can be used to describe real systems accurately. These symmetries determine the type of critical behavior (values of the critical exponents) known as the universality class. The fact that every system undergoing a continuous phase transition belongs to one of such universality classes and that the universality classes constitute relatively not numerous set is probably the most unusual feature of the phase transitions. The renormalization group theory predicts that the universality classes are determined by the spatial dimensionality d, dimension of the order parameter n or more generally its symmetry, and the range of interactions. In some systems the presence of some kinds of impurities may influence the critical exponents leading to a new universality class [23–27]. Besides the critical exponents also the scaling functions and some combinations of critical amplitudes like A^+/A^- or ξ_0^+/ξ_0^- are universal i.e. are the same for different sometimes quite dissimilar systems. The critical amplitudes alone are nonuniversal quantities and depend on a given system. In Table 1 we present the theoretical estimations of the most important (static) critical exponents and

Table 1. The theoretical estimations of the critical exponents and the universal amplitude ratios for three dimensional Ising ($n = 1$), XY ($n = 2$) and Heisenberg ($n = 3$) systems.

n	1	2	3
α	0.110(1)	$-0.0146(8)^*$	$-0.133(6)^*$
β	0.3265(3)	$0.348\,5(2)^*$	$0.3689(3)^*$
γ	1.2372(5)	1.3177(5)	1.3960(9)
δ	4.789(2)	$4.780(7)^*$	$4.783(3)^*$
η	0.0364(5)	0.0380(4)	0.0375(5)
ν	0.6301(4)	0.67155(27)	0.7112(5)
A^+/A^-	0.532(3)	1.062(4)	1.56(4)
ξ_0^+/ξ_0^-	1.956(7)	0.33	0.38
$\alpha A^+C^+/B^2$	0.0567(3)	0.127(6)	0.185(10)
References	[22]	[28]	[29]

The star (*) denotes the estimations obtained from the scaling laws $\alpha = 2 - 3\nu$, $\beta = \nu(1+\eta)/2$ and $\delta = (\beta+\gamma)/\gamma$

some universal amplitude ratios for three dimensional systems with n-vector order parameter and short range interactions. From the analysis of these data we can see that the change in the critical exponents from one class to another is not very impressive. Much greater variability is observed in the critical amplitude ratios and sometimes these ratios are better suited to identify the universality classes. Also the investigation of dynamic properties of the system as will be shown in the next section may be useful in solving this issue.

3. Critical Dynamics

We recall in this section the basic ideas which have contributed to the development of the modern theory of dynamic critical phenomena.

3.1. Critical Slowing Down

In description of the critical anomalies which are met in dynamic characteristics of the system like the linear response functions, we need an equation of motion describing the order parameter field. The most simple equation used in irreversible thermodynamics is that describing the rate of change in the quantity relaxing to its equilibrium state

$$\dot{\psi} = -L\frac{d\Phi}{d\psi}, \tag{22}$$

where the dot over ψ denotes the time derivative and L is a kinetic coefficient. The function $\Phi[\psi]$ is an increase in the corresponding thermodynamic potential related to the deviation of ψ from the equilibrium value ($\psi_{eq} = 0$) [30]. The probability of fluctuation ψ is proportional to

$$p_{eq} \propto \exp\{-\Phi[\psi]/k_B T\}. \tag{23}$$

If we assume that the probability distribution is Gaussian than we have

$$\Phi[\psi] = \frac{\psi^2}{2\chi}, \tag{24}$$

where

$$\chi = \frac{\langle \psi^2 \rangle}{k_B T} \tag{25}$$

is a susceptibility. The solution of (22) is given by

$$\psi(t) = \psi(0)e^{-t/\tau}, \tag{26}$$

where $\tau = \chi/L$ is known as the relaxation time of quantity ψ. In this section the symbol t refers to the time not to the reduced temperature and the distance to the critical point will be denoted by $(T - T_c)$. We have seen in the last section that an increase in the fluctuations near the critical point leads to the divergence of the susceptibility $\chi \propto (T - T_c)^{-\gamma}$. According the conventional theory of critical dynamics proposed by Van Hove [31] the kinetic coefficients stay finite at the critical point so the relaxation time goes to infinity at the critical point. We say that the system needs more and more time to get back to equilibrium. This phenomenon is known as the critical slowing down. The Van Hove's theory turns to be incorrect in most cases and the kinetic or transport coefficients diverge to infinity (or go to zero in some cases) but in no case does the kinetic coefficient diverge so strongly as the susceptibility [4] so the critical slowing down appears in all cases.

3.2. Dynamical Scaling

In the physics of dynamic phenomena another critical exponent known as dynamic critical exponent z must be defined. In the dynamic scaling hypothesis we assume that the characteristic frequency (known also as the critical frequency) of the order parameter mode $S_{\boldsymbol{k}}$ scales as

$$\omega_c(k) = k^z f(k\xi), \tag{27}$$

where k is the wave vector and f is the scaling function. The characteristic frequency is defined as the half width of the dynamic correlation function $C_S(k, \omega)$

$$\int_{-\omega_c(k)}^{\omega_c(k)} \frac{d\omega}{2\pi} C_S(k, \omega) = \frac{1}{2} C_S(k) \tag{28}$$

where

$$C_S(k, \omega) = \int d^d x \int dt\, e^{-i(\boldsymbol{k}\cdot\boldsymbol{x} - \omega t)} [\langle S(\boldsymbol{x}, t) S(\boldsymbol{0}, 0)\rangle - \langle S(\boldsymbol{x}, t)\rangle\langle S(\boldsymbol{0}, 0)\rangle]. \tag{29}$$

The characteristic frequency may be also defined [4] by

$$\omega_c(k) = \frac{2C_S(k)}{C_S(k,0)} = \frac{L_S(k)}{\chi_S(k)}. \tag{30}$$

$C(k)$ denotes here the Fourier transform of the static correlation function and $\chi_S(k) = C_S(k)/k_B$ is the susceptibility, $L_S(k)$ is the effective kinetic coefficient

$$\frac{1}{L_S(k)} = \frac{i\partial\chi_S^{-1}(k,\omega)}{\partial\omega}|_{\omega=0}, \tag{31}$$

where $\chi_S(k,\omega)$ is the linear response (dynamic susceptibility) to the infinitesimal field. We define the dynamic susceptibility $\chi_S(k,\omega)$ by the relation

$$\delta\langle S(k,\omega)\rangle_h = \chi_S(k,\omega)\, h(k,\omega). \tag{32}$$

Fourier transforms in wave vector and frequency are given by

$$S(\boldsymbol{x},t) = \int \frac{d^d k}{(2\pi)^d} \int \frac{d\omega}{2\pi}\, e^{i(\boldsymbol{k}\cdot\boldsymbol{x} - \omega t)} S(\boldsymbol{k},\omega).$$

The fluctuation-dissipation theorem for the classical systems says that the linear response and the correlation function are not independent:

$$C_S(k,\omega) = \frac{2k_B T}{\omega}\, \mathrm{Im}\,\chi_S(k,\omega). \tag{33}$$

If the correlation function has a Lorentzian peak centered about $\omega = 0$ the definitions (28) and (30) are equivalent. If there is a propagating mode in the system which is reflected in the correlation function $C(k,\omega)$ as a sharp peak at a finite frequency, the definition (30) is not appropriate [4].

Within the dynamical scaling hypothesis [32, 33] we assume that the linear response function is homogenous:

$$\chi_S(k,\omega;T-T_c) = b^{2-\eta}\chi_S\left(bk, b^z\omega; b^{1/\nu}(T-T_c)\right).$$

With a proper substitution for b we obtain

$$\chi_S(k,\omega;T-T_c) = (T-T_c)^{-\gamma} Y(k\xi, \frac{\omega\xi^z}{\Omega_0}), \tag{34}$$

where Y is a scaling function and Ω_0 is a constant setting the time scale in the system. It is assumed that the wave vector and the frequency are much smaller than the inverse of the microscopic length (e.g. the lattice constant) and the microscopic relaxation time.

In the simple model of relaxational dynamics described in last subsection we have

$$\omega_c(k \to 0) \to L/\chi \propto |T-T_c|^{-\gamma} \propto \xi^{-\gamma/\nu}, \tag{35}$$

as in this model the kinetic coefficient L does not depend on k for $k \to 0$. The scaling function f from Eq. (27) has to behave as

$$f(x) \propto x^{-z}, \tag{36}$$

in order that $\omega_c(k \to 0)$ was wave vector independent. It gives the relation $\omega_c \propto \xi^{-z}$ which can be compared with the Van Hove's dynamic exponent $z = \gamma/\nu = 2 - \eta$ where we have exploited the Fisher identity (19).

3.3. Mode Coupling and the Equations of Motion

Moreover, if we would like to take into account the fast movement associated with the other modes, we should add to Eq. (22) a stochastic Gaussian noise $\zeta(t)$ which mimics the thermal excitation

$$\dot{\psi} = -L\frac{d\Phi}{d\psi} + \zeta, \tag{37}$$

where the correlation function of the noise obeys the Einstein relation

$$\langle \zeta(t)\zeta(t') \rangle = 2k_B TL\delta(t-t'), \tag{38}$$

and $\delta(t)$ is the Dirac function. We assume that

$$\langle \zeta(t) \rangle = 0.$$

Both ψ and ζ should be regarded as the stochastic processes. Eq. (37) is known as a linear Langevin equation [34].

Generalizing Eq. (37) to include n-vector non uniform processes **S** and taking into consideration also (static) nonlinear couplings present in the thermodynamic potential as a term proportional to S^4, we obtain a model of dissipative dynamics known as the time-dependent Landau-Ginzburg model [35]

$$\dot{S}_i(\boldsymbol{x}) = -\Gamma_i \frac{\delta H}{\delta S_i(\boldsymbol{x})} + \zeta_i(\boldsymbol{x}), \tag{39}$$

where the potential (called the Landau-Ginzburg Hamiltonian or free energy)

$$H = \frac{1}{2}\int d^d x\{r_0 S^2 + (\nabla S)^2 + \frac{u}{2}S^4\}, \tag{40}$$

includes also nonlinear couplings between modes. In an external magnetic field $\boldsymbol{h}$ the term $-\int d^d x \boldsymbol{h} \cdot \boldsymbol{S}(\boldsymbol{x})$ should be added to (40). In Eq. (40) we used the following abbreviations:

$$S^2 = \sum_{i=1}^{n} S_i^2(\boldsymbol{x}),$$

$$(\nabla S)^2 = \sum_{i=1}^{n} (\nabla S_i(\boldsymbol{x}))^2,$$

$$S^4 = (S^2)^2.$$

The symbol $\frac{\delta H}{\delta S_i(\boldsymbol{x})}$ denotes a functional derivative [11, 36]. The noises fulfill the relations

$$\langle \zeta_i(\boldsymbol{x},t)\zeta_j(\boldsymbol{x}',t') \rangle = 2\Gamma_i \delta_{ij}\delta(\boldsymbol{x}-\boldsymbol{x}')\delta(t-t'). \tag{41}$$

Usually it is assumed that the kinetic coefficients for different components of the order-parameter are equal: $\Gamma_i = \Gamma$.

The equation of motion (39) does not contain the nonlinear mode-coupling terms which in a crucial way decide on whether the kinetic coefficients diverge or tend to zero when approaching the critical point. For an isotropic ferromagnet for example, Eq. (39) should be modified by a term describing the precession of the spins around the local magnetic field:

$$\dot{\boldsymbol{S}} = \lambda \boldsymbol{S} \times \frac{\delta H}{\delta \boldsymbol{S}} + D\nabla^2 \frac{\delta H}{\delta \boldsymbol{S}} + \zeta, \tag{42}$$

where the Onsager kinetic coefficient Γ was replaced by the term $-D\nabla^2$ which guarantees that the total spin $\boldsymbol{S}_{k=0}(t) = \int d^d x \boldsymbol{S}(\boldsymbol{x}, t)$ is a conserved quantity[5] (which does not change its value during the motion) analogously to microscopic models and to hydrodynamics [37]. The first term in Eq. (42) describes the precession around the local field $\boldsymbol{h}_{loc} = -\frac{\delta H}{\delta \boldsymbol{S}}$ and is known in literature as the *mode-coupling* term or the *streaming term*. It is a non dissipative term i.e. it does not change the total energy of the system when the noise and the damping force are absent:

$$\frac{dH}{dt} = \int d^d x \frac{\delta H}{\delta \boldsymbol{S}(\boldsymbol{x}, t)} \cdot \frac{\partial \boldsymbol{S}(\boldsymbol{x}, t)}{\partial t} = \int d^d x \frac{\delta H}{\delta \boldsymbol{S}(\boldsymbol{x}, t)} \cdot \left[\lambda \boldsymbol{S}(\boldsymbol{x}, t) \times \frac{\delta H}{\delta \boldsymbol{S}(\boldsymbol{x}, t)} \right] = 0. \tag{43}$$

When we investigate the critical dynamics we are not usually interested in complicated microscopic descriptions of the evolution of the system. Usually we tend to obtain the equations of motion for long-wavelength components of the so-called slow variables such as the conserved quantities, Goldstone modes and the order parameter. The fast variables are eliminated by a projection procedure on the subspace of slow variables [38, 39]. The reader can find the description of this procedure in the works of Mori et al. [40, 41]. The effective equations of motion for slow variables ϕ_α are reduced to nonlinear Langevin equations [40, 41]

$$\frac{d}{dt}\phi_\alpha(t) = V_\alpha(\{\phi_\alpha(t)\}) - \sum_\beta \Gamma_{\alpha\beta} \frac{\delta H(\{\phi_\alpha(t)\})}{\delta \phi^*_\beta(t)} + \zeta_\alpha(t), \tag{44}$$

where

$$H(\{\phi_\alpha(t)\}) = -k_B T \ln(P_{eq}(\{\phi_\alpha(t)\})), \tag{45}$$

and $P_{eq}(\{\phi_\alpha(t)\})$ is an equilibrium distribution function. The first term in Eq. (44) is the streaming term

$$V_\alpha(\{\phi_\alpha(t)\}) = -\lambda \sum_\beta \left[\frac{\delta}{\delta \phi_\beta} Q_{\alpha\beta}(\{\phi_\alpha\}) - Q_{\alpha\beta}(\{\phi_\alpha\}) \frac{\delta H(\{\phi_\alpha\})}{\delta \phi^*_\beta} \right], \tag{46}$$

with λ being a constant; and $Q_{\alpha\beta} = -Q_{\beta\alpha}$ are some functions constructed from the Poisson brackets or the commutators of slow variables $\{\phi_\alpha\}$. The second contribution describes the dumping and the last term is the stochastic force representing the effect of fast variables. The white noises have zero means and variations:

$$\langle \zeta_\alpha(t) \zeta_\beta(t') \rangle = 2\Gamma_{\alpha\beta}\delta(t - t').$$

[5] It is easy to see this performing a Fourier transform $\boldsymbol{S}(\boldsymbol{x}, t) = \frac{1}{\sqrt{V}} \sum_{k<\Lambda} e^{i\boldsymbol{k}\cdot\boldsymbol{x}} \boldsymbol{S}_{\boldsymbol{k}}(t)$. Then $D\nabla^2 \to -Dk^2$ and the damping coefficient for the total spin tends to zero for $k \to 0$. In this equation Λ is a cutoff parameter which is usually chosen in such a way that Λ^{-1} is much larger than the lattice constant and simultaneously much shorter than the correlation length. Models which describe the fluctuations in such a scale are known as mesoscopic models.

It can be shown that the equations of motion have the correct stationary state (the long-time limits of the one-time correlation functions are the same as the static quantities obtained from the equilibrium distribution function).

It should be noted that the Poisson brackets and the detailed form of the damping coefficients are not determined by the functional H, which implies that with each static universality class usually a few dynamic universality classes (determined by the static critical exponents as well the dynamic one) can be associated. In general, we can say that the dynamic universality class is determined by the number of slow variables and the structure of the Poisson brackets.

3.4. Dynamic Universality Classes

In this section we consider a few of the most important models used in the study of critical dynamics. We begin with the models describing the relaxational dynamics. The description of other models can be found in the excellent reviews [4, 42–47].

3.4.1. Model A

In this model there is no conserved quantity and the only slow variable is the order parameter of n components. It is described by simple equations [48, 49]:

$$\dot{S}_i(\boldsymbol{x}) = -\Gamma \frac{\delta H}{\delta S_i(\boldsymbol{x})} + \zeta_i(\boldsymbol{x}), \tag{47}$$

where H is the Ginzburg-Landau Hamiltonian of the form (40). The static exponents are determined in all dynamic models by the spatial dimensionality d and the order-parameter dimension. The dynamic critical exponent differs only slightly from that from the Van Hove's theory. The renormalization group theory gives the value:

$$z = 2 + c\eta, \tag{48}$$

where η is the correlation function exponent (12) and c is a function of d and n. One of the tools of this theory is the ϵ-expansion [3] of the critical exponents (and the scaling functions) where $\epsilon = 4 - d$. It was obtained that $c = 0.7261(1 - 1.687\epsilon) + O(\epsilon^2)$ [50, 51]. Another useful expansion is that in $1/n$ (exact for $n \to 0$), it gives the following estimate $c = 1/2$ for $d = 3$ [50]. According to Eq. (48) the renormalized (by interactions) kinetic coefficient Γ_R[6] goes slowly to zero [4] as $T \to T_c$ and it is actually finite to first order in ϵ.

3.4.2. Model B

It is a simple modification of Model A with the order parameter being a conserved quantity. In Eq. (47) the kinetic coefficient Γ is replaced by $-\lambda\nabla^2$ and the same replacement is made in Eq. (41). We find for this model

$$\omega_c(k) = \lambda k^2/\chi_S,$$

[6] The renormalized kinetic coefficient is usually defined by the relation $\Gamma_R = \omega_c\chi$ where χ is the susceptibility of the system.

as the transport coefficient λ is not renormalized by the nonlinear interactions present in the Hamiltonian [4]. Thus we obtain the classical result from the Van Hove's theory

$$z = z_{cl} = 4 - \eta.$$

The characteristic frequency $\omega_c(k)$ tends to zero for a given k as a result of the critical slowing down.

3.4.3. Model C

The nonconserved order parameter can be coupled to a conserved (noncritical) quantity such as the energy (or the magnetization in the case of uniaxial antiferromagnet). The static Landau-Ginzburg functional depends then on two quantities: the order parameter $\boldsymbol{S}$ and the additional conserved quantity m:

$$H = \frac{1}{2}\int d^d x\{r_0 S^2 + (\nabla S)^2 + \frac{u}{2}S^4 + \chi_m^{-1} m^2 + fmS^2\}, \tag{49}$$

where f is a new coupling constant and χ_m is the bare (i.e. without taking into considerations the effect of interactions which means for $f = 0$) susceptibility of m. Because m is a noncritical quantity it can be eliminated from statics by an integration over m [48]. This procedure leads to an effective Hamiltonian with the renormalized coupling constant u. Thus the static critical exponents are the same as for Hamiltonian (40).

The dynamics is described by two coupled equations:

$$\dot{S}_i(\boldsymbol{x}) = -\Gamma\frac{\delta H}{\delta S_i(\boldsymbol{x})} + \zeta_i(\boldsymbol{x}), \tag{50}$$

$$\dot{m}(\boldsymbol{x}) = \lambda_m \nabla^2 \frac{\delta H}{\delta m(\boldsymbol{x})} + \xi_m(\boldsymbol{x}), \tag{51}$$

where the additional noise obeys the relation

$$\langle \xi_m(\boldsymbol{x},t)\xi_m(\boldsymbol{x}',t')\rangle = -2\lambda_m \nabla^2 \delta(\boldsymbol{x}-\boldsymbol{x}')\delta(t-t'). \tag{52}$$

The coupling with the conserved quantity leads to a new dynamic critical exponent for the systems with positive specific heat critical exponent (which happens for the Ising like systems with $n = 1$)

$$z = 2 + \frac{\alpha}{\nu} \tag{53}$$

where α is the specific heat critical exponent. The renormalized kinetic coefficient goes to zero more rapidly than in model A, and then $\Gamma_R \propto (T - T_c)^\alpha$. For $\alpha < 0$ (or $n > 1$) the coupling with the conserved quantity is irrelevant and this model is equivalent to model A. It is worth noting that the exponent α/ν is rather small so it is difficult to distinguish the predictions of models A and C experimentally, but for the tricritical point[7] we have $\alpha_t = 1/2$ and $\nu_t = 1/2$ so the dynamic tricritical exponent $z_t = 3$ for model C is significantly different from model A where $z_t = 2$ [52, 53].

[7] At the tricritical point a change from the continuous to the first order transition occurs [57]. The tricritical exponents are classical ones for $d > 3$ with logarithmic corrections for $d = 3$.

3.4.4. Model D

In this model the conserved order parameter is coupled with the conserved noncritical quantity. The dynamics of this model is described by Eqs. (50) and (51) where $\Gamma = -\lambda_S \nabla^2$. The model's dynamics is reduced to that of model B (independently of the order parameter dimensionality n) with $z = z_{cl} = 4 - \eta$.

3.4.5. Models E and F

Let's consider a planar magnet described by the following equations [54] (model F)

$$\dot{\psi} = -2\Gamma \frac{\delta H}{\delta \psi^*} - ig\psi \frac{\delta H}{\delta m} + \theta, \tag{54}$$

$$\dot{m} = \lambda_m \nabla^2 \frac{\delta H}{\delta m} + 2g \,\mathrm{Im}(\psi^* \frac{\delta H}{\delta \psi^*}) + \xi_m, \tag{55}$$

where ψ is a complex order parameter representing $S_x - iS_y$ and m is the z-th component of magnetization (the z-axis is chosen to be perpendicular to the easy plane). The Landau-Ginzburg functional is given by

$$H = \frac{1}{2} \int d^d x \{ r_0 |\psi|^2 + |\nabla \psi|^2 + \frac{u}{2} |\psi|^4 + \chi_m^{-1} m^2 + fm |\psi|^2 - hm \}.$$

In easy plane ferromagnets the order parameter is not conserved but the z-th component of magnetization is the conserved quantity which is also the generator of rotations in the order parameter space. So there is a non-vanishing Poisson bracket

$$\{\psi, M\} = ig\psi, \tag{56}$$

where g is the mode-coupling and $M = \int d^d x \, m(\boldsymbol{x})$.

The static properties of this model are the same as those of model C but the dynamic behavior is different due to the nondissipative coupling g and the complex value of the damping coefficient Γ. It was shown by Halperin and Hohenberg [55] that below T_c there is a spin-wave of the frequency $c_{sw}k$. Model F is significantly simplified in the situation when the external magnetic field vanishes. In such a case the total magnetization $\langle M \rangle$ also vanishes and we have the symmetric planar model denoted as model E. In this model we put $f = 0$ and assume a real Γ.

The propagating mode below T_c permits determination of the dynamic exponent z only by means of the static exponents and the spatial dimension. For the antisymmetric planar model (F) which also describes the liquid helium transition we obtain

$$z = \frac{d}{2} + \frac{\tilde{\alpha}}{2\nu}, \tag{57}$$

where $\tilde{\alpha} \equiv \max(\alpha, 0)$ and α is the specific-heat exponent. For model E

$$z = \frac{d}{2}, \tag{58}$$

thus $z = 3/2$ for $d = 3$. In both models the kinetic coefficient Γ diverges for $T \to T_c^+$ (but not so strongly as the susceptibility, so the critical slowing down takes place).

3.4.6. Model G

The isotropic antiferromagnet is also the system with the mode coupling. We have there the nonconserved order parameter (the staggered magnetization) which is a three-dimensional vector. The second field describes the local magnetization. The equations of motion can be written as

$$\dot{\boldsymbol{N}} = -\Gamma \frac{\delta H}{\delta \boldsymbol{N}} + g\boldsymbol{N} \times \frac{\delta H}{\delta \boldsymbol{m}} + \boldsymbol{\theta}, \tag{59}$$

$$\dot{\boldsymbol{m}} = \lambda \nabla^2 \frac{\delta H}{\delta \boldsymbol{m}} + g\boldsymbol{N} \times \frac{\delta H}{\delta \boldsymbol{N}} + g\boldsymbol{m} \times \frac{\delta H}{\delta \boldsymbol{m}} + \zeta, \tag{60}$$

$$H = \frac{1}{2} \int d^d x \{r_0 N^2 + (\nabla N)^2 + \frac{u}{2} N^4 + \chi_m^{-1} m^2\}, \tag{61}$$

where $\boldsymbol{\theta}$ and ζ are white noises. There are non-vanishing Poisson brackets:

$$\{N_i, M_j\} = g\varepsilon_{ijk} N_k, \tag{62}$$

$$\{M_i, M_j\} = g\varepsilon_{ijk} M_k, \tag{63}$$

where $\boldsymbol{M} = \int d^d x \boldsymbol{m}(\boldsymbol{x})$ and ε_{ijk} is the antisymmetric tensor.

In this model we have also a propagating spin-wave mode and the dynamic critical exponent is the same as in model E: $z = d/2$ but some universal amplitude ratios are different [56].

3.4.7. Model J

The dynamics of this model is determined by spin precession and the conservation of the total magnetization:

$$\dot{\boldsymbol{S}} = \lambda \boldsymbol{S} \times \frac{\delta H}{\delta \boldsymbol{S}} + D\nabla^2 \frac{\delta H}{\delta \boldsymbol{S}} + \zeta,$$

where the Hamiltonian is given by (40). Below T_c we have the spin wave of the frequency $c_{sw}^{FM} k^2$ and the transport coefficient D diverges above T_c as

$$D \propto \xi_+^{(6-d-\eta)/2}, \tag{64}$$

revealing also the upper dynamic critical dimension $d_c^{dyn} = 6$ [58] which is the dimension above which the Van Hove theory applies. Below $d_c^{dyn} = 6$ the dynamic fluctuations become important and the kinetic coefficients diverge or vanish when approaching T_c. Thus in model J the dynamic critical dimension d_c^{dyn} differs from the static one d_c^{stat}, which equals four for the statics described by the Ginzburg–Landau model [3].

The dynamic critical exponent is determined only by the static exponent and the spatial dimension

$$z = \frac{1}{2}(d + 2 - \eta). \tag{65}$$

In three dimensions $z \simeq 5/2$. According to the renormalization group theory $\eta = 0$ for $d \geq 4$ so the critical exponent takes its classical value $z_{cl} = 4 - \eta$ for $d = 6$.

Table 2. Summary of the dynamic universality classes in magnets.

Model	Magnetic system	Order parameter dimension	Non-conserved fields	Conserv. fields	Poisson brackets	z
A	anizotropic magnets, uniaxial antiferro-magnets	n	S	none	none	$2+c\eta$
B	uniaxial ferromagn.	n	none	S	none	$4-\eta$
C	anisotropic magnets, uniaxial antiferro-magnets	1	S	m	none	$2+\frac{\alpha}{\nu}$
D	uniaxial ferro-magnets	n	none	S,m	none	$4-\eta$
E	easy plane magnets $h_z=0$	2	ψ	m	$\{\psi,m\}$	$\frac{d}{2}$
F	easy plane magnets $h_z\neq 0$	2	ψ	m	$\{\psi,m\}$	$\frac{d}{2}+\frac{\tilde{\alpha}}{2\nu}$
G	isotropic antiferro-magnets	3	N	m	$\{N,m\}$	$\frac{d}{2}$
J	isotropic ferro-magnets	3	none	S	$\{S,S\}$	$\frac{d+2-\eta}{2}$

3.4.8. Summary of the Universality Classes

In Table 2 the basic information about the dynamic universality classes is given. As shown [4] by the renormalization group theory, the addition of any number of nonconserved fields (which do not change the structure of the Poisson brackets) to the models specified in Table 2 does not change the critical dynamics in that sense that it does not change the critical exponents and other universal quantities. Sometimes it may be difficult to decide which dynamic universality class the real magnetic system belongs to. Many factors matter. For example in the real magnet also phonons contribute to the spin dynamics and model A with nonconserved energy may be a better description than model C. If however, the spin-lattice relaxation rate is low compared to the spin exchange frequency model C which is an idealization of thermally isolated spins is a better description of the system [48]. Moreover, in real spin systems there is always anisotropy. In this case one or more terms should be added to the Hamiltonian (40) and the crossover effects from the isotropic behavior ($n = 3$) to that described by anisotropic models ($n = 2$ or $n = 1$) should be studied. In such crossovers we generally observe the so-called effective[8] critical exponents [67, 68]. The universality classes which take into account the dipolar interactions are not included in Table 2.

4. Isothermal and Adiabatic Elastic Moduli

The sound velocity exhibits sharp dip near the critical temperature. Fig. 4 presents exemplary sound velocities for rare earth metals: Gd, Tb, Dy and Ho (Luthi et al. [60]). It is well known that the static isothermal and adiabatic elastic moduli are related to the corresponding sound velocities in the zero frequency limit. Let us assume that the elastic medium is

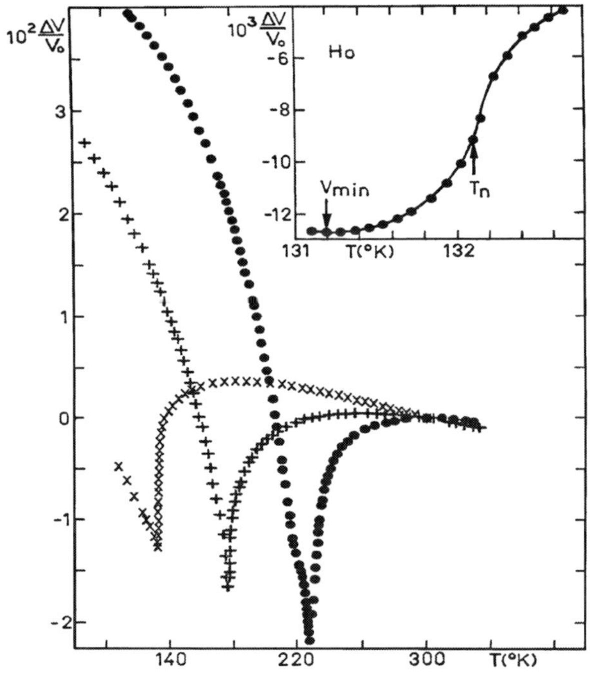

Figure 4. Temperature dependence of the sound velocity changes for rare earth metals. × Ho; + Dy; • Tb. The inset shows an expanded plot near T_N in Ho (Luthi at al. [60]).

isotropic and nondissipative. The equation of motion for an elastic wave has a simple form

[8] The effective exponent depends on the reduced temperature or magnetic field. It will be discussed in Section 6.

[61]

$$\rho_0 \ddot{\boldsymbol{u}} = (C_{11} - C_{44})\nabla(\nabla \cdot \boldsymbol{u}) + C_{44}\Delta \boldsymbol{u}, \tag{66}$$

where $\boldsymbol{u}$ is a local displacement vector, ρ_0 is the mass density of the system and C_{11} and C_{44} are elastic constants, Δ denotes the Laplacian and ∇ the Nabla operator. Decomposing $\boldsymbol{u}$ into a longitudinal part $\boldsymbol{u}_L$ for which $\nabla \times \boldsymbol{u}_L = 0$ and a transverse part $\boldsymbol{u}_T$ for which $(\nabla \cdot \boldsymbol{u}_T = 0)$, Eq. (66) splits into two independent wave equations:

$$\rho_0 \ddot{\boldsymbol{u}}_L = C_{11}\Delta \boldsymbol{u}_L, \qquad \rho_0 \ddot{\boldsymbol{u}}_T = C_{44}\Delta \boldsymbol{u}_T. \tag{67}$$

The solution of each equation is a planar wave $\boldsymbol{u} = \boldsymbol{u}_0 \exp i(\boldsymbol{k}\cdot\boldsymbol{x} - \omega t)$ with the wave vector $\boldsymbol{k}$ and the frequency ω related by the dispersion relation

$$\rho_0 \omega^2 = C_{eff} k^2, \tag{68}$$

where C_{eff} is an effective elastic constant for a given mode. The phase velocity $c = \omega/k$ is equal $\sqrt{C_{11}/\rho_0}$ for the longitudinal mode and $\sqrt{C_{44}/\rho_0}$ for the transverse modes. In the general case of anisotropic crystal, the sound velocity is given by a similar formula $c = \sqrt{C_{eff}/\rho_0}$ where the effective elastic constant is a linear combination of the elastic constants C_{ij}. We can take for C_{ij} both the isothermal as well as the adiabatic elastic constants depending on the conditions of propagation of the elastic mode. The adiabatic elastic constant is greater than the isothermal one. It is evident from the last equation that the static isothermal and adiabatic elastic moduli are related to the sound velocities in the zero frequency limit, so we can find the sound velocity singularities directly from the thermodynamics. In this section we will investigate the relations between adiabatic or isothermal moduli (or equivalently the sound velocities) and some correlation functions appearing in our model.

4.1. Model

All thermodynamic quantities can be obtained from the corresponding thermodynamic potential

$$F(T, P, h) = F^0(T, P, h) - \frac{k_B T}{V} \ln Z, \tag{69}$$

where T is temperature, P - pressure and h - an external magnetic field. $F^0(T, P, h)$ is the background part which is assumed to be smooth in the temperature and the magnetic field and

$$Z = \int \mathfrak{D}[S_\alpha, e_{\alpha\beta}, q] \exp(-H) \tag{70}$$

is the sum over the states which in our case is the sum over all paths $\{S_\alpha(\boldsymbol{x}), e_{\alpha\beta}(\boldsymbol{x}), q(\boldsymbol{x})\}$ which can be written as a functional integral. The fields $S_\alpha(\boldsymbol{x}), e_{\alpha\beta}(\boldsymbol{x})$ and $q(\boldsymbol{x})$ are the complete set of slow variables in our problem [55, 59]. In addition to the order parameter $S_\alpha(\boldsymbol{x})$, which for the magnetic phase transition is the local magnetization (or staggered one), we have the strain tensor: $e_{\alpha\beta}(\boldsymbol{x}) = \frac{1}{2}(\nabla_\alpha u_\beta + \nabla_\beta u_\alpha)$, connected with the displacement field $\boldsymbol{u}(\boldsymbol{x})$ [59] and the fluctuations of entropy per mass $q(\boldsymbol{x})$. The functional H determines the probability distribution of equilibrium fluctuations $p \propto \exp(-H)$ and for a magnetoelastic system of the Ising type $(n = 1)$ it can be written as

$$H = H_S + H_{el} + H_q + H_{int}, \tag{71}$$

where

$$H_S = \frac{1}{2}\int d^d x\{r_0 S^2 + (\nabla S)^2 + \frac{\widetilde{u}_o}{2} S^4\}$$

is the Landau-Ginzburg Hamiltonian for the order parameter fluctuations with $r_0 \propto T - T_c^0$, where T_c^0 is the mean field transition temperature. The elastic part

$$H_{el} = \frac{1}{2}\int d^d x\{C_{12}^0(\sum_\alpha e_{\alpha\alpha})^2 + 4C_{44}^0 \sum_{\alpha,\beta} e_{\alpha\beta}^2 + 2(P - P_0)e_{\alpha\alpha}\}$$

describes the elastic energy in the harmonic approximation (the bare elastic constants $C_{\alpha\beta}^0$ contain the factor $(k_B T)^{-1}$). We assume that the crystal is isotropic so only two elastic constants appear C_{12}^0 and C_{44}^0. P_0 is the pressure of a referential equilibrium state with respect to which we determine the strain. We assume that entropy fluctuations are Gaussian

$$H_q = \frac{1}{2C_V^0}\int d^d x q^2$$

where C_V^0 is bare specific heat. The last term in (71) describes the interactions

$$H_{int} = \int d^d x \left\{ g_0 \sum_\alpha e_{\alpha\alpha} S^2 + f_0 q S^2 + w_0 q \sum_\alpha e_{\alpha\alpha} \right\}$$

where the first term is the volume magnetostriction [6] with the coupling constant g_0. The second term is responsible for the divergence of the specific heat and last term mimics the mentioned coupling of sound mode to energy fluctuations proposed by Kawasaki [8].

The first step in analysis of such a system is the decomposition of a given elastic configuration into a uniform part and a phonon part which is a periodic function of the position [62]

$$e_{\alpha\beta}(x) = e_{\alpha\beta}^0 + \frac{1}{\sqrt{\rho_0 V}} \sum_{\boldsymbol{k} \neq \boldsymbol{0}, \lambda} k_\beta e_\alpha(\boldsymbol{k}, \lambda) Q_{\boldsymbol{k},\lambda} \exp(i\boldsymbol{k}\cdot\boldsymbol{x}), \tag{72}$$

where $Q_{\boldsymbol{k},\lambda}$ is the normal coordinate[9] of the sound mode with the polarization λ, wave vector $\boldsymbol{k}$ and the polarization vector $\mathbf{e}(\boldsymbol{k}, \lambda)$. $e_{\alpha\beta}^0$ is the uniform deformation. For simplicity we will assume that the mass density is equal unity. In the new variables the elastic Hamiltonian takes the form

$$H_{el} = H_{el}(e_{\alpha\beta}^0) + \frac{1}{2} \sum_{\boldsymbol{k} \neq \boldsymbol{0}, \lambda} k^2 c_0^2(\widehat{\boldsymbol{k}}, \lambda) \left|Q_{\boldsymbol{k},\lambda}\right|^2, \tag{73}$$

where $c_0(\widehat{\boldsymbol{k}}, \lambda)$ is the bare sound velocity for polarization λ and the versor $\widehat{\boldsymbol{k}}$. Analogously for the interaction Hamiltonian we obtain

$$H_{int} = H_{int}(e_{\alpha\beta}^0) + \sum_{\boldsymbol{k}} \left\{ f_0 q_{\boldsymbol{k}} S_{-\boldsymbol{k}}^2 + \sum_\lambda [\boldsymbol{k}\cdot\boldsymbol{e}(\boldsymbol{k},\lambda)] Q_{\boldsymbol{k},\lambda}(g_0 S_{-\boldsymbol{k}}^2 + w_0 q_{-\boldsymbol{k}}) \right\}, \tag{74}$$

[9] The factor i was incorporated into the variable Q.

where $S_{\boldsymbol{k}}^2 = \frac{1}{\sqrt{V}} \sum_{\boldsymbol{k}_1} S_{\boldsymbol{k}_1} S_{\boldsymbol{k}-\boldsymbol{k}_1}$ is the Fourier transform of the square of the order parameter. For the isotropic system only the longitudinal sound modes are coupled to the order parameter and the entropy fluctuations and $\boldsymbol{k} \cdot \boldsymbol{e}(\boldsymbol{k}, \lambda = L) = k$. As a result the transverse modes do not show any critical anomaly in this model. This is what one observes normally in experiment [60] at least for high-symmetry propagation directions. It is clearly mani-

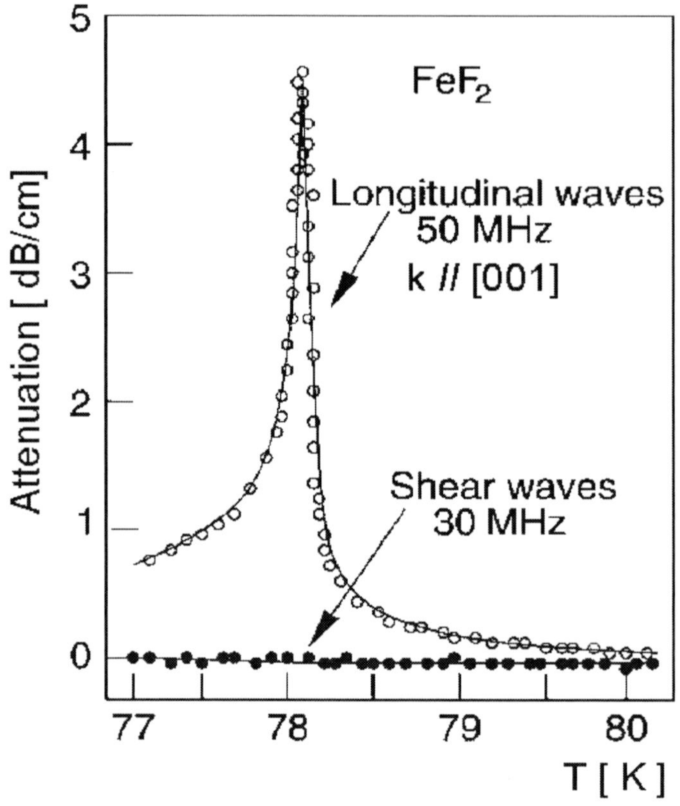

Figure 5. Ultrasonic attenuation of longitudinal and shear waves propagating along the tetragonal axis (symmetry direction) near the Neel temperature (Ikushima and Feigelson [119])

fested in Fig. 5 where the results for longitudinal and shear sound attenuation for FeF_2 in the vicinity of Neel temperature are shown (Ikushima and Feigelson [119]).

The next step is the integration over the homogenous deformations and the transverse modes

$$\exp[-H(S_{\boldsymbol{k}}, Q_{\boldsymbol{k},L}, q_{\boldsymbol{k}})] = \int \mathfrak{D}[e^0_{\alpha\beta}, Q_{\boldsymbol{k},T}] \exp[-H(S_{\boldsymbol{k}}, Q_{\boldsymbol{k},\lambda}, q_{\boldsymbol{k}}, e^0_{\alpha\beta}), \tag{75}$$

where index T refers to the transverse modes. From this point on we can forget about the transverse modes and homogenous deformations. The effect of homogenous deformations is only the renormalization of the parameters $\widetilde{u}$, f_0 and C_V^0 [72]. It leads in principle to a non-analyticity (with respect to the wave vector) in the couplings $u(k), f(k), C_V(k)$ but because in magnetic systems the coupling constants g and w are usually very small we can argue that the first order phase transition expected for the Ising systems with positive specific-heat exponent [63] can be seen only extremely close to the critical point and in the experimentally accessible temperature range the transition is continuous, which is in perfect agreement with experimental observations. The problem was thoroughly investigated in the 1970s and the reasonable conclusion is to neglect the additional contributions generated by the homogenous deformations [63–65, 68]. So our effective Hamiltonian expressed by the Fourier components of the fields is

$$H = \frac{1}{2}\sum_{\boldsymbol{k}} \left\{ (r_0 + k^2)\left|S_{\boldsymbol{k}}\right|^2 + k^2 c_{0L}^2 \left|Q_{\boldsymbol{k}}\right|^2 + \frac{1}{C_V^0}\left|q_{\boldsymbol{k}}\right|^2 \right\} + H_{int}, \tag{76}$$

with

$$\begin{aligned} H_{int} = w_0 \sum_{\boldsymbol{k}} k Q_{\boldsymbol{k}} q_{-\boldsymbol{k}} + \sum_{\boldsymbol{k},\boldsymbol{k}_1,\boldsymbol{k}_2} \left(f_0 q_{\boldsymbol{k}} + g_0 k Q_{\boldsymbol{k}}\right) S_{\boldsymbol{k}_1} S_{-\boldsymbol{k}-\boldsymbol{k}_1} + \\ + \frac{u_0}{2V} \sum_{\boldsymbol{k},\boldsymbol{k}_1,\boldsymbol{k}_2} S_{\boldsymbol{k}} S_{\boldsymbol{k}_1} S_{\boldsymbol{k}_2} S_{-\boldsymbol{k}-\boldsymbol{k}_1-\boldsymbol{k}_2} \end{aligned} \tag{77}$$

where $c_{0L} = \sqrt{C_{11}^0}$ is the bare sound velocity of the longitudinal modes and the normal coordinate Q refers only to the longitudinal modes.

4.2. Isothermal Sound Velocity

The isothermal elastic constant or equivalently the isothermal sound velocity of the longitudinal modes can be determined from the corresponding correlation function

$$\langle Q_{\boldsymbol{k}} Q_{-\boldsymbol{k}} \rangle = \frac{1}{c_{is}^2 k^2}, \tag{78}$$

where $k \neq 0$ is assumed. It is easy to calculate this correlation function by a simple separation of variables in the Hamiltonian

$$\begin{aligned} q_{\boldsymbol{k}} &= q'_{\boldsymbol{k}} - w_0 k C_V^0 Q_{\boldsymbol{k}} - f_0 C_V^0 (S^2)_{\boldsymbol{k}}, \\ Q_{\boldsymbol{k}} &= Q'_{\boldsymbol{k}} - (g_0 - w_0 f_0 C_V^0) c_{0r}^{-2} k^{-1} (S^2)_{\boldsymbol{k}}, \end{aligned} \tag{79}$$

where $c_{0r}^2 = c_0^2(1 - r^2)$ and $r^2 = w_0^2 C_V^0 c_0^{-2}$. In these new variables the Hamiltonian takes a form

$$H = \frac{1}{2}\sum_{\boldsymbol{k}} \left\{ k^2 c_{0r}^2 \left|Q'_{\boldsymbol{k}}\right|^2 + \frac{1}{C_V^0}\left|q'_{\boldsymbol{k}}\right|^2 \right\} + H_{eff}^T(S), \tag{80}$$

with the effective spin Hamiltonian of the Landau-Ginzburg form

$$H_{eff}^T(S) = \frac{1}{2}\left\{ \sum_{\boldsymbol{k}} (r_0 + k^2)\left|S_{\boldsymbol{k}}\right|^2 + \frac{u_0^T}{V} \sum_{\boldsymbol{k},\boldsymbol{k}_1,\boldsymbol{k}_2} S_{\boldsymbol{k}} S_{\boldsymbol{k}_1} S_{\boldsymbol{k}_2} S_{-\boldsymbol{k}-\boldsymbol{k}_1-\boldsymbol{k}_2} \right\}, \tag{81}$$

where $u_0^T = u_0 - v_T^{ph} - v_T^q$, $v_T^{ph} = \overline{g}^2 c_{0r}^{-2}$, $\overline{g}_0 = (g_0 - w_0 f_0 C_V^0)$ and $v_T^q = f_0^2 C_V^0$. The non-analyticity mentioned earlier is neglected here.

With such Hamiltonian we can write

$$\langle Q_{\boldsymbol{k}} Q_{-\boldsymbol{k}} \rangle = \langle Q'_{\boldsymbol{k}} Q'_{-\boldsymbol{k}} \rangle + v_T^{ph} c_{0r}^{-2} k^{-2} \langle S_{\boldsymbol{k}}^2 S_{-\boldsymbol{k}}^2 \rangle_{H_{eff}^T}. \tag{82}$$

or

$$c_{is}^2 = \frac{c_0^2(1 - r^2)}{1 + v_T^{ph} \langle S_{\boldsymbol{k}}^2 S_{-\boldsymbol{k}}^2 \rangle_{H_{eff}^T}}. \tag{83}$$

The index H_{eff}^T at the second average in Eq. (82) means that this average does not contain the elastic and entropic variables. The scaling behavior of $\langle S_{\boldsymbol{k}}^2 S_{-\boldsymbol{k}}^2 \rangle_{H_{eff}^T}$ is well known [11, 36, 68]. This function behaves as the specific-heat

$$\langle S_{\boldsymbol{k}}^2 S_{-\boldsymbol{k}}^2 \rangle_{H_{eff}^T(S)} \propto A t^{-\alpha} \Phi(k\xi) - B, \tag{84}$$

where A and B are the some nonuniversal constants and Φ is a scaling function (usually we assume that $\Phi(0) = 1$); ξ is the correlation length and t is the reduced temperature. In ultrasonic experiments the wavelength is much greater than the correlation length so we can take $k\xi = 0$. The specific-heat exponent α is positive for the Ising universality class and equal to about 0.11 so in this case the denominator in Eq. (83) tends to infinity and the isothermal sound velocity must go to zero

$$c_{is} \propto t^{\alpha/2} \searrow 0. \tag{85}$$

as we approach the critical temperature. We can say that the isothermal sound mode is softening at the critical point of Ising type systems. Otherwise, for the Heisenberg universality class $n = 3$, we have $\alpha < 0$ and the isothermal sound velocity stays finite at T_c.

The experimental observation of the relation (85) is extremely difficult for the two reasons. The first is that the critical exponent of the sound velocity, $\alpha/2$, is very small of an order of 0.05 and we must be very close to the critical temperature in order to observe a significant changes in the sound velocity. The second is that the coupling constant $v_T^{ph} = (g_0 - w_0 f_0 C_V^0)^2 c_{0r}^{-2}$, which precedes the singular term in the denominator of Eq.(83), depends on the coupling constants g_0 and w_0 which are usually very small in magnets (contrary to e.g. the structural phase transitions [53]). Because v_T^{ph} is a small quantity it is reasonable to expand the expression (83) obtaining

$$c_{is}^2 \simeq c_{0r}^2 - A_T t^{-\alpha}. \tag{86}$$

This expression is a very good approximation to the experimentally observed measurements of isothermal sound velocity [7, 60, 115] (isothermal elastic moduli C_{11}^T). The expressions (85) and (86) were given independently by Dengler and Schwabl [70] and by the author [68, 71].

4.3. Adiabatic Sound Velocity

Another static quantity of interest is the adiabatic elastic constant. In our case it is the modulus C_{11}^{ad} or the related quantity c_{ad}. From the theory of fluctuations [30] we know that the adiabatic compressibility is given by the correlation function of pressure fluctuations.

The pressure fluctuations are defined as the quantity which is orthogonal to the entropy fluctuations. The orthogonality is understood as vanishing of the corresponding correlation function $\langle \mathcal{P}_{\boldsymbol{k}} q_{-\boldsymbol{k}} \rangle = 0$ (where $\mathcal{P}_{\boldsymbol{k}}$ is a fluctuation of pressure). Looking at the Hamiltonian (76) we see that the variables $Q_{\boldsymbol{k}}$ and the entropy fluctuations $q_{\boldsymbol{k}}$ are not orthogonal as a result of the coupling between these quantities in H_{int}. In order to get a quantity which is orthogonal to $q_{\boldsymbol{k}}$ we have to perform the Schmidt orthogonalization procedure (choosing as the first variable the entropy fluctuations)

$$\mathcal{P}_{\boldsymbol{k}} = Q_{\boldsymbol{k}} - \langle Q_{\boldsymbol{k}} q_{-\boldsymbol{k}} \rangle \frac{q_{\boldsymbol{k}}}{\langle q_{\boldsymbol{k}} q_{-\boldsymbol{k}} \rangle}. \tag{87}$$

or in other words we must subtract from the acoustic variable $Q_{\boldsymbol{k}}$ a part linear in $q_{\boldsymbol{k}}$. Immediately we get that

$$\langle \mathcal{P}_{\boldsymbol{k}} \mathcal{P}_{-\boldsymbol{k}} \rangle = \frac{1}{c_{ad}^2 k^2} = \langle Q_{\boldsymbol{k}} Q_{-\boldsymbol{k}} \rangle - \frac{\langle Q_{\boldsymbol{k}} q_{-\boldsymbol{k}} \rangle^2}{\langle q_{\boldsymbol{k}} q_{-\boldsymbol{k}} \rangle}. \tag{88}$$

By a shift of variables $Q_{\boldsymbol{k}}$ and $q_{\boldsymbol{k}}$ we can separate these variables in the Hamiltonian obtaining

$$H = \frac{1}{2} \sum_{\boldsymbol{k}} \left\{ k^2 c_0^2 \left| Q_{\boldsymbol{k}}'' \right|^2 + \frac{1}{\overline{C}_V^0} \left| q_{\boldsymbol{k}}'' \right|^2 \right\} + H_{eff}^{ad}(S),$$

where q'' and $Q_{\boldsymbol{k}}''$ are the shifted variables, $\overline{C}_V^0 = C_V^0 (1 - r^2)^{-1}$ and

$$H_{eff}^{ad}(S) = \frac{1}{2} \left\{ \sum_{\boldsymbol{k}} (r_0 + k^2) \left| S_{\boldsymbol{k}} \right|^2 + \frac{u_0^{ad}}{V} \sum_{\boldsymbol{k}, \boldsymbol{k}_1, \boldsymbol{k}_2} S_{\boldsymbol{k}} S_{\boldsymbol{k}_1} S_{\boldsymbol{k}_2} S_{-\boldsymbol{k}-\boldsymbol{k}_1-\boldsymbol{k}_2} \right\}, \tag{89}$$

is effective adiabatic Hamiltonian with $u_0^{ad} = \widetilde{u}_0 - v_{ad}^{ph} - v_{ad}^q$, $v_{ad}^{ph} = g_0{}^2 c_0^{-2}$ and $v_{ad}^q = \overline{f}_0^2 \overline{C}_V^0$, where $f_0 = (f_0 - w_0 g_0 c_0^{-2})$.

Now we can find the correlation functions in (88). A simple algebra shows that

$$c_{ad}^2 = c_0^2 \frac{1 + v_{ad}^q \langle S_{\boldsymbol{k}}^2 S_{-\boldsymbol{k}}^2 \rangle_{H_{eff}^{ad}}}{1 + v_+^{ad} \langle S_{\boldsymbol{k}}^2 S_{-\boldsymbol{k}}^2 \rangle_{H_{eff}^{ad}}}, \tag{90}$$

where $v_+^{ad} = v_{ad}^q + v_{ad}^{ph}$.

Straightforward calculations show that

$$v_{ad}^{ph} + v_{ad}^q = v_T^{ph} + v_T^q \equiv v_+, \tag{91}$$

therefore $u_0^T = u_0^{ad} \equiv u_0$ and the effective spin Hamiltonians H_{eff}^{ad} and H_{eff}^T and identical. As a consequence, the correlation function $\langle S_{\boldsymbol{k}}^2 S_{-\boldsymbol{k}}^2 \rangle_{H_{eff}^{ad}}$ is identical with the function $\langle S_{\boldsymbol{k}}^2 S_{-\boldsymbol{k}}^2 \rangle_{H_{eff}^T}$ which shows a specific-heat singularity as was discussed earlier (84). The expression (90) can be given in more transparent form

$$c_{ad}^2 = c_0^2 (1 - \frac{v_{ad}^{ph}}{v_+}) + c_0^2 \frac{v_{ad}^{ph}}{v_+} \frac{1}{1 + v_+ \langle S_{\boldsymbol{k}}^2 S_{-\boldsymbol{k}}^2 \rangle}, \tag{92}$$

where we have omitted the Hamiltonian index. It is seen from this equation that there is a constant term and a correction which tends to zero at T_c

$$c_{ad}^2 = c_0^2(1 - \frac{v_{ad}^{ph}}{v_+}) + A_{ad}t^{\alpha}. \tag{93}$$

The critical amplitude of the singular term $A_{ad} = c_0^2 \frac{v_{ad}^{ph}}{v_+^2} A^{-1}$, where A is the critical amplitude of the specific heat from Eq. (84), is very small for magnets because $v_{ad}^{ph} \ll v_{ad}^q \simeq v_+$, which explains small sound velocity changes near the magnetic phase transition [6].

It should be noted that this result for the adiabatic sound velocity obtained by the author [72] differs from that obtained by Drossel and Schwabl [73] who obtained for the adiabatic sound velocity a result similar to Eq. (83) for the isothermal velocity (only v_T^{ph} should be replaced by v_{ad}^{ph} in this equation). The reason for this discrepancy is a different choice of the pressure variable. Drossel and Schwabl took for the pressure a variable which is orthogonal to entropy only in the Gaussian approximation and it leads to non-vanishing correlation function of pressure-entropy. The correlation function of such „pressure" containing a non-zero entropy component is similar to that obtained for isothermal sound.

On the other hand, the expression (92) shows a close analogy to the adiabatic sound velocity in liquid He^4, obtained by Pankert and Dohm [74, 75]. A similar result was also obtained by Folk and Moser [76] for binary liquids.

5. Phenomenological Theory of Sound Attenuation and Dispersion

The sound wave propagation through the medium disturbs the existing balance as a consequence of the temperature (or pressure) changes in a wave of successive compressions and dilatations. Let the molecular equilibrium of the system be described by a parameter ψ called the reaction coordinate (or the degree of advance) [78] which can correspond to the extent of the chemical reaction or to the temperature of some internal degrees of freedom. For gases such internal degrees of freedom are the rotational or vibrational modes of many-atomic molecules. The parameter ψ does not follow the temperature and pressure changes and this delay is described by the relaxation equation

$$\dot{\psi} = -\frac{\psi - \bar{\psi}}{\tau}, \tag{94}$$

where τ is the relaxation time characterizing the rate at which the coordinate y approaches the equilibrium value $\bar{\psi}(p, T)$ determined by temporary pressure and temperature in the ultrasonic wave. The lag between the oscillations of the temperature and pressure, and the excitation of a given mode leads to the dynamic hysteresis, to dissipation of energy and dispersion of the sound wave.

The Eq. (94) is the simplest equation of the irreversible thermodynamics. Historically the method of irreversible thermodynamics was first applied to the sound dynamics by Herzfeld and Rice [79] in 1928. They postulated that the medium through which the sound waves propagates is characterized by two temperatures: one of them is called the external temperature and determines the energy distribution of translational degrees of freedom of molecules and the other one is the internal temperature connected with the energy distribution in internal degrees of freedom (e.g. the rotational or vibrational modes in a gas of many-atom molecules). Herzfeld and Rice assumed that the rate at which the internal temperature changes is proportional to the difference between these temperatures and the coefficient of proportionality is the inverse of the relaxation time. They noticed that each process in which the energy is transferred with some delay from translational motion (the sound wave) to other (internal) degrees of freedom, is connected with a dissipation of acoustic energy or in other words to the attenuation of the sound wave. As a result we obtain a complex effective elastic constant (and sound velocity) in the dispersion relation $\hat{C}_{eff}$, where for the single relaxational process we obtain

$$\hat{C}_{eff} = C_{eff}^{\infty} - \frac{\Delta'}{1 - i\omega\tau}, \tag{95}$$

The constant C_{eff}^{∞} is the high frequency limit of (95), where the reaction coordinate does not follow the stress changes. The symbol τ stands for the relaxation time and Δ' is a parameter describing the coupling of the sound to the relaxing variable known as the relaxation strength. In the ultrasonic experiments the sound frequency is a real-valued quantity and for the propagation wave vector we assume a complex value $k = k_r + i\alpha$, where α is the sound attenuation coefficient. A single relaxational process results in a frequency dependent sound velocity

$$c^2(\omega) = \frac{\omega^2}{k_r^2} = c_{\infty}^2 - \frac{\Delta}{1 + \omega^2\tau^2} = c^2(0) + \frac{\Delta\omega^2\tau^2}{1 + \omega^2\tau^2} \tag{96}$$

and the sound attenuation

$$\alpha(\omega) = \frac{\Delta}{2c_{\infty}^3}\frac{\omega^2\tau}{1 + \omega^2\tau^2} = B\frac{\omega^2\tau}{1 + \omega^2\tau^2}, \tag{97}$$

where $c_{\infty} = \sqrt{C_{eff}^{\infty}/\rho_0}$ is the high-frequency limit of (96) and $\Delta = \Delta'/\rho_0$ and $B = \Delta/2c_{\infty}^3$. It was assumed that the frequency dependence of the sound velocity is weak.

Figure (6) shows these dependencies in the logarithmic scale for frequency. It should be noted that the sound velocity increases from the value $c(0)$ for $\omega\tau = 0$ to the value c_{∞} for $\omega\tau \to \infty$. The velocity $c(0)$ corresponds to the situation when the temperature and the pressure of the sound wave change so slowly that system remains at the thermodynamic equilibrium (the reaction coordinate has the same phase as the pressure applied). The sound attenuation coefficient $\alpha(\omega)$ increases from zero for low frequencies to the „saturation" value B/τ for very high frequencies. In the low-frequency regime the attenuation coefficient is proportional to the square of frequency and to the relaxation time: $\alpha(\omega) = B\omega^2\tau$.

For many relaxational processes with the relaxation times τ_j and the relaxation strengths Δ_j the equations (96) and (97) change into

$$c^2(\omega) = c_{\infty}^2 - \sum_j \frac{\Delta_j}{1 + \omega^2\tau_j^2} \tag{98}$$

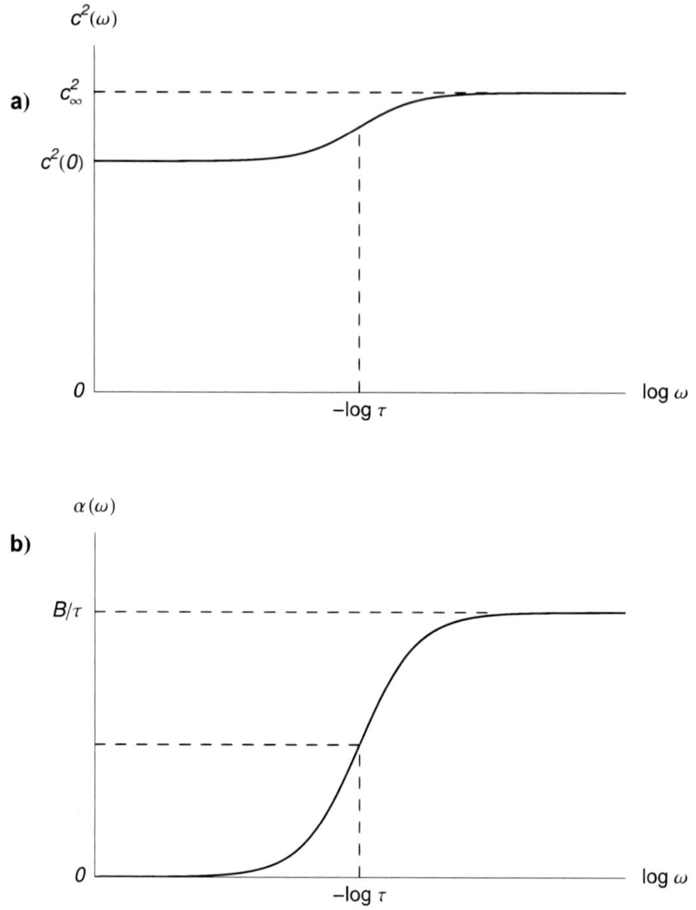

Figure 6. The sound velocity (a) and the attenuation coefficient (b) for a single relaxational process; $c^2(0) = c^2_\infty - \Delta, \quad \alpha(\infty) = B/\tau.$

and

$$\alpha(\omega) = \omega^2 \sum_j \frac{B_j \tau_j}{1 + \omega^2 \tau_j^2}. \tag{99}$$

In Fig. 7 the dependences described by Eqs. (98) and (99) are shown for two relaxational processes with the relaxation times τ_1 and τ_2. In the classical theory, the relaxational processes do not interact and if the relaxation times are well separated from each other one can see something like a staircase (with slightly rounded stairs). The height of the j-th stair for the sound velocity is Δ_j, and $\Delta_j / 2c^3_\infty \tau_j$ for the sound attenuation.

At the magnetic phase transition we have a quasi continuum[10] of the relaxation times. They are attributed to the internal degrees of freedom which are the Fourier components of the order parameter[11]. Their relaxation times are typically i.e. far from the critical point,

[10]The index j at τ_j in the case of phase transitions denotes the wave vector which is a quasi continuous variable for the finite volume of the crystal.

[11]To be precise they are usually the Fourier components of the square of the order parameter field for the magnetostrictive coupling.

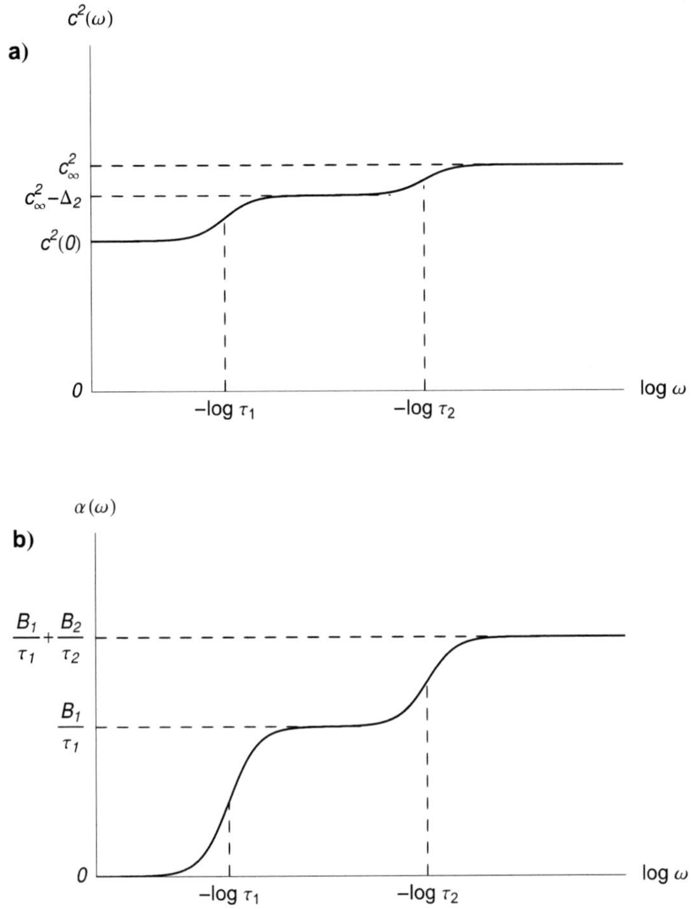

Figure 7. The sound velocity (a) and the attenuation coefficient (b) for two relaxational processes with the relaxation times τ_1 and τ_2.

very short of the order of 10^{-12} s, so their inverses are much lower than the ultrasound frequencies used in the study of the acoustic properties of solids. They are typically in the range from 1 MHz to 1000 MHz. Due to the critical slowing down the relaxation times are getting longer and longer and some of them may become comparable with the period of the ultrasonic wave. The longest of the relaxation times may diverge even to infinity. It is illustrated in Fig. (7). Also the relaxation strengths increase when approaching the critical point. It is easily seen if we consider the contribution to the acoustic linear response function in the Gaussian[12] approximation for the order parameter fluctuations. In this approximation the complex sound velocity has a simple form [80]

$$\hat{c}^2(\omega) = c_\infty^2 - g^2 \int \frac{d^3p}{(\xi^{-2}+p^2)^2[1-i\omega\tau(p)]}, \tag{100}$$

where $\tau(p) = [2\Gamma(\xi^{-2}+p^2)]^{-1}$ is a relaxation time of the product of two Fourier

[12]The Gaussian approximation assumes that the Fourier components of the order parameter do not interact

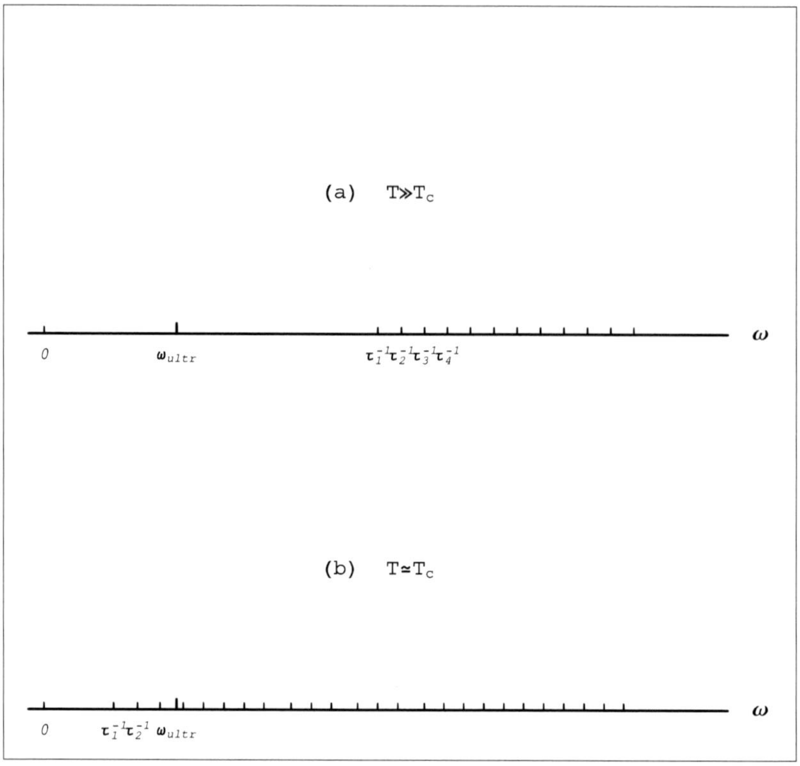

Figure 8. The effect of the critical slowing down on the relaxation times of the order parameter.

components of the order parameter $S_{\boldsymbol{p}}S_{-\boldsymbol{p}}$, and the integration is over the wave vectors inside a sphere $|\boldsymbol{p}| \leq \Lambda$. The coefficients g^2 and Γ are some constants and Λ is a cutoff wave-vector. $\xi^{-2} \propto T - T_c$ is the correlation length in the Gaussian approximation. It is evident that the summation over j in Eqs. (98) and (99) is replaced by the integration over the wave vector $\boldsymbol{p}$ in Eq. (100). The relaxation strength Δ_j corresponds to $(\xi^{-2} + p^2)^{-2}$ which is a contribution of the mode $S_{\boldsymbol{p}}$ to the specific heat in the Gaussian approximation [35]. The correspondence $\tau_j \longleftrightarrow \tau(p)$ is also obvious so Eq. (100) is a continuous version of Eqs. (98) and (99) written in the complex form. From the expression for $\tau(p)$ we find that the relaxation times for small wave vectors will diverge roughly as ξ^2 when $T \rightarrow T_c$. It is a quite strong divergence which is the principal cause of the divergence of the attenuation coefficient in the so-called hydrodynamic regime i.e. for $\omega\tau(p) \ll 1$. In the hydrodynamic regime we have $\alpha(\omega) \propto \int_{\boldsymbol{p}} \Delta(p)\tau(p)$ and the sound attenuation strongly increases when we come near the critical temperature. It is visualized in Fig. 9. Very far from the critical temperature all modes are in the hydrodynamic regime, as illustrated in Fig. 9a and the attenuation is very small. As we approach the critical temperature, the attenuation starts to increase due to an increase in $\tau(p)$ and $\Delta(p)$. Only when the slowest modes' relaxation times become longer than the period of the sound wave or in other words the inverses of the relaxation times „get across" to the left of the ultrasonic frequency, only then the

and there is only the first term in the spin Hamiltonian (89).

slowest modes with $\omega_{\rm ultr}\tau(p) > 1$ get into the „saturated" state and the curve of the sound attenuation coefficient starts to level off. It is illustrated in Fig. 9b where for the sake of the figure transparency, only the frequencies close to the ultrasonic frequency are shown. Not all the relaxational modes could get across $\omega_{\rm ultr}$, because the relaxation times $\tau(p)$ depend not only on temperature (through ξ^2) but also on the square of the wave vector which does not change for $T \to T_c$. For $p^2 > \xi^{-2}$ the relaxation time is not very sensitive to temperature changes, similarly as the relaxation strength $\Delta(p)$. The crossover of the slowest modes (and simultaneously those giving the largest contribution to $\alpha(\omega, T)$) into the saturated state and the smaller sensibility (to the temperature change) of the other modes leads ultimately to the levelling off the sound attenuation coefficient.

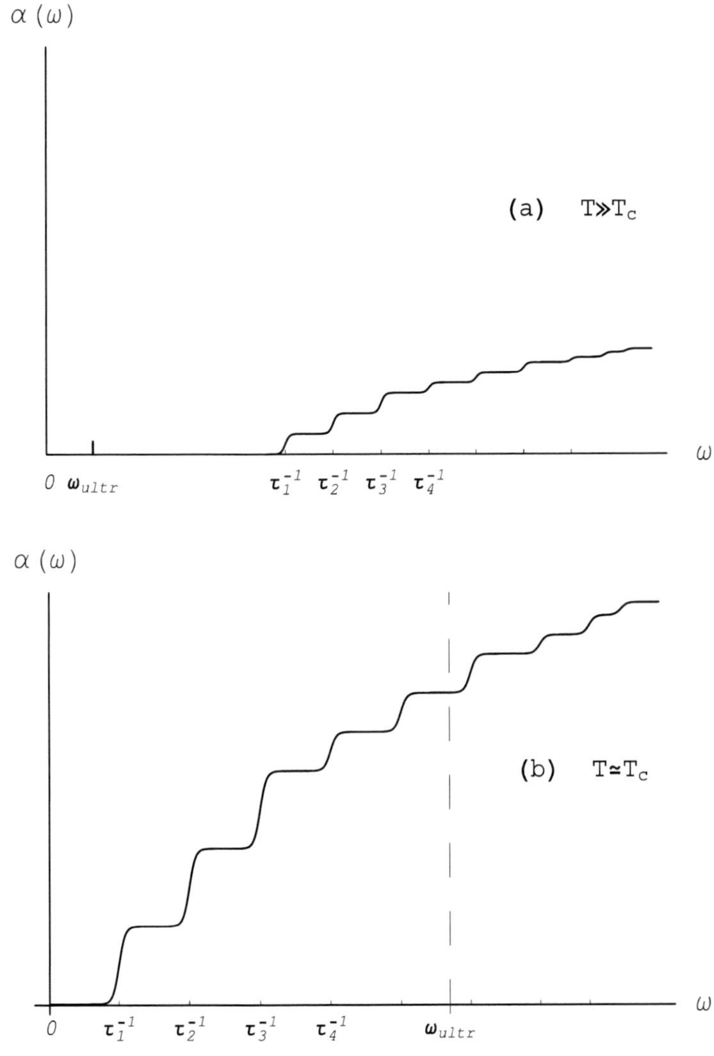

Figure 9. It demonstrates in intuitive way how the sound attenuation changes when approaching the critical temperature. In Fig. 9(b) the frequency was rescaled in order to show more precisely the situation around the ultrasonic frequency.

For the sound velocity in the hydrodynamic regime we have $c^2(\omega) = c_\infty^2 - \int_p \Delta(p)$. Thus we observe a weaker singularity as the integral $\int_p \Delta(p)$ is less singular than the corresponding integral $\int_p \Delta(p)\tau(p)$ for the sound attenuation coefficient. It is well known that it is a singularity of the specific heat type. As shown in Fig. 10, some of the relaxational modes get into the saturated state as the critical temperature is approached. At the first sight it would seem that the sound velocity should increase (for a given frequency) as we get closer T_c because more and more „stairs" (contributions) from saturated modes are added to $c^2(0)$. It is however, not so as equation (96) only describes the way the sound velocity

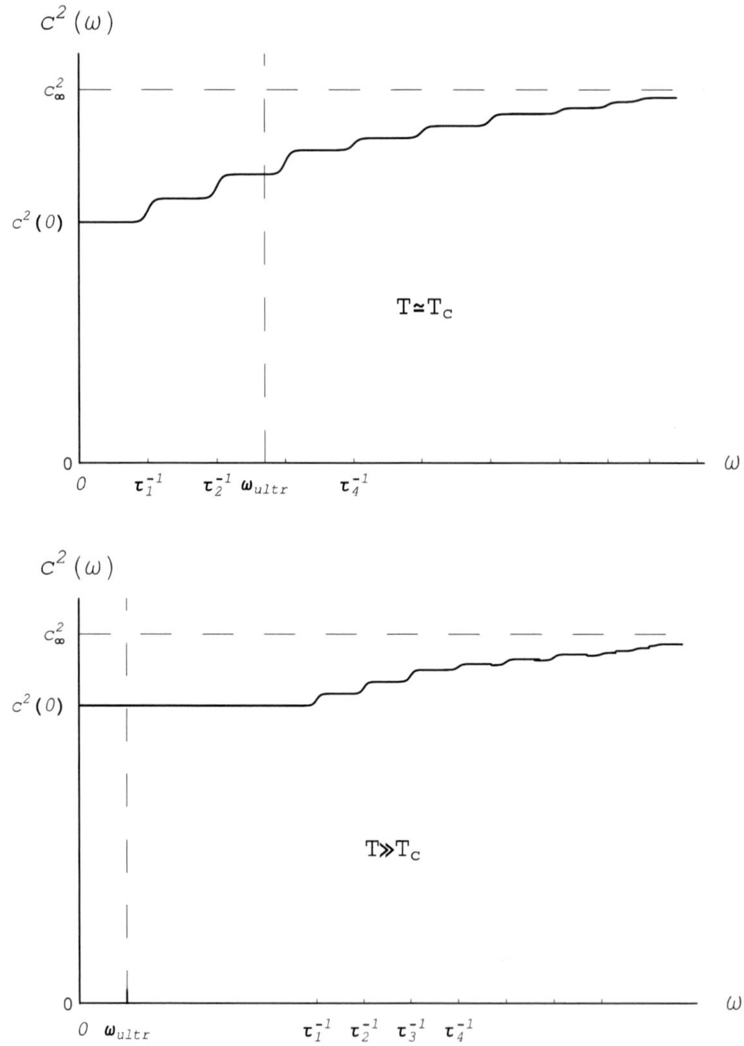

Figure 10. The sound velocity changes when approaching the critical temperature.

changes relatively to $c^2(0)$ or c_∞^2. It gives no clue about the temperature dependence of these parameters near T_c. We know from the previous section that the sound velocity in the limit of zero frequency is a thermodynamic quantity which is singular near the critical point so it is not a good reference level to measure the critical sound velocity changes. Such a good reference level turns out to be c_∞^2. It is a quantity corresponding to the infinite frequency so it is measured far from the critical point which corresponds to $\xi^{-1} = k = \omega = 0$ as we know from the phase transition theory. Such uncritical quantity does not depend strongly on temperature and can to a good approximation be taken as a constant near T_c. Having decided which quantity c_∞^2 or $c^2(0)$ is a constant near the phase transition we see that the sound velocity can only decrease with $T \to T_c$. Of course this decrease is eventually stopped by the crossover of the slowest modes to the saturated state like for the sound attenuation. The higher the sound frequency the larger is the sound velocity as more modes will get across this frequency. It should be also noted that the difference $c_\infty^2 - c^2(0)$ increases as $T \to T_c$ because the sum of the relaxational strengths increases as the specific heat.

The presented here phenomenological theory of sound attenuation and dispersion has only a qualitative character as it does not take into account the interaction between the relaxational processes (modes). It is well known [3, 4] that these very interactions between the modes are the origin of the nontrivial singularities encountered in many physical quantities near the critical point. We need a more elaborated theory of the dynamic phenomena to obtain a precise form of the sound attenuation coefficient and dispersion. It will be presented in the next section.

6. Model of Critical Sound Propagation

In order to build a detailed theory of sound propagation near the critical point we need the equations of motion. We use the phenomenological hydrodynamic description in terms of nonlinear Langevin equations for slow variables described in Sect. 3. Our choice of the slow variables depends mainly on the nature of the physical system we want to describe (e.g. whether it is a ferro- or antiferromagnet, isotropic or anisotropic system, etc.) and on the quantity of interest. It depends additionally on the frequency range (the time scale used to investigate the system) and in some cases we should add also some fast variables important in the time scale considered to the system of slow variables. In practice in each universality class we would need a different system of equations. Fortunately, some of the aspects of critical attenuation and dispersion presented in this chapter can be easily generalized over other magnetic dynamic universality classes.

6.1. Anisotropic Magnet with the Spin-Lattice Relaxation

As mentioned in the Introduction, in magnetic materials which are also insulators we observe a weak divergence of the sound attenuation coefficient. The critical exponent characterizing this initial increase (in the hydrodynamic region) is usually very small of an order of 0.2. It was postulated by Kawasaki [8] that the origin of such a weak divergence (or even the lack of any singularity) is connected with the linear coupling of the sound mode with the spin energy density which decays through spin-lattice relaxation process. With such assumptions the sound attenuation can be written as a quantity proportional to the square of the specific heat

$$\alpha(\omega, T) \propto \omega^2 C^2 \propto \omega^2 t^{-2\alpha}.$$

The other group of magnets, these which are conductors of the electric current such as for example the rare earth metals, we observe a much stronger increase in $\alpha(\omega, T)$ for $T \to T_c$, with the sound attenuation critical exponent $\rho_s > 1$ [81–83]. It was recognized [60] that for such a strong singularity another coupling which involves one acoustic mode and two fluctuations of the order parameter is responsible. As we noted in the Introduction, both types of the coupling arise from the dependence of the exchange integral on the distance between the interacting magnetic ions (we call this general interaction a volume magnetostriction) and both are present in all magnets. Only when the interactions with the next nearest neighbors are neglected and only for some sound propagation directions the other part of the coupling typical of metals can be neglected. Therefore, a general model which involves both couplings was proposed [9]. The Hamiltonian comprising all important interactions is of the form Eq. (71) and the dynamics is given by the Langevin equations

$$\dot{S}_{\boldsymbol{k}} = -\Gamma \frac{\delta H}{\delta S_{-\boldsymbol{k}}} + \zeta_{\boldsymbol{k}}, \tag{101}$$

$$\ddot{Q}_{\boldsymbol{k}} = -\frac{\delta H}{\delta Q_{-\boldsymbol{k}}} - \Theta k^2 \dot{Q}_{\boldsymbol{k}} + \xi_{\boldsymbol{k}}, \tag{102}$$

$$\dot{q}_{\boldsymbol{k}} = -\gamma \frac{\delta H}{\delta q_{-\boldsymbol{k}}} + \varphi_{\boldsymbol{k}}, \tag{103}$$

where $S_{\boldsymbol{k}}$, Q_k and $q_{\boldsymbol{k}}$ denote similarly as in Sect. 4 the Fourier components of the order parameter, longitudinal acoustic mode and entropy. $\zeta_{\boldsymbol{k}}$, $\xi_{\boldsymbol{k}}$ and $\varphi_{\boldsymbol{k}}$ are white Gaussian noises simulating the thermal agitation forces. They have zero means and their variances are connected with the bare damping terms by relations

$$\left\langle \zeta_{\boldsymbol{k}}(t) \zeta_{-\boldsymbol{k}}(t') \right\rangle = 2\Gamma \delta(t - t'), \tag{104}$$

$$\left\langle \xi_{\boldsymbol{k}}(t) \xi_{-\boldsymbol{k}}(t') \right\rangle = 2\Theta k^2 \delta(t - t'), \tag{105}$$

$$\left\langle \varphi_{\boldsymbol{k}}(t) \varphi_{-\boldsymbol{k}}(t') \right\rangle = 2\gamma \delta(t - t'), \tag{106}$$

where now t denotes a time not the reduced temperature[13]. Equation (101) describes the relaxation of the order parameter fluctuation which is a non-conserved quantity and this equation corresponds to the model A dynamic universality class [4, 48, 49]. This class comprises e.g. the anisotropic magnets. Eq. (102) describes the longitudinal sound mode and the damping coefficient Θk^2 is responsible for all other interactions of the sound mode except those with the long-wavelength fluctuations of the order parameter. The last equation (103) characterizes the non-conserved quantity $q(\boldsymbol{x})$ which decays by the spin-lattice relaxation. The rate of the spin-lattice relaxation is γ / C_V^0. In this equation the heat conduction process is neglected. It describes an idealized situation when the lattice is characterized by infinite thermal capacity or infinite thermal conductivity. The problem of the heat conduction will be touched in Sect. 6.2. It can be shown that Eqs. (102) and (103) do not change the dynamic critical exponent z which corresponds to the model A universality class $z = 2 + c\eta$.

[13]Unfortunately, it is well established manner to denote time and the reduced temperature by the same symbol t but it is easy to find out from the context whether t refers to time or to the temperature. We will need the first meaning only in Sect. 6.1 for the definition of the dynamic model and in Sect. 6.2 in formulation of the functional representation of the equations of motion.

6.1.1. Functional Form of the Equations of Motion

There are a few ways of constructing the perturbation expansion for the dynamic model [44, 45, 85, 86]. One of the most frequently used is the path-integral formalism called also the functional form of the equations of motion which uses the dynamic functional known also as Lagrangian or Onsager-Machlup functional [85]. This functional determines the probability of the whole trajectory $\{S_{\boldsymbol{k}}, Q_k, q_{\boldsymbol{k}}\}_{t\in[-t_0,t_0]}$ in some time interval. Let us forget for a moment about the fields Q and q and the indexes $\boldsymbol{k}$. The stochastic process e.g. $\zeta(t)$ is characterized by the set of the probability densities

$$P_1(\zeta_1, t_1),$$

$$P_2(\zeta_1, t_1; \zeta_2, t_2),$$

$$\dots\dots\dots\dots\dots$$

$$P_n(\zeta_1, t_1; \zeta_2, t_2; \dots; \zeta_n, t_n),$$

with P_i having the sense of the probability density to find the system in the state ζ_1 at the time moment t_1, ζ_2 at time t_2 etc. The probability density $P_\infty[\zeta]$ for the whole trajectory $\{\zeta(t)\}_{t\in[-t_0,t_0]}$ is obtained in the limit where the differences between the successive times go to zero. It is a function of the infinite number of variables $\zeta(t)$, for all times in the interval $[-t_0, t_0]$. For the Gaussian noise it has the form

$$P_\infty[\zeta]\mathcal{D}\zeta = \frac{1}{Z}\exp\left\{-\int_{-t_0}^{t_0} dt\left[\frac{\zeta^2(t)}{4\Gamma}\right]\right\}\mathcal{D}\zeta \equiv \tag{107}$$

$$\equiv \frac{1}{Z}\lim_{\Delta t\to 0}\exp\left\{-\sum_{\sigma}\Delta t\left[\frac{\zeta_\sigma^2}{4\Gamma}\right]\right\}\prod_{\sigma=0}^{N}(\frac{\Delta t}{4\Gamma})^{\frac{1}{2}}d\zeta_\sigma,$$

where the time interval was divided into $N = 2t_0/\Delta t$ sub-intervals, and Z is a normalization factor. The second part of this expression defines the functional differential $\mathcal{D}\zeta$. We can interpret $P_\infty[\zeta]\mathcal{D}\zeta$ as the probability density of a given trajectory passing through the infinite set of time windows at times $t_\sigma = -t_0 + \sigma\Delta t$ of the width $d\zeta_\sigma$.

If we want to determine the time-dependent correlation function which depends on S not on the noise, it is favorable to have the functional $P_\infty[S]$, instead of $P_\infty[\zeta]$. The transformation which allow us to change the variables in $P_\infty[S]$ is the equation of motion (101) which in the discrete form is given by

$$\frac{1}{\Delta t}(S_\sigma - S_{\sigma-1}) + \Gamma\frac{\delta H}{\delta S_\sigma} = \zeta_\sigma. \tag{108}$$

By eliminating ζ_σ we obtain

$$P_\infty[S] = \frac{1}{Z}\exp\left\{-\int_{-t_0}^{t_0} dt\left[\frac{\left|\dot{S} + \Gamma\frac{\delta H}{\delta S}\right|^2}{4\Gamma}\right]\right\}, \tag{109}$$

where the Jacobian of the transformation, $\frac{1}{2}\Gamma\frac{\delta^2 H}{\delta S^2}$, was omitted as it is compensated in the perturbation expansion by some acausal terms which are also usually omitted [44, 85, 86]. The exponent in (109) is known as the Onsager-Machlup functional. Its form is not very useful for the two reasons. The first is that it involves the interactions of the high order and the second is that for the conserved quantities we have $\Gamma \to Dk^2$ so the functional becomes infinite for $k \to 0$. In order to avoid such problems we perform a Gaussian transformation

$$P_\infty[S] = \frac{1}{Z}\int \mathcal{D}[i\tilde{S}] \exp \mathcal{J}\{\tilde{S}, S\}, \tag{110}$$

with

$$\mathcal{J}\{\tilde{S}, S\} = \int_{-t_0}^{t_0} dt\, \mathcal{L}(\tilde{S}, S) = \int_{-t_0}^{t_0} dt \left[\tilde{S}\Gamma\tilde{S} - \tilde{S}\left(\dot{S} + \Gamma\frac{\delta H}{\delta S}\right)\right], \tag{111}$$

where a new functional $\mathcal{L}(\tilde{S}, S)$ known as the dynamic functional or Lagrangian (by analogy to the quantum field theory) is introduced. It is a function of an artificial imaginary field $\tilde{S}$ known as the response field because an additional term related to the external magnetic field hS in the Hamiltonian, gives the contribution $\Gamma h\tilde{S}$ in the Lagrangian and the linear response function which is defined as the derivative $\langle S(t)\rangle$ over the external field $h(t')$ is given by

$$\frac{\delta\langle S(t)\rangle}{\delta h(t')} = \Gamma\langle S(t)\tilde{S}(t')\rangle. \tag{112}$$

The angular brackets denote nonequilibrium average

$$\langle O(S, \tilde{S})\rangle = \frac{1}{Z}\int \mathcal{D}[i\tilde{S}]\mathcal{D}[S]\; O[S, \tilde{S}]\; \exp \mathcal{J}\{\tilde{S}, S\}. \tag{113}$$

In this formalism all the correlation and response functions are obtained as path integrals weighted with the favorite exponential factor $\exp(\mathcal{J})$ which permits expressing the dynamics in a form analogous to that in statics [44, 45]. The dynamic diagram technique is a simple generalization of the diagram technique used in statics. We have two kinds of propagators. In the Fourier representation the free response propagator $G_0^S(k, \omega)$ has the form

$$\Gamma\langle S_{\boldsymbol{k},\omega}\tilde{S}_{-\boldsymbol{k}',\omega'}\rangle_0 = \delta_{\boldsymbol{k},\boldsymbol{k}'}\delta(\omega+\omega')\frac{\Gamma}{-i\omega + \Gamma\left(r_0 + k^2\right)} = \delta_{\boldsymbol{k},\boldsymbol{k}'}\delta(\omega+\omega')G_0^S(k,\omega), \tag{114}$$

and the free two-point correlation function $K_0^S(k, \omega))$ is given by

$$\langle S_{\boldsymbol{k},\omega}S_{-\boldsymbol{k}',\omega'}\rangle_0 = \delta_{\boldsymbol{k},\boldsymbol{k}'}\delta(\omega+\omega')\frac{2\Gamma}{\omega^2 + \Gamma^2\left(r_0 + k^2\right)^2} = \delta_{\boldsymbol{k},\boldsymbol{k}'}\delta(\omega+\omega')K_0^S(k,\omega). \tag{115}$$

The lower index 0 denotes that the average is taken over the free Lagrangian with $u_0 = g_0 = f_0 = w_0 = 0$. The upper index indicates the field which the propagator is referred to. Another important advantage of this formalism is the possibility to carry out the Gaussian transformations decoupling different modes in the Lagrangian and a simple generalization of the renormalization group into the dynamics.

6.1.2. The Acoustic Response Function

In the studies of the sound propagation we are interested primarily in the acoustic response function

$$\langle Q_{\boldsymbol{k},\omega}\tilde{Q}_{-\boldsymbol{k}',\omega'}\rangle = \delta_{\boldsymbol{k},\boldsymbol{k}'}\delta(\omega+\omega')G^{Q}(k,\omega). \tag{116}$$

The imaginary part of $G^{Q}(k,\omega)$ determines the sound damping and the real part - the sound dispersion. The perturbation expansion for $G^{Q}(k,\omega)$ can be represented by the Dyson equation

$$G^{-1}(k,\omega) = G_0^{-1}(k,\omega) - \Sigma(k,\omega), \tag{117}$$

where the index Q was omitted and

$$G_0^{-1}(k,\omega) = -\omega^2 - i\Theta k^2\omega + c_0^2 k^2,$$

with c_0 being the bare longitudinal sound velocity. The self-energy $\Sigma(k,\omega)$ is an infinite sum of one-particle irreducible Feynman diagrams [44, 45, 85, 86]. Performing the dynamic Gaussian transformations [9, 10] and extracting an irreducible, with respect to the acoustic propagators as well as to the entropy propagators, part of the four-spin response function we are able to preserve also relevant nonasymptotic effects in the critical sound attenuation coefficient. The acoustic self-energy after these transformations can be written as [9]

$$\Sigma(k,\omega) = \frac{c_0^2 k^2\left[(v_T^{ph} - i\tilde{\omega}v_{ad}^{ph})\Pi^{(A)}(k,\omega) - r^2\right]}{1 + v_T^{ph}\Pi^{(A)}(k,\omega) - i\tilde{\omega}\left[1 + v_+\Pi^{(A)}(k,\omega)\right]}, \tag{118}$$

where the coefficients v_T^{ph}, v_{ad}^{ph}, v_+ and r^2 are defined in Sect. 4, $\tilde{\omega} = \omega C_V^0/\gamma$ is the ratio of the sound frequency to the bare spin-lattice relaxation frequency. The function

$$\Pi^{(A)}(k,\omega) = 2\Gamma\int_{-\infty}^{\infty} dt\, e^{i\omega t}\left\langle S_{\boldsymbol{k}}^2(t)\widetilde{S^2}_{-\boldsymbol{k}}(0)\right\rangle_{\mathcal{L}_S^{(A)}(S,\tilde{S})} \tag{119}$$

is the response function of the square of the order parameter. $\widetilde{S^2}_{\boldsymbol{k}}(t) = \frac{1}{\sqrt{V}}\sum_{\boldsymbol{k}_1}\tilde{S}_{\boldsymbol{k}_1}(t)S_{\boldsymbol{k}-\boldsymbol{k}_1}(t)$ is the reaction field coupled to $S_{\boldsymbol{k}}^2$. The index $\mathcal{L}_S^{(A)}(S,\tilde{S})$ denotes the effective spin Lagrangian of the model A. The function $\Pi^{(A)}(k,\omega)$ depends on the spin variables only and does not contain any irrelevant parameters, so it can be relatively easily calculated by the perturbation technique [68, 70, 87]. In the limit $\omega = 0$ the function $\Pi^{(A)}$ becomes the static correlation function $\langle S_{\boldsymbol{k}}^2 S_{-\boldsymbol{k}}^2\rangle_{H^{(A)}}$ which is well known from the Sect.4. For $k = 0$ the last quantity is proportional to the specific heat so we can interpret $\Pi^{(A)}(k,\omega)$ as the frequency and wave-vector dependent specific heat of the system. It is worth noting at this point that the structure of the self-energy described by Eq.(118) does not depend on the order-parameter dimension nor on the detailed form of the equation of motions so it has a quite universal character. The dynamic universality class, i.e. the fact whether we study an isotropic antiferromagnet or a planar ferromagnet, influences only the form of the function $\Pi^{(A)}(k,\omega)$. Also the fact whether the system is below or above the critical temperature influences only the shape of the function $\Pi^{(A)}(k,\omega)$. Of essential importance is the fact if the entropy is a conserved quantity or not. The case of conserved entropy[14] will be discussed later. As already mentioned, the sound attenuation coefficient and the sound dispersion are easily obtained from self energy:

$$\alpha(\omega) = \frac{1}{2c(\omega)} \operatorname{Im} \frac{\Sigma(k,\omega)}{\omega}, \tag{120}$$

$$c^2(\omega) - c^2(0) = \operatorname{Re}\left[\Sigma(k,\omega) - \Sigma(k,0)\right]. \tag{121}$$

In the last equation the noncritical contribution to the sound attenuation $\alpha_{nc} = \Theta\omega^2/2c^3$ was omitted. Also the temperature dependence of these quantities was not displayed. The self-energy depends on the temperature through the four spin response function $\Pi^{(A)}$ which in the static limit turns into the specific-heat. The scaling theory predicts that

$$\Pi^{(A)}(k,\omega;t) = At^{-\alpha}\Phi'(k\xi,\omega\tau_c) + B, \tag{122}$$

where A and B are some constants and t denotes the reduced temperature, α and ν are the critical exponents of the specific-heat and the correlation length, respectively. $\tau_c = \frac{1}{2\Gamma}t^{-z\nu}$ is a characteristic relaxation time of the order parameter fluctuations, which diverges to infinity at the critical point. The dynamic critical exponent z determines the dynamic universality class. We will focus on the high-temperature phase of the Ising type magnets ($n = 1$) which are usually described by the universality class of model A with $z = z_A = 2 + c\eta$ where c is a constant of an order of unity and η is the critical exponent of the correlation function [4].

For the ultrasonic frequencies the wavelength is much longer than the correlation length, $k\xi \ll 1$, so we usually put $k\xi = 0$ in the function Φ'. The scaling function Φ' can be obtained by the exponentiation procedure [3] or by integration of the recursion relations of the renormalization group [68]. In the leading order in the expansion parameter $\epsilon = 4 - d$ it is given by

$$\Phi'(y) = \left[1 + (y/2)^2\right]^{-\alpha/2z\nu} \left\{ \frac{\nu}{\alpha} + \frac{i}{y}\left[i\left(1 - iy/2\right)\arctan(y/2) + \right.\right.$$
$$\left.\left. -\frac{1}{2}\ln\left(1 + (y/2)^2\right)\right]\right\} K_4, \tag{123}$$

where $y = \omega\tau_c$ is the reduced frequency and K_4 is a constant. Substituting (122) into Eq. ((120)) and neglecting the irrelevant terms we obtain

$$\frac{\alpha(\omega,t)c_0^3}{\omega^2} = \frac{W_1(\omega)t^{-(\alpha+z\nu)}\operatorname{Im}(\Phi(y)/y) + W_2 t^{-2\alpha}\left|\Phi(y)\right|^2}{\left|1 - i\tilde{\omega}[1 + t^{-\alpha}\Phi(y)]\right|^2}, \tag{124}$$

where: $W_1(\omega) = \frac{1}{2v_+\Gamma}(\overline{g}_0^2 + \tilde{\omega}^2 g_0^2)$, $W_2 = \frac{\overline{g}_0^2 C_V^0}{v_+\gamma}$ and $\Phi = Av_+\Phi'$.

Eqs. (120) and (121) show many different types of behavior depending on the relative size of the reduced temperature, frequency and the bare relaxation times: for the order parameter $\tau_c^0 = 1/(2\Gamma)$ and the spin lattice one $\tau_{SL}^0 = C_V^0/\gamma$.

[14]It is well known that in all irreversible processes, entropy must increase. However, the terms describing the entropy production are of higher order and can be shown to be irrelevant (in the language of renormalization group theory) parameters in the equation of motion.

6.1.3. Low-Frequency Regime

Sound Attenuation Let's assume at the beginning that the sound frequency is very low, so that $\omega\tau^0_{SL} \ll 1$ is satisfied. It is true if the sound frequency is much lower than the spin-lattice relaxation rate. Than the denominator of Eq.(124) can be approximated by unity and the term proportional to $\tilde{\omega}^2$ can be neglected so we obtain

$$\frac{\alpha(\omega,t)c_0^3}{\omega^2} = \frac{\overline{g}_0^2}{v_+}\left\{\tau_c^0 t^{-(\alpha+z\nu)}\mathrm{Im}(\Phi(y)/y) + \tau_{SL}^0 t^{-2\alpha}\left|\Phi(y)\right|^2\right\}. \tag{125}$$

We can see two competing terms with different exponents but also with different amplitudes which are proportional to the relaxation times τ_c^0 and τ_{SL}^0. It is worth noting here that the effective coupling constant $\overline{g}_0$ is identical for both terms, so the second term is present in the sound attenuation even if the initial coupling constant „entropy-sound" w_0 is equal zero. Asymptotically, i.e. for very small reduced temperature the first term dominates because of the larger the critical exponent $\rho_s = z\nu + \alpha$, as long as the sound frequency is sufficiently low.

$$\alpha(\omega,t) \propto \omega^2 t^{-(z\nu+\alpha)} g_1(y). \tag{126}$$

The scaling function for the sound attenuation coefficient $g_1(y) = \mathrm{Im}(\Phi(y)/y)$ was introduced in Eq. (126). This strong singularity with the sound attenuation critical exponent

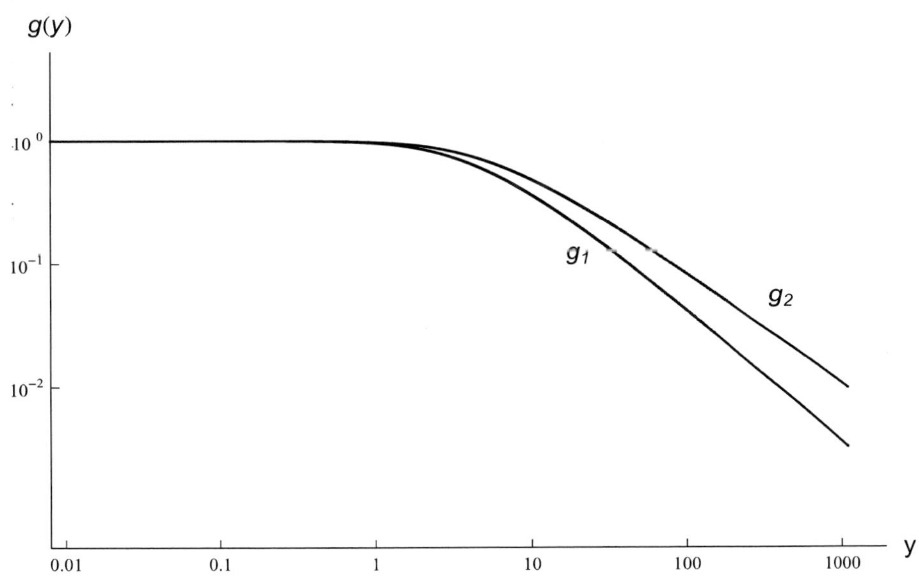

Figure 11. The scaling functions of the sound attenuation coefficient for the low frequency, g_1, and high frequency, g_2, regimes. The normalization $g_1(0) = g_2(0) = 1$ was used.

$\rho_s = z\nu + \alpha$ was predicted for all systems with the magnetoelastic coupling of the type QS^2 by Murata [83] and Iro and Schwabl [87]. For the Ising type systems $\alpha \simeq 0.110$, $\nu \simeq 0.630$ and $z \simeq 2.013$ so $\rho_s \simeq 1.38$. The scaling function $g_1(y)$ (after the normalization to unity in the hydrodynamic region) is shown in Fig. 11.

The reduced frequency can vary from the values much lower to those much higher than unity. As mentioned above, the range of the reduced frequency $y \ll 1$ is known as the hydrodynamic region and that for which $y \gg 1$ is called the critical region. In the hydrodynamic region $g_1(y) \simeq g_1(0) = \text{const}$ so $\alpha(\omega, t) \propto \omega^2 t^{-\rho_s}$. In the critical region $g_1(y \to \infty) \propto y^{-\rho_s/z\nu}$ thus $\alpha(\omega, t)$ reach a saturation value which is independent of temperature: $\alpha_{sat}(\omega) \propto \omega^{1-\alpha/z\nu}$. In Fig. 12 the temperature dependence of the sound attenuation coefficient given by Eq. (126) is shown for a few ultrasonic frequencies. The crossover from the strong increase region (hydrodynamic region) for not very small t to the saturation region (critical region) is seen. Note that in the double logarithmic scale used in this figure the power law behavior is seen as a straight line.

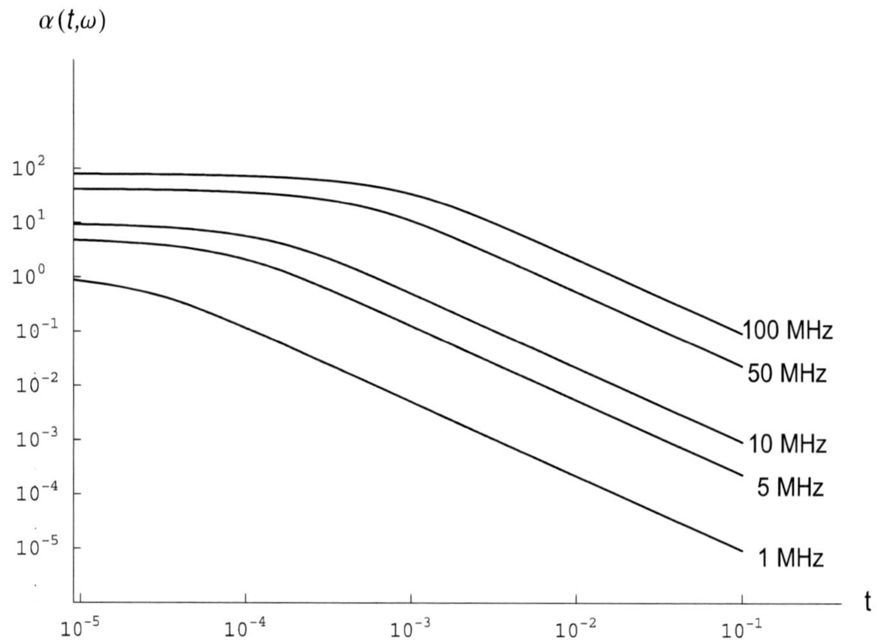

Figure 12. The sound attenuation coefficient as a function of the reduced temperature and frequency in the regime described by Eq. (126).

For $\tau_{SL}^0 \gg \tau_c^0$ despite the smaller exponent, the second term in (125) dominates as long as the reduced temperature is much higher than the crossover temperature $t_{\text{cross}} = \left(\frac{vK_4\tau_c^0}{4a^2\tau_{SL}^0}\right)^{\frac{1}{z\nu-\alpha}}$ where $a = v_+/v_+^*$ and $v_+^* K_4 = \alpha/\nu + O(\epsilon^2)$ is a fixed point value of this coupling in the renormalization group analysis. In this regime known as the Kawasaki regime a weaker singularity is expected:

$$\alpha(\omega, t) \propto \omega^2 t^{-2\alpha}, \tag{127}$$

where for the scaling function suitable for this case we have taken a constant because we have here $y < \omega\tau_c^0 \ll 1$. The critical sound attenuation exponent $\rho_s = 2\alpha \simeq 0.22$ $(n = 1)$ is relatively small in comparison with the one for the Murata-Iro-Schwabl (MIS) regime described previously. It should be noted that this analysis implies that the crucial factor which decides which singularity dominates in this low frequency region is the ratio of the relaxation times τ_{SL}^0/τ_c^0 not the ratio of the coupling constants w_0/g_0.

One may ask now why the measured sound attenuation exponents are usually small in insulators and large in metallic magnets. It seems that it is connected with the fact that in metals due to the coupling to the conduction electrons the spin-lattice relaxation times are generally shorter than in the insulators. For some dielectric magnets like MnF_2, $RbMnF_3$, Fe_5O_{12} and $RbMnF_3$ the spin-lattice relaxation time is known from the ultrasonic measurements. It is $\tau_{SL} = 3 \times 10^{-9}$ s for MnF_2 [88], whereas the critical relaxation time for order parameter fluctuations τ_c is of an order of 10^{-11} s for this compound. For $RbMnF_3$ a relatively short spin-lattice relaxation time was measured near the critical point [88] varying from 2×10^{-10} s to 4×10^{-10} s whereas τ_c measured in the inelastic neutron scattering varied from 0.08×10^{-10} s do 3×10^{-10} s [89]. For both these compounds the inequality $\tau_{SL}^0 \gg \tau_c^0$ held in the experimental temperature range which allowed the experimental observation of the Kawasaki singularity in the sound attenuation coefficient. Much longer spin-lattice relaxation time of an order of 10^{-8} s was observed for $Y_3Fe_5O_{12}$ [90]. The experimental values for MnF_2 and $RbMnF_3$ agree (to the order of magnitude) with the theoretical estimations of for insulating antiferromagnets of reported by Huber [91] and Itoh [92].

While for insulators the ultrasonic measurements seem to be a good tool in determining the spin-lattice relaxation time, for conducting magnets this method is less suitable as the high sound attenuation critical exponents measured in these magnets suggest that the sound attenuation is dominated by the first term in Eq.(125) which is connected with τ_c rather than τ_{SL}. Unfortunately, there are only very few methods permitting studies of the spin-lattice relaxation in metals. Recently, Vaterlaus et al. [94] were the first who measured τ_{SL} in rare earth metals. They used a pioneering technique of time resolved spin-polarized photoemission. Applying strong 10 ns laser heating pulses followed by 60 ps weak probe pulses they determined τ_{SL} in gadolinium. This result $\tau_{SL} = 100 \pm 80\,\mathrm{ps}$ (averaged in the temperature interval $45 < T < 225\,\mathrm{K}$) is in satisfactory agreement with a theoretical estimation by Hübner and Bennemann [95] who obtained $\tau_{SL} = 48\,\mathrm{ps}$ for Gd. Furthermore Bloembergen obtained $\tau_{SL} = 4 \times 10^{-11}\,\mathrm{s}$ for nickel by extrapolating the magnetic resonance data to the Curie temperature. These results confirm the expectations that the spin-lattice relaxation times in metals can be even a few orders of magnitude shorter than in insulators and that this is the reason why the strong Murata-Iro-Schwabl singularity dominates in metallic magnets. However, it would be of a great interest to get more experimental data on the spin-lattice relaxation time in metals.

The sound velocity It is useful to write Eqs. (117) and (118) in terms of the complex sound velocity $\hat{c}^2(\omega) = \frac{1}{k^2}\left[G^{-1}(k,\omega) - \omega^2\right]$, where

$$\hat{c}^2(\omega) = c_0^2 \frac{(1-r^2) - i\tilde{\omega}\left[1 + v_{ad}^q \Pi^{(A)}(k,\omega)\right]}{1 + v_T^{ph}\Pi^{(A)}(k,\omega) - i\tilde{\omega}\left[1 + v_+\Pi^{(A)}(k,\omega)\right]}. \tag{128}$$

In the limit $\omega \to 0$ the last equation becomes the static relation

$$\hat{c}^2(0) = c^2(0) = c_0^2 \frac{(1-r^2)}{1 + v_T^{ph}\left\langle S_{\mathbf{k}}^2 S_{-\mathbf{k}}^2 \right\rangle}, \tag{129}$$

which is exactly the isothermal sound velocity (83) obtained in Sect.4, if the effective spin Hamiltonians used in the calculations of the average $\left\langle S_{\mathbf{k}}^2 S_{-\mathbf{k}}^2 \right\rangle$ are identical. We have thus obtained a proof of the internal consistency of this theory.

In the low-frequency regime (but for the finite frequencies) Eq. (128) becomes the „isothermal" relation

$$\hat{c}^2(\omega) = c_0^2 \frac{(1-r^2)}{1+v_T^{ph}\Pi^{(A)}(k,\omega)} = c_0^2 \frac{(1-r^2)}{1+\frac{v_T^{ph}}{v_+}t^{-\alpha}\Phi(y)}, \tag{130}$$

whose structure is similar to Eq. (129). The complex sound velocity (130) depends on frequency through y. As a measure of dispersion we take $c^2(\omega) - c^2(0)$. In the low-frequency regime it is given by

$$c^2(\omega) - c^2(0) \propto t^{-\alpha} f_1(y), \tag{131}$$

where $f_1(y) = \mathrm{Re}\,\Phi(0) - \mathrm{Re}\,\Phi(y)$ is a new scaling function shown in Fig.13. In the hydrodynamic limit ($y \to 0$) this function behaves as $f_1(y) \propto y^2$ and then we can write $c^2(\omega) - c^2(0) \propto t^{-(2z\nu+\alpha)}\omega^2$. Thus the sound dispersion is characterized by a high critical exponent $2z\nu + \alpha$ and a quadratic sound frequency dependence. In the critical range $f_1(y)$ behaves as $1 - y^{-\alpha/z\nu}$, so it reaches a constant value for $T = T_c$.

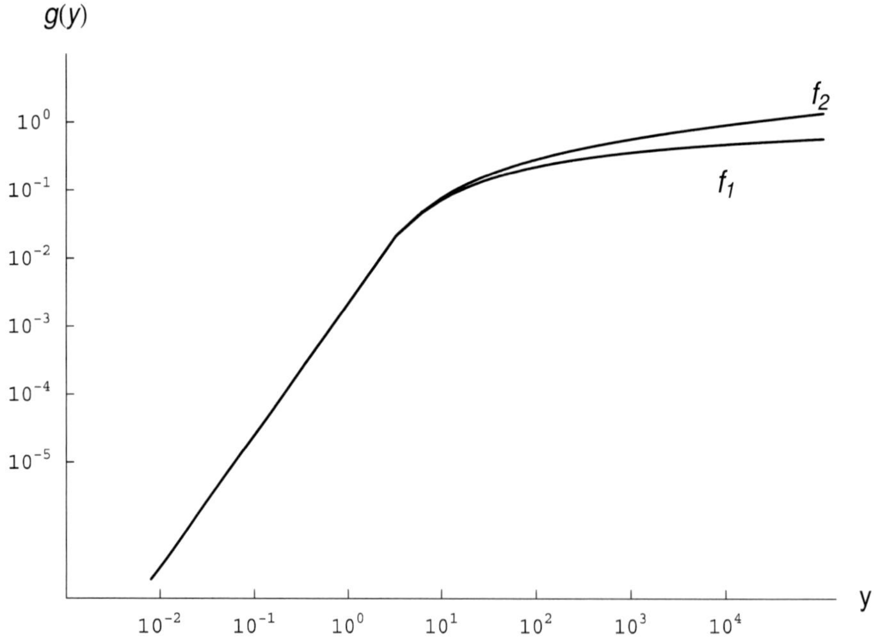

Figure 13. The plot of the scaling functions for the sound dispersion for low-frequency, f_1, and high-frequency, f_2, regimes. The normalization $f_1(0) = f_2(0)$ and $f_1(\infty) = 1$ was used.

The experiments on the critical sound velocity [60] reveal that the sound changes are usually very small so the coupling constants v_T^{ph} and v_{ad}^{ph} are expected also to be very small. That is the reason why the term $\frac{v_T^{ph}}{v_+}t^{-\alpha}\Phi(y)$ in the denominator of Eq. (124) was neglected in the paragraph concerning the sound attenuation. However, it is worth noting here that this term leads to qualitatively different behavior of the sound attenuation coefficient in the strong-coupling limit i.e. for such strong magnetoelastic couplings that $\frac{v_T^{ph}}{v_+}t^{-\alpha}\Phi(y)$ can be greater than one [53, 68]. For such strong-coupling regime the sound velocity tends to zero as $t^{\alpha/2}$ and the sound attenuation exponent $\rho_s = z\nu + \alpha/2$. The behavior of sound characteristics in the strong-coupling limit is very similar to the that of the sound velocity and attenuation in liquids near a critical point [68]. Although it is hard to be expected in magnets (because of the smallness of the coupling constants) it is believed that it takes place at some structural phase transitions as for example the order-disorder transition in ammonium halides [53]. In the tricritical points in NH_4Cl and NH_4Br at which the high tricritical specific-heat exponent $\alpha_t = 0.5$ favors the revealing of the term $\frac{v_T^{ph}}{v_+}t^{-\alpha}\Phi(y)$ the tricritical sound attenuation exponents 1.2 and 1.1 were found [96], to be compared with the theoretical strong-coupling tricritical value $\rho_s^t = z_t\nu_t + \alpha_t/2 = 1.25$.

6.1.4. High-Frequency Regime

Sound attenuation For the sound frequencies much higher than the spin-lattice relaxation rate ($\omega\tau_{SL}^0 \gg 1$), the denominator of Eq. (124) becomes singular and the term proportional to $\tilde{\omega}^2$ dominates the numerator so $\alpha(\omega, t)$ can be written as

$$\alpha(\omega, t) = \left(\frac{g_0^2}{c_0^3 v_+}\right)\omega\frac{t^{-\alpha}\mathrm{Im}\Phi(y)}{|1 + t^{-\alpha}\Phi(y)|^2} \tag{132}$$

or in the form analogous to the Eq. (126) as

$$\alpha(\omega, t) \propto \omega^2 t^{-(z\nu - \alpha)} g_2(y), \tag{133}$$

where the scaling function $g_2(y) = \frac{\mathrm{Im}\Phi(y)}{y|\Phi(y)|^2} = -\frac{\mathrm{Im}\Phi^{-1}(y)}{y}$ is shown in Fig. 11.

The new regime with the critical sound attenuation exponent $z\nu - \alpha \simeq 1.16$, is obtained. This exponent is about 0.22 smaller than that in the MIS regime. The coupling constant is also different: g_0^2 instead of $\overline{g}_0^2$. The scaling function in the critical range behaves as $g_2(y) \propto y^{-\rho_s/z\nu}$, so the value at which the attenuation saturates is $\alpha_{sat}(\omega) \propto \omega^{1+\alpha/z\nu}$ a little different from $\alpha_{sat}(\omega) \propto \omega^{1-\alpha/z\nu}$ for the one in the low-frequency regime.

The sound attenuation exponent for the high-frequency regime is given by the same formula ($\rho_s = z\nu - \alpha$) as for binary liquids [70, 97, 98], where of course the numerical value of ρ_s is different from that for the Ising magnet considered here, because of different dynamic universality class ($z \simeq 3.06$ in the critical mixtures). It is related to the structure of the expression for in the high-frequency region. Eq. (132) resembles the well known expression for the sound attenuation coefficient introduced by Ferrell and Bhattacharjee

$$\alpha(\omega, t) \propto -\omega\,\mathrm{Im}[C_{FB}(\omega)]^{-1} \simeq \frac{\omega\,\mathrm{Im}\,C_{FB}(\omega)}{(\mathrm{Re}\,C_{FB}(\omega))^2}, \tag{134}$$

where $C_{FB}(\omega)$ is a phenomenological frequency dependent specific heat[15] [99, 100]. This expression applies also to the λ phase transition in the liquid helium [74, 75, 99, 100], to the

[15]The correspondence between (132) and (134) is obtained if we interpret $C_V^0[1 + v_+\Pi^{(A)}(\omega)]$ as the Ferrell-Bhattacharjee specific heat. Note however that it is not exactly the specific heat even in $\omega \longrightarrow 0$ limit as v_+ differs slightly from v_T^q.

binary mixtures [76] and to some extend to the liquid crystals [101]. The essential factor for the change of the attenuation critical exponent from $z\nu + \alpha$ to $z\nu - \alpha$ is the divergence of $\mathrm{Re}\, C_p^{FB}$ near a critical temperature. In liquid helium the order parameter dimensionality is two and the specific heat exponent is very close to zero. For the Heisenberg systems ($n = 3$) this exponent is negative so there is no change in the sound attenuation exponent in the high-frequency regime. However, the singularity in the denominator of (124) will influence the nonasymptotic behavior of the sound attenuation coefficient. It may also happen that the background part of the specific heat will be much higher than its singular part and the high-frequency exponent $\rho_s = z\nu - \alpha$, will not be observed for some Ising type systems in experimentally accessible temperature range.

The sound velocity In the high-frequency regime the sound frequency takes a familiar „adiabatic" form

$$\hat{c}^2(\omega) = c_0^2 \frac{1 + v_{ad}^q \Pi^{(A)}(k,\omega)}{1 + v_+ \Pi^{(A)}(k,\omega)} = c_0^2 (1 - \frac{v_{ad}^{ph}}{v_+}) + c_0^2 \frac{v_{ad}^{ph}}{v_+} \frac{1}{1 + t^{-\alpha}\Phi(y)}. \tag{135}$$

In the limit $y \to 0$ the Eqs. (92) and (93) are recovered. The above should be expected because for the long spin-lattice relaxation times the fluctuations of temperature of the spin system do not decay by a fast process of energy relaxation to the lattice. The sound dispersion is also easily obtained

$$c^2(\omega) - c^2(0) \propto t^{\alpha} f_2(y), \tag{136}$$

where the scaling function $f_2(y) = \mathrm{Re}\,\Phi^{-1}(y) - \mathrm{Re}\,\Phi^{-1}(0)$ is shown in Fig. 13. It behaves as $f_2(y) \propto y^2$ in the hydrodynamic range and as $f_2(y) \propto y^{\alpha/z\nu}$ in the critical range so in the hydrodynamic region $c^2(\omega) - c^2(0) \propto t^{-(2z\nu - \alpha)}\omega^2$, and the critical exponent for the sound dispersion is lowered by 2α in comparison to the low-frequency regime.

6.2. Low Temperature Phase and Other Dynamic Models

In the low-temperature phase only the four-spin response function $\Pi(k,\omega)$ has to be obtained. The theory becomes more complicated as there is spontaneous polarization $\langle S \rangle$ and introducing $S = \langle S \rangle + \delta S$ where δS is the spin fluctuation we can write this function as [70]

$$\Pi = 2\langle(\delta S \delta S)(\Gamma \widetilde{S\delta S})\rangle + 4\langle S\rangle\langle(\delta S)(\Gamma \widetilde{S\delta S})\rangle + 2\langle S\rangle\langle(\delta S \delta S)(\Gamma \widetilde{S})\rangle + 2\langle S\rangle^2\langle \delta S \Gamma \widetilde{S}\rangle \tag{137}$$

The last term is the analogue of the Landau-Khalatnikov sound damping [104] which is the only one which contributes to in the mean-field theory. The first term in (137) is known as a fluctuation contribution to the sound attenuation and the other two terms are sometimes called the mixing contribution [5]. Considered separately, the three contributions are characterized by different critical exponents and for example the Landau-Khalatnikov term diverges with the critical exponent equal to $2(\gamma - \beta)$ [5]. As was noted by Halperin and Hohenberg [4] in the scaling region there should be cancellations between different contributions and the critical sound attenuation exponent should be the same as in the disordered phase. It was shown explicitly by Dengler and Schwabl [70] that this was the case. They calculated also the scaling function $g_1^-(y)$ in the MIS low frequency regime to the second order in ϵ. The upper index indicates the low-temperature phase. Interestingly, this scaling function showing the same asymptotic properties (for $y \to 0$ and $y \to \infty$) as $g_1^+(y)$ exhibits a characteristic maximum below the transition temperature at $y \simeq 1$ as shown in Fig. 14. This maximum is due to the Landau-Khalatnikov term. The authors were also able

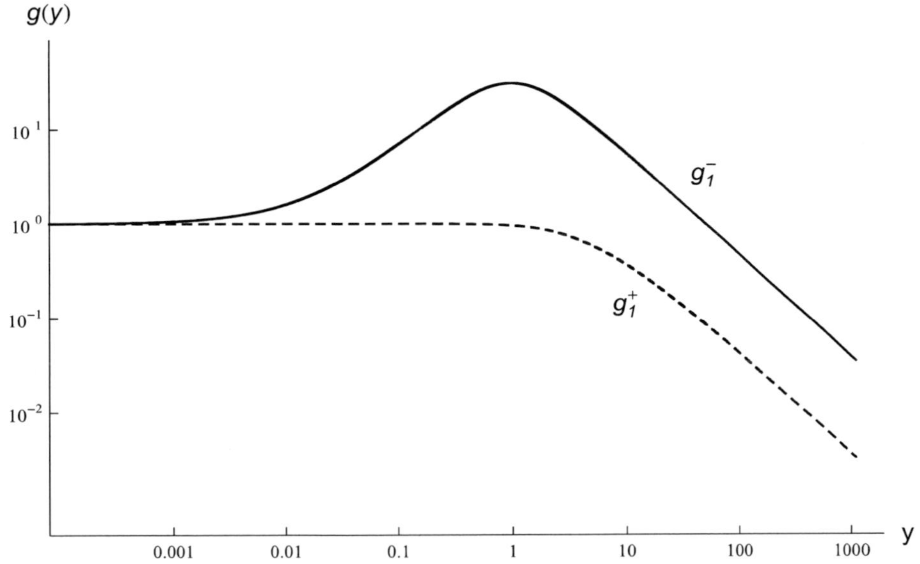

Figure 14. The scaling functions of the sound attenuation coefficient for the low temperature phase, g_1^- (continuous line), in comparison with that for the high temperature phase, g_1^+ (dashed line). The normalization $g_1^+(0) = g_1^-(0) = 1$ was used.

to calculate the universal amplitude ratios for ultrasonic attenuation coefficient above and below the critical point [70]. For uniaxial magnets investigated in this section the critical amplitude ratio

$$\alpha^+/\alpha^- = \frac{\varepsilon}{72} 2^{z\nu+\alpha}(1 + \frac{29}{27}\epsilon) + O(\epsilon^2) \tag{138}$$

is small, of an order of 0.05 as a reminiscent of the LK theory (for which this ratio is zero).

Another point is the calculations of the scaling functions in other universality classes. There a substantial progress has been made due to the renormalization group theory. Referring the reader to the original works [70, 105, 106] we focus only shortly on the model C with conserved energy field as it is a limiting case for the model with spin-lattice relaxation considered here. In the limit of very slow spin-lattice relaxation the spin energy is a conserved quantity. Assuming that the fluctuations of q decay due to diffusion the kinetic coefficient Γ in Eq. (103) should be replaced by κk^2 where k is the wave vector and κ is the thermal conductivity. Analysis of this model [107] gives the following expressions for the sound attenuation and dispersion

$$\alpha(\omega, t) =\propto \omega^2 t^{-\rho_s} g_C(y), \tag{139}$$

$$c^2(\omega) - c^2(0) \propto t^{\alpha} f_C(y), \tag{140}$$

with the sound attenuation exponent $\rho_s = z\nu - \alpha = 2\nu \simeq 1.26$, and $g_C(y) = \mathrm{Im}\,\Psi(y)/y$ and $f_C(y) = \mathrm{Re}\,\Psi(y)$ as the scaling functions, where

$$\Psi(y) = \left[1 + \left(\frac{y}{2}\right)^2\right]^{\frac{\alpha}{2z\nu}} \left\{1 + \frac{\alpha}{\nu}\left[\frac{1 - \frac{iy}{2}}{y}\arctan\frac{y}{2} + i\frac{\ln(1+(\frac{y}{2})^2)}{2y}\right]\right\}. \tag{141}$$

The scaling functions in the one-loop approximation differ from g_2 and f_2 only by the value of the dynamic critical exponent ($z \simeq 2.175$ for the model C). We can also express the complex sound velocity as

$$\widehat{c}(\omega)^2 = c_0^2(1 - \frac{v_{ad}^{ph}}{v_+}) + \frac{\mathrm{const}}{C_{FB}(\omega)}. \tag{142}$$

with a Ferrell–Bhattacharjee function[16] $C_{FB}(\omega)$ which diverges as the static specific heat near a critical point. In the limit $y \to 0$, the adiabatic formula (90) for the sound velocity is recovered again. The sound attenuation behavior in model C is very similar to that in the high-frequency regime of model A and Eqs. (139) and (140) appear as the high-frequency regime expressions of model A (133) and (131) in which the crossover to another dynamic class has taken place (at least in the one-loop approximation).

The formula 142 seems to have quite general character (magnets, liquid helium, binary mixtures and liquid crystals). However, for the liquid-gas critical point we have different expression $\widehat{c}(\omega)^2 = \mathrm{const}/C_{FB}(\omega)$ which induces different behavior of the sound velocity and different sound attenuation exponent $\rho_s = z\nu + \alpha/2$ [70]. The difference originates from the different role played by the order parameter in this system [75]. In magnets, liquid helium, binary mixtures the order parameter couples to heat and sound modes by two static couplings f_0 and g_0, whereas in the liquid-gas system there is only one static coupling between the heat (which is the order parameter) and the sound modes.

7. Eksperiment

There are many experimental works on the ultrasonic propagation near a critical point. It is beyond the scope of this review to discuss them all and we refer the reader to the excellent reviews devoted the experimental results [6, 7, 60, 90] focusing only on some questionable problems. First of all it is worth considering the question of how do the measured sound attenuation critical exponents compare with the theoretical estimations: $\rho_s \simeq 1.26$ for the model with conserved energy and the Kawasaki singularity $\rho_s \simeq 0.22$, MIS singularity $\rho_s \simeq 1.38$ as well as for the high-frequency singularity ($\rho_s \simeq 1.16$) in the Ising type model (n=1) with nonconserved energy (spin-lattice relaxation). In Table 5 we present also the theoretical estimations for the other universality classes in the MIS regime. In real systems e.g. for isotropic Heisenberg ($n = 3$) magnets the cubic anisotropy destroys the conservation of the order parameter leading to the purely dissipative dynamics described by model A. Therefore, in Table 5 this possibility is also taken into account. Other anisotropies

[16]The Ferrell–Bhattacharjee function can be expressed ($C_{FB}(\omega) = C_V^0/\Gamma_{m_2m_2}(\omega)$) in terms of a vertex function of an idealised phonon-free model [107].

can induce the crossover to the Ising universality class ($n = 1$). In real systems it is not clear a priori whether the total spin energy is conserved or not. It is connected with the strength of the spin-lattice interactions and in some cases model A may be a better description of the magnetic system and in other cases model C will be more suitable.

7.1. Insulators

For the most of insulators the exponent ρ_s takes small vales from zero for europium oxide (EuO) to 0.75 for FeF_2. In Table 3 the values of ρ_s defined for the hydrodynamic range are presented for magnetic insulators. A conspicuous exception is the chromium oxide (Cr_2O_3) for which $\rho_s = 1.3$. This value is very close to those observed in metals (see Table 4). This compound will be discussed later. First we discuss typical magnets from this group. The small values of the critical sound attenuation exponents in insulators are commonly interpreted as an evidence that the Kawasaki regime is realized there. Two questions arise here: how the exponent 0.77, observed in FeF_2 corresponds to the value $\rho_s^{KAW} = 2\alpha_I \simeq 0.22$ obtained for the Ising like magnets (here α_I denotes the specific heat exponent for the Ising systems). What is the connection of $\rho_s \simeq 0.32$ for the isotropic antiferromagnet $RbMnF_3$ with the negative value $2\alpha_H \simeq -0.27$ obtained from $\rho_s^{KAW} = 2\alpha_I$ by replacing α_I with α_H? It could be explained by the inaccuracy in determining the critical temperature which can influence the values of the critical exponents measured [90].

Table 3. Critical sound attenuation exponents for insulators.

Compound	Exponent ρ_s	Anisotropy parameter	Frequency range (MHz)	Reduced temperature range	Refer.
MnF_2	$0.14(1)$	$1.4 \cdot 10^{-2}$	$10 - 110$	$10^{-4}-10^{-1}$	[113, 114] [60, 115]
$RbMnF_3$	$0.32(2)$	$5 \cdot 10^{-6}$	$30 - 150$	$10^{-4}-6 \cdot 10^{-2}$	[116]
EuO	0	$4 \cdot 10^{-4}$	$50,\ 170$	$10^{-4}-10^{-1}$	[60, 117]
$Y_3Fe_5O_{12}$	$0.5(1)$	10^{-5}	$5,\ 30$	$2 \cdot 10^{-4}$-$3 \cdot 10^{-2}$	[90]
$Gd_3Fe_5O_{12}$	$0{,}42(10)$	10^{-5}	$5,\ 30$	$2 \cdot 10^{-4}$-$5 \cdot 10^{-1}$	[90]
FeF_2	$0.77(7)$	0.6	$10 - 70$	$3 \cdot 10^{-4}$-$2 \cdot 10^{-2}$	[119]
Cr_2O_3	$1.3(1)$	$3 \cdot 10^{-4}$	$100 - 1500$	$3 \cdot 10^{-5}-10^{-3}$	[122]

In our opinion the cause of this apparent contradiction rests with the fact that in experiment only an effective critical exponent is measured from the slope of the sound attenuation curve vs. the reduced temperature (shown in the double-logarithmic scale) in the hydrodynamic range. The effective sound attenuation exponent can be defined [67, 68] as

$$\rho_s(t) = -\frac{\partial \ln \left[\alpha(\omega, t)\right]_{\omega \to 0}}{\partial \ln t}. \tag{143}$$

It was shown in the last section that the Kawasaki singularity is proportional to the square of the specific heat so the asymptotic value for this exponent in Ising systems is $\rho_s^{Kaw} = 2\alpha_I$. It should be noted that this is only an asymptotic approximation because the specific heat behaves near T_c as

$$C = At^{-\alpha} + B, \tag{144}$$

The constant term B can be neglected for Ising systems only infinitesimally close to T_c. In the Heisenberg model where $A < 0$ this constant is necessary to assure the positivity of the specific heat. Only the two terms assure a peak in the specific heat [35]. In real systems we should take into account also the first correction to scaling [111, 112, 123] writing

$$C = At^{-\alpha}(1 + Et^{\Delta_1}) + B + Ft, \tag{145}$$

where Δ_1 is the exponent of the first correction to scaling equal about 0.52 [22] for the one-component order parameter systems. A regular term proportional to t is also added. The effective attenuation exponent can be found from (144) in the Kawasaki regime as

$$\rho_s(t) = 2\alpha \frac{At^{-\alpha}}{At^{-\alpha} + B}. \tag{146}$$

For the Heisenberg magnets ($n = 3$) we have $\alpha < 0$, $A < 0$ and $B > 0$ ($|A| < B$) so the effective exponent is positive because the product αA is positive. Moreover, its absolute value can be higher than $2|\alpha_H| \simeq 0.27$ which could explain the experimental data in ferrites $Gd_3Fe_5O_{12}$ and $Y_3Fe_5O_{12}$.

7.1.1. FeF_2

In this strongly anisotropic antiferromagnet, both the specific heat exponent and its amplitude are positive and the constant B is negative [123]. In Figure 15 the experimental data [119] in the symmetric phase are fitted to the general formula (145) and in Fig. 16 the effective attenuation exponent is shown for $B/A = -1.59$, $E = 1.2$, $F/A = 0.1$, $\alpha = 0.11$ and $\Delta_1 = 0.52$ (Pawlak and Fechner [118]). The solid line in Fig. (15) is the plot of the square of specific heat. The dashed line represents a simple power law behavior $\alpha(\omega, t) \propto t^{-\rho_s}$ with the exponent $\rho_s = 0.73$. Remembering about stochastic scatter of data it is not hard to imagine that the exponent ~ 0.73 could be obtained in the temperature range $10^{-2} - 10^{-3}$. The lower value of this exponent in the low-temperature phase of FeF_2 $\rho_s \simeq 0.5$ [119] should be connected to the greater amplitude A^- because for the Ising systems we have $A^+/A^- \simeq 0.53$. The greater amplitude A^- implies a lower value of the ratio B/A^- so also the smaller deviation of ρ_s from the value $2\alpha_I$.

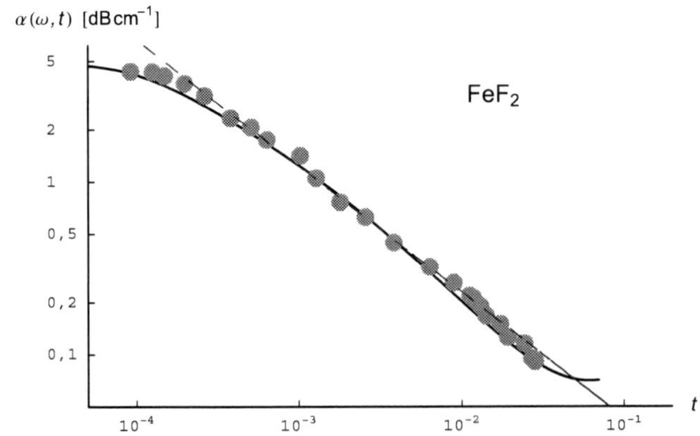

Figure 15. The ultrasonic attenuation vs the reduced temperature for $f = 50$ MHz in FeF_2 along $[0, 0, 1]$ direction (the data are from [119]). The solid curve is the plot of Eq. (145). The dashed line is the fit to the single power law with exponent $\rho_s = 0.73$ (Pawlak and Fechner [118]).

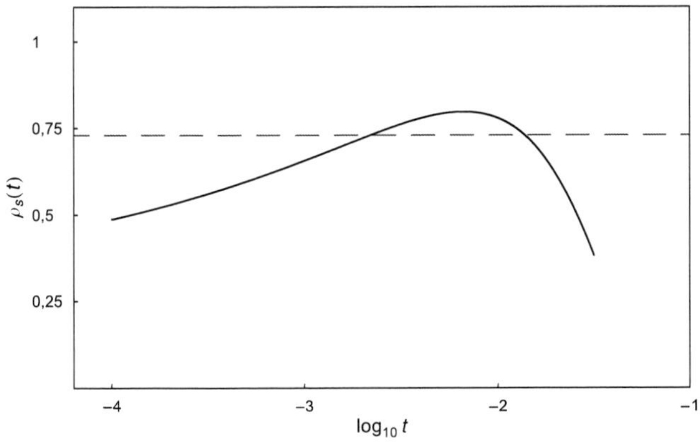

Figure 16. The effective sound attenuation exponent in the Kawasaki regime for the specific heat characterized by Eq.(145) (Pawlak and Fechner [118]). The dashed line corresponds to "experimental" value $\rho_s = 0.73$.

7.1.2. $RbMnF_3$

Because $RbMnF_3$ is the isotropic antiferromagnet with very small anisotropy its specific-heat exponent is negative [120]. The analysis is restricted to relatively low frequencies at which the saturation effects (appearing when the reduced frequency y is comparable to unity) can be neglected in the explored temperature range. In Fig. (17) the experimental data [92] in the high-temperature phase are fitted to the expression (145). The Heisenberg critical exponents $\alpha = -0.133$, $\Delta_1 = 0.5$ [22] and $A = -0.372$, $B = 0.292$, $E = -0.0273$, $F = 0.25$ are used (Pawlak and Fechner [118]). The critical amplitudes for the high temperature specific heat are consistent with the experimental estimations obtained by Marinelli et al. [120]. In Fig. (18) the effective sound attenuation exponent versus the reduced temperature is shown. It is seen that it is positive for the Heisenberg type antiferromagnet $RbMnF_3$ and its average value in the experimental range of reduced temperature is about $0.25 - 0.35$ as is observed in experiment (see Table 4). That explains also the positive sound attenuation exponents in Heisenberg magnets like $Y_3Fe_5O_{13}$ and

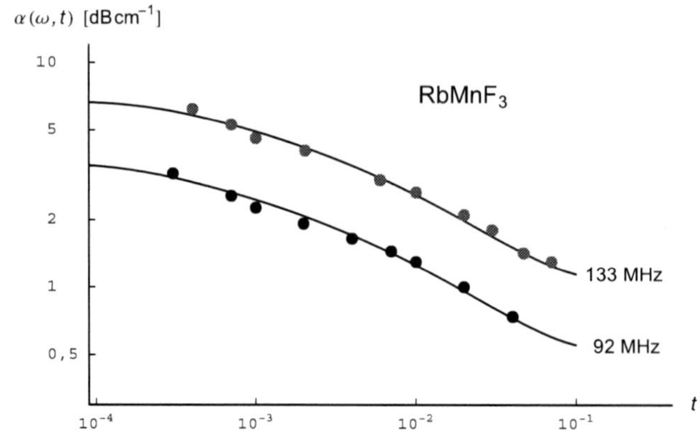

Figure 17. The high temperature ultrasonic attenuation vs the reduced temperature for low frequencies: $f = 92$ and 133 MHz in $RbMnF_3$ along [1,0,0] direction. The solid curve is the plot of Eq. (145) with $\alpha = -0.133$, $\Delta_1 = 0.5$ and $A = -0.372$, $B = 0.292$, $E = -0.0273$, $F = 0.25$ (Pawlak and Fechner [118]).

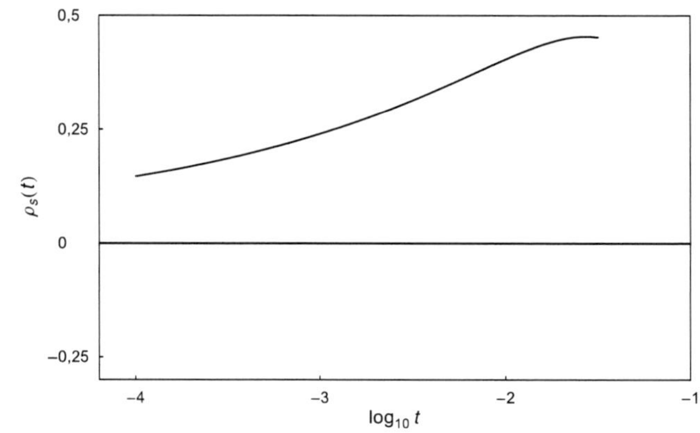

Figure 18. The effective sound attenuation critical exponent for $RbMnF_3$ (Pawlak and Fechner [118]).

$Gd_3Fe_5O_{13}$.

7.1.3. EuO

Another source of concern is the lack of the critical attenuation observed in isotropic ferromagnet EuO ($\rho_s = 0$) although a singularity in the specific heat is observed in this compound [124]. It is however connected with a very long spin-lattice relaxation time which is of an order of milliseconds ($\tau_{SL}^{-1} \approx 1.5 \cdot 10^6$ s^{-1} [125]). The lack of singularity in the sound attenuation for EuO was commonly explained with the aid of simple formula (97) which for this case was usually written as

$$\alpha(\omega, t) \propto C_p \frac{\omega^2 \tau}{1 + \omega^2 \tau^2}, \tag{147}$$

where $\tau \propto C_p$ and C_p is the specific heat. For $\omega\tau \gg 1$ this relation leads to $\alpha(\omega, t) \propto$ const. For the frequency $f = 50$ MHz and $\tau = \tau_{SL} \simeq 10^{-6}$ s we have $\omega\tau \simeq 300$

so the assumption $\omega\tau \gg 1$ is well satisfied. However, there is a small problem with the formula (147) based on only one relaxation time. Namely, the sound velocity change $\Delta c \equiv c(\omega) - c_\infty$ on the grounds of Eq. (96) is given by

$$\Delta c \simeq \frac{C_p}{2c_\infty(1+\omega^2\tau^2)}. \tag{148}$$

For $\omega\tau \gg 1$ Δc it is very small and depends on the frequency as ω^{-2}. It is very difficult to reconcile with a small anomaly observed in the sound velocity in EuO which practically does not depend on the sound frequency [125]. The lack of the ω^{-2} dependence seems to be unquestionable. So the lack of the critical attenuation and the anomaly in the sound velocity cannot be simultaneously explained. It can be easily done with the aid of formula (118) if we assume that

$$\omega^2\tau_{SL}^2\frac{\tau_c}{\tau_{SL}} \ll \frac{v_T^{ph}}{v_{ad}^{ph}} \ll \omega^2\tau_{SL}^2. \tag{149}$$

Then the imaginary part of (118) gives the equation analogous to (147)

$$\alpha(\omega,t) \propto \frac{\omega^2\tau_{SL}^0 \operatorname{Re}\Pi^{(A)}(\omega;t)(1+v_+\operatorname{Re}\Pi^{(A)}(\omega;t))}{1+\omega^2(\tau_{SL}^0(1+v_+\operatorname{Re}\Pi^{(A)}(\omega;t)))^2}, \tag{150}$$

where $\Pi^{(A)}(k,\omega;t) = t^{-\alpha}\Phi'(\omega\tau_c) + B$ is the discussed four spin response function which turns into the static specific heat in the $\omega\tau_c \to 0$ limit. The role of the relaxation time τ in Eq. (147) is played here by the product $\tau_{SL}^0(1+v_+\operatorname{Re}\Pi^{(A)}(\omega;t))$ which for $\omega\tau_c \ll 0$ behaves as the specific heat. The real part of Eq. (118) becomes the adiabatic sound velocity described by Eq. (135) with the singular term proportional to the inverse of the specific heat. Thus both aspects of the sound propagation in EuO can be explained with the aid of formula (118).

Because the spin-lattice relaxation time in EuO is extremely long (much longer than τ_c) the inequality (149) is probably satisfied for very wide range of frequencies. It should be however expected that for very low as well as for very high frequencies anomalous sound attenuation should be observed. Actually, it was mentioned [125] that the critical attenuation was observed in vibrating reeds experiments EuO for the very low frequency range 0.4–3 kHz.

7.1.4. MnF_2

The critical sound attenuation in magnets was for the first time observed in the antiferromagnet MnF_2 [126]. As shown by the experimental investigation of the specific heat [127] the magnet shows a crossover from the isotropic Heisenberg behavior ($n = 3$) to the Ising type behavior at t of an order of 10^{-2}. The sound velocity and attenuation coefficient of manganese fluoride have been studied by many authors [60, 113–115, 122, 126, 128–130] and it is the best known (from the experimental point of view) magnetic compound so far. In Fig. 19 the longitudinal sound attenuation dependence on the reduced temperature is shown (Pawlak [121]). The experimental data are taken from Ikushima's work [113]. The continuous curves represent the expression (124) for the frequencies $f = 10,\ 30,\ 50$ and 70 MHz.

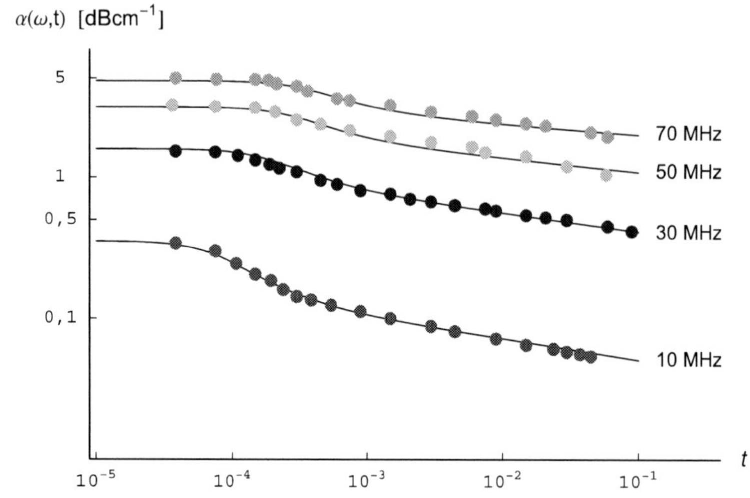

Figure 19. The temperature dependence of the attenuation coefficient for longitudinal waves in $[1,1,0]$ direction for MnF_2 ($T > T_N$) (Pawlak [121]). The continuous curves represent the imaginary part of Eq. (118).

The acoustic self energy is taken in the most general form (118) for the model with the spin-lattice relaxation, with $\tau_{SL}^0 = 3\cdot 10^{-9}$ s, $\tau_c^0 = 10^{-13}$ s, $\frac{v_{ad}^{ph}}{v_T^{ph}} = 1.9$, $(A\nu K_4/\alpha B) = -1.03$. It can be seen from this plot that for low frequencies e.g. for $f = 10\,\mathrm{MHz}$, $\alpha(\omega,t)$ exhibits a typical Kawasaki behavior with a small slope of the curve and only for the reduced temperatures of an order of 10^{-4} the curve starts to climb up and soon saturates. It is connected with the fact that $\tau_{SL}^0 \gg \tau_c^0$ in MnF_2, and although the critical relaxation time for the order parameter fluctuations increases much faster than the spin-lattice relaxation time the MIS term exceeds the Kawasaki term only for t close to 10^{-4}. The crossover to the critical range ($y \gg 1$) is observed when ω^{-1} becomes comparable with τ_c.

Fig. 20 shows the contributions of different terms to the total sound attenuation coefficient (Pawlak [121]). The background noncritical term is also included. It should be noted that every contribution saturates at roughly the same temperature at which $\omega\tau_c \sim 1$. For higher frequencies the MIS term cannot exceed the Kawasaki term (α_{KAW}) because the saturation of both terms takes place before.

Fig. 21 and 22 present analogous plots for the longitudinal sound waves along the $[1,0,0]$ direction for $T > T_N$. The experimental points are from [114]. This time the range of the ultrasonic frequencies is wider (10-110 MHz). Generally, it is expected that in the systems without full isotropic (elastic) symmetry the effective coupling constants will depend on the direction of propagation and this time $\frac{v_{ad}^{ph}}{v_T^{ph}} = 0.8$ is assumed. The share of the individual terms in the total attenuation is seen in Fig. 22 for the highest frequency $f = 110$ MHz. This time α_{KAW} significantly exceeds α_{MIS} term and competes with α_{HF} for the small reduced temperatures. The background term is very high now (about 50% of the total attenuation) and suppresses the relative magnitude of the attenuation peak.

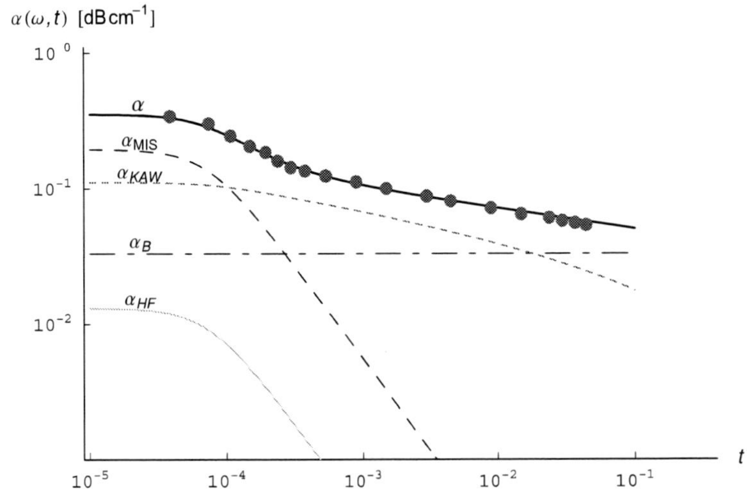

Figure 20. The contribution of different terms to the total attenuation of the longitudinal wave along the $[1, 1, 0]$ direction in MnF_2 for $f = 10$ MHz (Pawlak [121]). The MIS, Kawasaki, high-frequency and the background terms are denoted as α_{MIS}, α_{KAW}, α_{HF} and α_B, respectively.

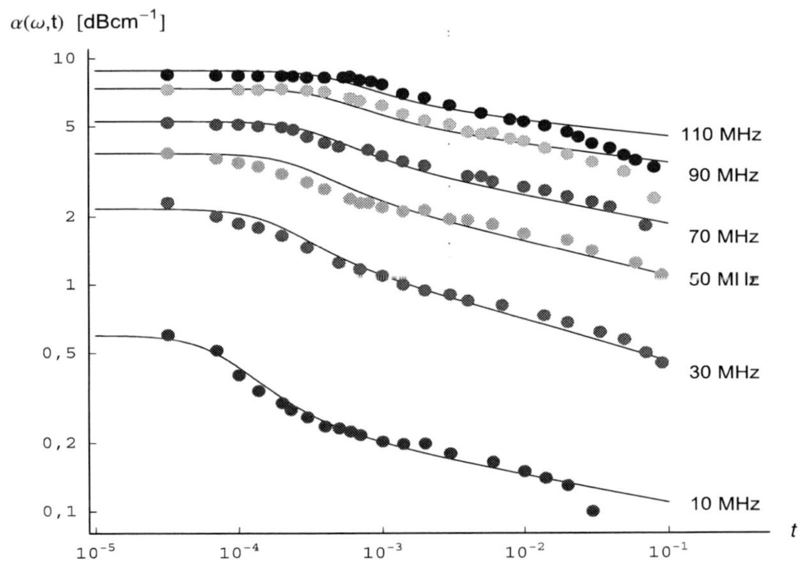

Figure 21. The temperature dependence of the attenuation coefficient for longitudinal waves in $[1, 0, 0]$ direction for MnF_2 ($T > T_N$). The continuous curves represent the imaginary part of Eq. (118). The experimental point are taken from [114].

7.1.5. Cr_2O_3

From Table 3 it is seen that this antiferromagnet does not match the other insulators because of its high exponent $\rho_s = 1.3$. The experiment [112] shows that this system is characterized by Ising specific-heat exponent for $t < 3 \cdot 10^{-3}$, despite only small anisotropy. As follows from our analysis, high sound attenuation exponent ρ_s can appear in three cases. One is

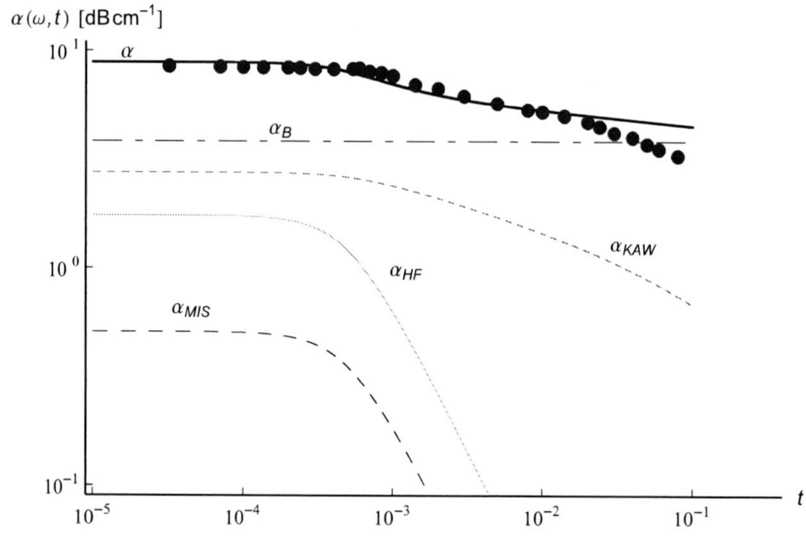

Figure 22. The contribution of different terms to the total attenuation of the longitudinal wave along the $[1, 0, 0]$ direction in MnF_2 for $f = 110$ MHz. The MIS, Kawasaki, high-frequency and the background terms are denoted as α_{MIS}, α_{KAW}, α_{HF} and α_B, respectively.

typical of metals where $\tau^0_{SL} \sim \tau^0_c$ and then due to the critical slowing down the MIS behavior is observed with the exponent $\rho_s = z\nu + \alpha$ where $z = z_A$ and the nonconserved energy is assumed. For this universality class another high-frequency regime is also characterized by large attenuation exponent $\rho_s = z\nu - \alpha$ and it is the second case. The third case is that the energy is conserved and the attenuation exponent is given by $\rho_s = z_C\nu - \alpha$ with $z_C = 2 + \alpha/\nu$ for the model C. In model A we have $\rho_s \simeq 1.38$ for the MIS regime and $\rho_s \simeq 1.16$ for the high-frequency regime. The experimental value 1.3 measured for Cr_2O_3 is closest to the value for model C $\rho_s \simeq 1.26$.

It should be noted that the sound attenuation was measured for very high frequencies, 100–1500 MHz, in this compound. Assuming even very short spin-lattice relaxation time of an order of one nanosecond we have $1 \leq \omega\tau_{SL} \leq 15$ which indicates the crucial role of the denominator in Eq. (124) which rather excludes the MIS regime. Further measurements of the sound attenuation coefficient in the low-frequency range as well as an experimental estimation of the spin-lattice relaxation time are desirable for this magnet in order to recognize the source of its exceptionality.

7.2. Metals

In Table 4 the critical exponents for conducting magnets are given. They are equal or greater than one, which confirms the hypothesis that the spin-lattice relaxation time are shorter than in insulators and the Kawasaki regime is less important in these compounds. Taking into account that the ultrasonic frequencies used were not very high i.e. such that $\omega\tau^0_{SL} \sim 1$ could be met, we can suppose that in the investigated frequency interval the MIS term dominates with the exponent $\rho_s = z\nu + \alpha$. What is the source of so large differences in the attenuation exponents measured in metals? Of course, they follow from the fact that these

Table 4. Critical sound attenuation exponents for magnetic metals.

Metal	Exponent ρ_s	Anisotropy parameter	Frequency range (MHz)	Reduced temperature range	Refer.
Gd	$1.2(1)$ $1.63(10)$ $1.8(2)$ $1.15(10)$	$5{\cdot}10^{-4}$	$30-180$ $10-70$ 5 $5-30$	$10^{-3}-10^{-1}$ $10^{-3}-10^{-1}$ $3.4{\cdot}10^{-3}-2.4{\cdot}10^{-2}$ $10^{-4}-10^{-1}$	[132] [60] [133] [90]
Tb	$1.24(10)$	-0.4	$30-170$	$7{\cdot}10^{-3}-7{\cdot}10^{-2}$	[134]
Dy	$1.37(10)$ $1.26(10)$	-0.3	$30-170$	$3{\cdot}10^{-3}-10^{-1}$ $3{\cdot}10^{-3}-10^{-1}$	[134] [60]
Ho	$1.0(1)$	$-7{\cdot}10^{-2}$	$30-170$	$3{\cdot}10^{-4}-10^{-1}$	[134]
MnP	$1.1(1)$ $1.1(1)$		$30-210$ $10-520$	$10^{-4}-10^{-2}$ $10^{-4}-10^{-2}$	[135] [136]
Ni	$1.4(2)$	$\sim 10^{-4}$	$20,\ 60$	$10^{-6}-3{\cdot}10^{-3}$	[137]

compounds belong to different static (n) and dynamic universality classes as explains Table 5 [4].

7.2.1. Ni

The attenuation exponent observed in this isotropic ferromagnet $\rho_s = 1.4 \pm 0.2$ suggests that the earlier mentioned cubic anisotropy terms are important in this ferromagnet, which indicates the crossover to the relaxational dynamics with $z \simeq 2$. Below T_C a temperature interval was observed in which $\rho_s \simeq 0.3$ [137]. Such a small value of this exponent suggests that the Kawasaki regime is important here for not very small reduced temperature.

7.2.2. Gd

The most interesting metal is the ferromagnetic gadolinium. Is it possible to explain such great differences in the sound attenuation exponent shown in Table 4? It seems that yes.

Table 5. Theoretical estimations of ρ_s in the MIS regime for different universality classes.

n	Magnetic system	Model	ρ_s
1	anisotropic magnets, uniaxial antiferromagnets	A	1.38
1	uniaxial ferromagnets	B	2.50
1	anisotropic magnets, uniaxial antiferromagnets	C	1.48
1	uniaxial ferromagnets	D	2.50
2	easy plane magnets, $h_z = 0$	A E	1.35 1.00
3	isotropic antiferromagnets	A G	1.31 0.94
3	isotropic ferromagnets	A J	1.31 1.65

Firstly, the four experiments mentioned in Table 4 have been performed for different samples and the structure and the symmetry of the crystal depend on the number and character of impurities. Secondly, in Gd besides the anisotropy the dipole interactions are very important. It follows from the renormalization group analysis [138] that in such system there are four fixed points: Heisenberg (H), anisotropic Ising (I), anisotropic dipolar (AD) and isotropic dipolar (ID) fixed points. Depending on the ratio of the anisotropy to the parameter describing the strength of the dipolar interactions we have a series of the crossovers between these fixed points which can be seen as transitions from one set of critical exponents to another. One of the possible series is H $\rightarrow$ I $\rightarrow$ AD and the second one is H $\rightarrow$ ID $\rightarrow$ AD. For the sequence H $\rightarrow$ I $\rightarrow$ AD the behavior of the isotropic ferromagnet with $\rho_s = 1.65$ will be observed for high reduced temperatures. The exponents reported in [60] and [133] probably refer to this temperature range. For smaller reduced temperatures the Ising behavior with relaxational dynamics and $\rho_s = 1.38$ can be expected in such a sample (or in other sample in different temperature interval). Eventually, the system will be found in the temperature range controlled by the anisotropic dipolar fixed point where also the relaxational dynamics and the mean-field critical exponents with $\rho_s = 1$ are expected. Because the upper critical dimension for such system is $d = 3$ [19, 20] fractal powers of logarithms also appear [70, 139, 140]. For the sequence H $\rightarrow$ ID $\rightarrow$ AD we should observe first $\rho_s = 1.65$ (for Heisenberg fixed point) then a crossover to the isotropic dipolar fixed point ($\rho_s = 1.31$) is expected with the static exponents only slightly different from the Heisenberg exponents. However, the dipolar interactions do not conserve the order parameter so $z_{ID} \simeq 2$. Smaller dynamic critical exponent implies smaller sound attenuation exponent $\rho_s = 1.31$ in this regime. The asymptotically stable fixed point for this sequence is again the anisotropic dipolar fixed point with $\rho_s = 1$ and the logarithmic factors [70, 139, 140]. As follows from this discussion gadolinium is such a complex system that it is very difficult to describe it with the aid of one fixed point (and one value of ρ_s). Instead a sequence of crossovers between different exponents is expected and the effective exponents are measured in different temperature ranges and for different samples.

7.3. Conclusion

This chapter deals with the theoretical and experimental studies of ultrasonic wave attenuation and velocity in magnets. It begins with a short description of the basic concepts of the statics of the phase transitions such as critical exponents, universality etc. Then a short introduction to the critical dynamics is presented, in which dynamical scaling, critical slowing down and main universality classes are described. It was shown that the adiabatic longitudinal sound velocity remains finite at the magnetic phase transition temperature. Additionally, an extensive discussion of the phenomenological theory of sound attenuation and dispersion is given. The effect of spin-lattice relaxation on the sound propagation is investigated on the basis of the stochastic model. Three asymptotic regimes for sound attenuation are discussed. Two of them: MIS and Kawasaki regime refer to the low-frequency range. The additional regime refers to the high-frequency range and corresponds to the adiabatic sound propagation. It transforms into the sound propagation in model C (with conserved energy) in the limit of vanishing spin-lattice relaxation frequency. An overview of experimental and theoretical sound attenuation exponents both for magnetic insulators as well as magnetic metals is given. The concept of the effective sound attenuation exponent is discussed and illustrated on the example of FeF_2 and $RbMnF_3$. The crossovers between different regimes are shown for the antiferromagnet MnF_2.

However some unsolved questions still remain. In future the experimental measurements should cover a wider frequency range. For example the measurements for frequencies lower than 10 MHz in MnF_2 would be extremely important in verification of the theory. The same applies to Cr_2O_3 for which only high frequency measurements are accessible so far. The critical sound propagation in low-dimensional systems is a very interesting topic both from theoretical and experimental point of view. Another challenge is the construction of critical sound attenuation in an external magnetic field. For the ferromagnets the magnetic field h is coupled to the order parameter so the non-zero magnetic field destroys the phase transition and the scaling relation for the attenuation takes a form

$$\alpha(t,\omega,h) = \omega^2 t^{-\rho_s} f(\omega t^{-z\nu}, h t^{-\Delta})$$

or

$$\alpha(t,\omega,h) = \omega^2 h^{-\rho_s/\Delta} g(\omega h^{-z\nu/\Delta}, t h^{-1/\Delta})$$

where $\Delta = \gamma + \beta$ is the gap exponent and f, g are scaling functions [141]. Additional variable in the scaling function induces the existence of six asymptotic regions in the space (ω, t, h) instead of two (hydrodynamic and critical ones) in the space (ω, t). The scaling functions need to be determined theoretically as well as experimentally. So far only some mean-field results have been obtained [142–145] and the measurements far from critical temperature were carried out in MnP [146]. In particular the sound attenuation and velocity exactly at the Curie temperature would be of importance. In antiferromagnets the magnetic field does not destroy the continuous transition and only shifts the Neel temperature in a similar way as the transverse field shifts the Curie temperature in anisotropic ferromagnets [147]. The effects of the external field on the critical sound propagation in antiferromagnets need a further theoretical and experimental studies.

References

[1] Moran T. J., Luthi B., *J. Phys. Chem. Solids* **31**, 1735 (1970).

[2] Wilson K.G. *Rev. Mod. Phys.* **55**, 583 (1983).

[3] Wilson K.G., Kogut J., *Phys. Rep. C* **12**, 75 (1974).

[4] Hohenberg P.C. and Halperin B.I., *Rev. Mod. Phys.* **49**, 435 (1977).

[5] Fossum J.O., *J. Phys. C* **18**, 5531 (1985).

[6] Lüthi B., [in:] G.K. Horton, A.A. Marudin (eds.), *Dynamical Properties of Solids*, V. **3**, North-Holland, Amsterdam, 1980.

[7] B. Lüthi, Physical Acoustics in the Solid State, chap. 6, pp. 93–108, Springer, Berlin, 2005.

[8] Kawasaki K., *Phys. Lett. A* **29**, 406 (1968).

[9] Pawlak A., *Eur. Phys. J. B* **4**, 179 (1998).

[10] Pawlak A., *Acta Phys. Pol. A* **98**, 23 (2000).

[11] Binney J.J., Dowrick N.J., Fisher A.J., Newman M.E., *The Theory of Critical Phenomena: An Introduction to the Renormalization Group*, Oxford University Press, New York, NY, 1992.

[12] Vicsek T. , Czirok A., Ben-Jacob E., Cohen I., Shochet O., *Phys. Rev. Lett.* **75**, 1226 (1995).

[13] Toner J., Tu Y., *Phys. Rev. Lett.* **75**, 4326 (1995).

[14] Goldstone J. *Nuovo Cimento* **19**, 154 (1961).

[15] Brezin E., Wallace D.J., *Phys. Rev. B* **7**, 1967 (1973).

[16] Stanley H.E., *Introduction to Phase Transitions and Critical Phenomena*, Oxford, NY, 1972.

[17] Widom B., *J. Chem. Phys.* **43**, 3898 (1965).

[18] Kadanoff L.P., *Physics* **2**, 263 (1966).

[19] Wegner F.J., *Phys. Rev. B* **6**, 1891 (1972).

[20] Wegner F.J., Riedel E., *Phys. Rev. B* **7**, 248 (1973).

[21] Fisher M.E., *Rev. Mod. Phys* **70**, 653 (1998).

[22] Pelissetto A., Vicari E., *Phys. Rep.* **368**, 549 (2002).

[23] Fisher M.E., *Rev. Mod. Phys* **46**, 597 (1974).

[24] Fisher M.E., Aharony A., *Phys. Rev. Lett.* **30**, 559 (1973).

[25] Harris A.B., *J. Phys. C* **7**, 1671 (1974).

[26] Pawlak A., Fechner B., *Phys. Rev. B* **40**, 9324 (1989).

[27] Pawlak A., *Z. Phys. B* **79**, 279 (1990).

[28] Campostrini M., Hasenbusch M., Pelissetto A., Rossi P., Vicari E., *Phys. Rev. B* **63**, 214503 (2001).

[29] Campostrini M., Hasenbusch M., Pelissetto A., Rossi P., Vicari E., *Phys. Rev. B* **65**, 144520 (2002).

[30] Landau L.D., Lifszyc E. M., *Statistical Physics*, Adison-Wesley, Reading, MA, 1969.

[31] Van Hove L., *Phys. Rev.* **93**, 1374 (1954).

[32] Hohenberg P.C., Halperin B.I., *Phys. Rev. Lett.* **19**, 900 (1967).

[33] Ferrell R.A., Menyhard N., Schmidt H., Schwabl F., Szepfalusy P., *Phys. Rev. Lett.* **18**, 891 (1967).

[34] Langevin P., *Comptes Rendus Acad. Sci. Paris* **146**, 530 (1908).

[35] Ma S.-k., *Modern Theory of Critical Phenomena*, Benjamin/Cummings, Reading, MA, 1976.

[36] Amit D.A., *Field Theory, the Renormalisation Group and Critical Phenomena*, McGraw-Hill, New York 1978.

[37] Landau L.D., Lifszyc E.M., *Fluid Mechanics*, Pergamon, London 1959.

[38] Zwanzig R., *J. Chem. Phys.* **33**, 1388 (1960).

[39] Mori H., *Prog. Theor. Phys.* **33**, 423 (1965).

[40] Mori H., Fujisaka H., *Prog. Theor. Phys.* **49**, 764 (1973).

[41] Mori H., Fujisaka H., Shigematsu H., *Prog. Theor. Phys.* **51**, 109 (1974).

[42] Janssen H. K., *Z. Phys. B* **23**, 377 (1976).

[43] Bausch R., Janssen H. K., Wagner H., *Z. Phys. B* **24**, 113 (1976).

[44] Janssen H.K., [in:] C.P. Enz (ed.), *Proceedings of the International Conference on Dynamic Critical Phenomena*, Springer, Berlin 1979.

[45] De Dominicis C., Peliti L., *Phys. Rev. B* **18**, 353 (1978).

[46] Frey E., Schwabl F., *Adv. Phys.* **43**, 577 (1994).

[47] Taüber U. C., *Critical Dynamics*, http://www.phys.vt.edu/˜tauber/

[48] Halperin B.I., Hohenberg P.C., Ma S., *Phys. Rev. B* **10**, 139 (1974).

[49] Halperin B.I., Hohenberg P.C., Ma S., *Phys. Rev. B* **13**, 4119 (1976).

[50] Halperin B.I., Hohenberg P.C., Ma S., *Phys. Rev. Lett.* **29**, 1548 (1972).

[51] De Dominicis C., Brezin E., Zinn-Justin J., *Phys. Rev. B* **12,** 4945 (1975).

[52] Siggia E.D., Nelson D.R., *Phys. Rev.* **B** 15, 1427 (1977).

[53] Pawlak A., *Phys. Rev. B* **44**, 5296 (1991).

[54] Halperin B.I., Hohenberg P.C., Siggia E.D., *Phys. Rev. Lett.* **32**, 1289 (1974).

[55] Halperin B.I., Hohenberg P.C., *Phys. Rev. B* **188**, 898 (1969).

[56] Halperin B.I., Hohenberg P.C., Siggia E.D., *Phys. Rev. B* **13**, 1299 (1976).

[57] Riedel E. K., Wegner F. J. , *Phys. Rev. Lett.* **29**, 349 (1972).

[58] Ma S.-k., Mazenko G.F., *Phys. Rev. B* **11**, 4077 (1975).

[59] Chaikin P.M., Lubensky T.C., *Principles of Condensed Matter Physics*, Cambridge University Press, Cambridge 1995.

[60] Lüthi B., Moran T.J., Pollina R.J., *J. Phys. Chem. Solids* **31**, 1741 (1970).

[61] Cheeke J.D.N., *Fundamentals and Applications of Ultrasonic Waves,* CRC Press, Boca Raton 2002.

[62] Larkin A.I., Pikin S.A., *Ż.E.T.F.* **56**, 1664 (1969).

[63] Sak J., *Phys. Rev. B* **10**, 3957 (1974).

[64] Bergman D.J., Halperin B.I., *Phys. Rev. B* **13**, 2145 (1976).

[65] Bruno J., Sak J., *Phys. Rev. B* **22**, 3302 (1980).

[66] Landau L.D., Lifszyc E.M., *Theory of Elasticity*, PWN, Warszawa, PL, 1968.

[67] Riedel E.K., Wegner F.J., *Phys. Rev. B* **9**, 294 (1974).

[68] Pawlak A., *J. Phys. CM: Condens. Matter* **1**, 7989 (1989).

[69] Wegner F., [in:] C. Domb, M.S. Green (eds.), *Phase Transitions and Critical Phenomena*, V. 6, Academic Press, New York, 1976.

[70] Dengler R., Schwabl F., *Z. Phys. B* **69**, 327 (1987).

[71] Pawlak A., *The effecs of spin-phonon couplings on the critical sound propagation*, PhD thesis, A. Mickiewicz University, Poznań, 1987.

[72] Pawlak A., *phys. stat. sol.* (b) **236**, 474 (2003).

[73] Drossel B., Schwabl F., *Z. Phys. B* **91**, 93 (1993).

[74] Pankert J., Dohm V., *Phys. Rev. B* **40**, 10842 (1989).

[75] Pankert J., Dohm V., *Phys. Rev. B* **40**, 10856 (1989).

[76] Folk R., Moser G., *Phys. Rev. E* **58**, 6246 (1998).

[77] Fisher M.E., *Phys. Rev.* **176**, 257 (1968).

[78] Lamb J., [in:] W.P. Mason (ed.), *Physical Acoustics*, V. **1**, Academic Press, New York 1964.

[79] Herzfeld K.F., Rice F.O., *Phys. Rev.* **31**, 691 (1928).

[80] Bhattacharjee J.K., *Phys. Rev. B* **25**, 3404 (1982).

[81] Laramore G.E., Kadanoff L.P., *Phys. Rev.* **187**, 619 (1969).

[82] Schwabl F., *Phys. Rev. B* **7**, 2038 (1973).

[83] Murata K.K., *Phys. Rev. B* **13**, 4015 (1976).

[84] Lüthi B., Rehwald W., [in:] K.H. Müller, H. Thomas (ed.), *Topics in Current Physics*, Springer, Berlin 1981.

[85] Graham R., [in:] G. Höhler (ed.), *Springer Tracts in Modern Physics*, V. 66, Springer, Berlin 1973.

[86] Folk R.,Iro H., Schwabl F., *Phys. Rev. B* **20**, 1229 (1979).

[87] Iro H., Schwabl F., *Solid State Commun.* **46**, 205 (1983).

[88] Moran T.J., Lüthi B., *Phys. Rev. B* **4**, 122 (1970).

[89] Lau H.Y., Corliss L.M., Delapalme A., Hastings J.M., Nathans R., Tucciarone A., *J. Appl. Phys.* **41**, 1384 (1970).

[90] Kamilov I.K., Aliev H.K., *Usp. Fiz. Nauk* **168**, 953 (1998).

[91] Huber D.L., *Phys. Rev. B* **3**, 836 (1971).

[92] Itoh Y., *J. Phys. Soc. Jap.* **38**, 336 (1975).

[93] Bloembergen N., *Phys. Rev.* **78**, 572 (1950).

[94] Vaterlaus A., Beutler T., Meier F., *Phys. Rev. Lett.* **67**, 3314 (1991).

[95] Hübner W., Bennemann K.H., *Phys. Rev. B* **53**, 3422 (1996).

[96] Leung R. C., Zahradnik C., Garland C. W., *Phys. Rev. B*, **19**, 2612 (1979).

[97] Ferrell R. A., Bhattacharjee J. K., *Phys. Rev. A* **31**, 1788 (1985).

[98] Kroll D.M., Ruhland J.M., *Phys. Rev. A* **23**, 371 (1981).

[99] Ferrell R.A., Bhattacharjee J.K., *Phys. Rev. Lett.* **44**, 403 (1980).

[100] Ferrell R.A., Bhattacharjee J.K., *Phys. Rev. B* **23**, 2434 (1981).

[101] Bhattacharjee J.K., Ferrell R.A., *Phys. Rev. E* **56**, 5549 (1997).

[102] Martin P.C., Parodi O., Pershan P.S., *Phys. Rev. A* **6**, 2401 (1972).

[103] Folk R., Moser G., *J. Low. Temp. Phys.* **99**, 11 (1995).

[104] Landau L. D., Khalatnikov I. M., *Dokl. Akad. Nauk SSSR* **96**, 469 (1954); reprinted in *Collected Papers of L. D. Landau*, ed. D. ter Haar (Pergamon, London, 1965).

[105] Dengler R., PhD thesis, Technische Universität München 1987.

[106] Schorgg A. M., Schwabl F., *Phys. Rev. B* **46**, 8828 (1992).

[107] Pawlak A., *Phys. Rev. B* **68**, 094416 (2003).

[108] Dengler R., Schwabl F., *Europhys. Lett.* **4**, 1233 (1987).

[109] Folk R., Moser G., *Phys. Rev. E* **57**, 683, 705 (1998).

[110] Folk R., Moser G., *Phys. Rev. E* **58**, 6246 (1998).

[111] Marinelli M., Mercuri F., Belanger D.P., *Phys. Rev. B* **51**, 8897 (1995).

[112] Marinelli M., Mercuri F., Zammit U., Pizzoferrato R., Scuderi F., Dadarlat D., *Phys. Rev. B* **49**, 4356, 9523 (1994).

[113] Ikushima A., *J. Phys. Chem. Solids* **31**, 939 (1970).

[114] Ikushima A., *J. Phys. Chem. Solids* **31**, 283 (1970).

[115] Kawasaki K., Ikushima A., *Phys. Rev. B* **1**, 3143 (1970).

[116] Golding B., *Phys. Rev. Lett.* **20**, 5 (1968).

[117] Lüthi B., Pollina R.J., *Phys. Rev. Lett.* **22**, 717 (1969).

[118] Pawlak A., Fechner B., phys. *phys. stat. sol.* (c) **3**, 208, (2006).

[119] Ikushima A., Feigelson R., *J. Phys. Chem. Solids* **32**, 417 (1971).

[120] Marinelli M., Mercuri F., Foglietta S., Belanger D. P., *Phys. Rev. B* **54**, 4087 (1996).

[121] Pawlak A., *phys. stat. sol.* (c) **3**, 204, (2006).

[122] Bachellerie A., Frenois C.H., *J. Phys. (Paris)* **35**, 30 (1974).

[123] Bagnulus C., Bervillier C., *Phys. Rev. B* **32**, 7209 (1985).

[124] Kornblit A., Ahlers G., Buehler E., *Phys. Lett. A* **43**, 531 (1973).

[125] Golding B., Barmatz M., Buehler E., Salamon M.B., *Phys. Rev. Lett.* **30**, 968 (1973).

[126] Neighbours J.R., Oliver R.W., Stillwell R.W., *Phys. Rev. Lett.* **11**, 125 (1963).

[127] BelangerD.P.,NordbladP.,KingA.R.,JaccarinoV.,LundgrenL., Beckman O., *J. Magn. Magn. Mater.* **31–34**, 1095 (1983).

[128] Evans G., *Phys. Lett. A* **27**, 451 (1968).

[129] Leisure R.G., Moss R.W., *Phys. Rev.* **188**, 840 (1969).

[130] Ikushima A., *Phys. Lett. A* **29**, 417 (1969).

[131] Bachellerie A., Joffrin J., Levelut A., *Phys. Rev. B* **30**, 617 (1973).

[132] Lüthi B., Pollina R.J., *Phys. Rev.* **167**, 488 (1968).

[133] Long M., Stern R., *Phys. Rev. B* **4**, 4094 (1971).

[134] Pollina R.J., Lüthi B., *Phys. Rev.* **177**, 841 (1969).

[135] Suzuki M., Komatsubara T., *J. Phys. C* **18**, 5459 (1982).

[136] Golding B., *Phys. Rev. Lett.* **34**, 1102 (1975).

[137] Golding B., Barmaz M., *Phys. Rev. Lett.* **23**, 223 (1969).

[138] Henneberger S., Frey E., Maier P.G., Schwabl F., Kalvius G.M., *Phys. Rev. B* **60**, 9630 (1999).

[139] Meissner G., Pirc R., *Solid State Commun.* **33**, 253 (1980).

[140] Nattermann T., *phys. stat. sol. (b)* **85**, 291 (1978).

[141] Pawlak A., *Acta Phys. Pol. A* **115** , 229 (2009).

[142] Tachiki M. , Maekawa S., *Prog. Theor. Phys.* **51**, 1 (1974).

[143] Maekawa S., Treder R. A., Tachiki M., Lee M. C., Levy M., *Phys. Rev. B* **13**, 1284 (1976).

[144] Erdem R., *Phys. Lett. A* **312**, 238 (2003).

[145] Erdem R., Keskin M., *Phys. Lett. A* **326**, 27 (2004).

[146] Komatsubara T., Ishizaki A., Kusaka S., Hirahara E., *Solid State Commun.* **14**, 741 (1974).

[147] Pawlak A., Fechner B., *Acta Phys. Pol. A* **115** , 232 (2009).

In: Horizons in World Physics, Volume 268
Editors: M. Everett and L. Pedroza, pp. 65-95 ISBN 978-1-60692-861-5

Chapter 2

MAGNETISM IN PURE AND DOPED MANGANESE CLUSTERS

Mukul Kabir[1,*] ***Abhijit Mookerjee***[1] ***and D.G. Kanhere***[2]
[1] S. N. Bose National Centre for Basic Sciences, JD Block, Sector III,
Salt Lake City, Kolkata 700 098, India
[2] Department of Physics and Centre for Modeling and Simulation,
University of Pune, Pune - 411 007, India

Abstract

In this chapter, we report electronic and magnetic structure of pure and (As-) doped manganese clusters from density functional theory using generalized gradient approximation for the exchange-correlation energy. Ferromagnetic to ferrimagnetic transition takes place at $n = 5$ for pure manganese clusters, Mn_n, and remarkable lowering of magnetic moment is found for Mn_{13} and Mn_{19} due to their closed icosahedral growth pattern and results show excellent agreement with experiment. On the other hand, for As-doped manganese clusters, Mn_nAs, ferromagnetic coupling is found only in Mn_2As and Mn_4As and inclusion of a single As stabilizes manganese clusters. Exchange coupling in the Mn_nAs clusters are anomalous and behave quite differently from the Ruderman-Kittel-Kasuya-Yosida like predictions. Finally, possible relevance of the observed magnetic behaviour is discussed in the context of Mn-doped GaAs semiconductor ferromagnetism.

Key Words: Cluster, magnetism, semiconductor ferromagnetism.

1. Introduction

The terms "clusters" and "microclusters" are usually used to describe aggregates of number of atoms, starting with the diatomic molecule and reaching an upper bound of several hundred thousands of atoms, which are too small to resemble small pieces of crystals. They bridge the gap between the atom and crystals and serve as the dome of basic physics at reduced dimension. The structural, electronic and magnetic properties of these aggregates

*E-mail address: mukul@bose.res.in. (Corresponding author)

differ widely from their corresponding bulk material and these properties can change dramatically with the addition of just one or few atoms to it. Surface rearrangements can take place on crystals with adatoms, but these changes are less drastic than the changes occur when one or few atoms are added to smaller clusters.

Particularly, transition metal (TM) clusters and their magnetic properties are of more interest and are unique. How the magnetic properties behave in the reduced dimension and how it evolve with cluster size to the bulk? — are the great fundamental questions with a potential technological importance. Several unexpected magnetic ordering have already been reported in clusters: (1) *Non-zero magnetic moment in the clusters of nonmagnetic bulk material* — Cox and co-workers found that the bare rhodium clusters display non-zero magnetic moment for less than 60 atoms in it [1], which is an indication of either ferromagnetic or ferrimagnetic ordering even though the bulk rhodium is Pauli paramagnet at all temperature. In Rh_n clusters magnetism show strong size dependence: Magnetic moment per atom decreases as the cluster size increases and become non-magnetic above 60 atoms[1]. (2) *Enhancement of magnetic moment in the clusters which is already ferromagnetic in bulk.* For Fe_n clusters, it was found that the magnetic moment per atom oscillates with the size of the cluster, slowly approaching its bulk value[2], and finally (3) *Finite magnetic moment observed in the clusters which is antiferromagnetic as bulk* – The Cr_n [3]and Mn_n [4, 5] have shown such property. It has been found that Cr_n clusters have magnetic moment 0.5-1 μ_B per atom. Knickelbein observed, Mn_5 - Mn_{99} clusters posses finite magnetic moment, which is otherwise antiferromagnetic as bulk. All these magnetic measurements on free clusters have been done by using Stern-Gerlach molecular deflection experiments at "low" temperatures.

As we will be discussing the pure as well as doped manganese clusters in this chapter, we discuss some few aspects of manganese. Among all the $3d$ transition metal elements, manganese is unique as an atom, clusters or crystals. It has $3d^5\ 4s^2$ electronic configuration and has high (2.14 eV) $3d^5\ 4s^2 \rightarrow 3d^6\ 4s^1$ promotion energy. This large promotion energy reduces the degree of $s-d$ hybridization as atoms are brought together and leads to weaker bonding. For example, Mn_2 dimer is a weakly bound van der Walls dimer with very low binding energy, ranging $0.1 \pm 0.1 - 0.56 \pm 0.26$ eV per atom[6]. If an additional atom is added and the process keeps on going to form different sized clusters, the binding energy increases monotonically but the improvement is not much and remains the lowest among all the $3d$-transition metal clusters. In the solid phase, manganese exists in four allotropic forms exhibiting a complex phase diagram [7]. The stable form is known as $\alpha-$Mn, which has a very complex lattice structure with as much as 58 atoms in the unit cell and has lowest cohesive energy (2.92 eV). For an isolated manganese atom, according to Hund's rule, the half-filled localized $3d$ electrons give rise to a magnetic moment of 5 μ_B.

We discuss electronic and magnetic properties of Mn_n ($n \leq 20$) and Mn_nAs ($n \leq 10$) clusters from the density functional theory. The pure manganese clusters undergo a ferromagnetic $\rightarrow$ ferrimagnetic transition at $n = 5$. In the recent Stern-Gerlach experiments a few interesting observations were made [4, 5]. (1) Although the experimental uncertainty in the measurement decreases with the increase in cluster size, Knickelbein found an extraordinarily large uncertainty in the measured magnetic moment of Mn_7, 0.72 ± 0.42 μ_B/atom[5]. (2) A relative decrease in the magnetic moment was observed at $n = 13$ and 19. We have discussed these issues in a recent paper[8]. Parks *et al.* found that free Mn_n

clusters are nonreactive towards H_2 for small sizes $n \leq 15$. However, this reaction rate show an abrupt increase at $n = 16$ [9] and it was argued to be attributed from a nonmetal-metal transition at $n = 16$. All these issues are discussed in a great detail here.

Next we will move on to see how a single dopant affect the electronic and magnetic properties. Recently we observed [10] that if we dope a single As-atom into a Mn_n cluster that binding energy of the resultant Mn_nAs cluster improves substantially due to their strong $p - d$ hybridization, i.e. As-atom stabilizes the Mn_n clusters. Mn_2As and Mn_4As are the only clusters that we found to have ferromagnetic Mn-Mn coupling and for all other sizes this coupling is turned out to be ferrimagnetic. It was also found that the exchange coupling in these doped-clusters are anomalous and behave differently from the Ruderman-Kittel-Kasuya-Yosida like predictions. All these issues will be discussed in the context of semiconductor ferromagnetism. Before going to the results, we briefly discuss the projector augmented wave formalism of pseudopotential what we have been used throughout.

2. The Projector-Augmented-Wave Formalism

The most widely used electronic structure methods can be divided into two classes. First one is the linear method [11] developed from the augmented-plane-wave method [12, 13] and Koringa-Kohn-Rostocar method [14, 15] and the second one is the norm-conserving pseudopotentials developed by Hamann, Schlüter and Chiang [16]. In that scheme, inside some core radius, the all electron (AE) wave function is replaced by a soft nodeless pseudo (PS) wave function, with the restriction to the PS wave function that within the chosen core radius the norm of the PS wave function have to be the same with the AE wave function and outside the core radius both the wave functions are just identical. However, the charge distribution and moments of AE wave function are well reproduced by the PS wave function only when the core radius is taken around the outer most maxima of AE wave function. Therefore, elements with strongly localized orbitals pseudopotentials require a large plane wave basis set. This was improved by Vanderbilt [17], where the norm-conservation constraint was relaxed and a localized atom centered augmentation charges were introduced to make up the charge deficit. But the success is partly hampered by rather difficult construction of the pseudopotential. Later Blöchl [18] has developed the projector-augmented- wave (PAW) method, which combines the linear augmented plane wave method with the plane wave pseudopotential approach, which finally turns computationally elegant, transferable and accurate method for electronic structure calculation. Further Kresse and Joubert [19] modified this PAW method and implemented in their existing Veina ab-initio pseudopotential package (VASP). Here in this section we will discuss briefly the idea of the method.

2.1. Wave Functions

The exact Kohn-Sham density functional is given by,

$$E = \sum_n f_n \langle \Psi_n | -\frac{1}{2}\nabla^2 | \Psi_n \rangle + E_H[n + n_Z] + E_{xc}[n], \tag{1}$$

where $E_H[n + n_Z]$ is the hartree energy of the electronic charge density n and the point charge densities of the nuclei n_Z, $E_{xc}[n]$ is the electronic exchange-correlation energy and

f_n are the orbital occupation number. $|\Psi_n\rangle$ is the all-electron wave function. This physically relevant wave functions $|\Psi_n\rangle$ in the Hilbert space exhibit strong oscillations and make numerical treatment difficult. Transformation of this wave functions $|\Psi_n\rangle$ into a new pseudo wave functions $|\tilde{\Psi}_n\rangle$ in the PS Hilbert space,

$$|\Psi_n\rangle = \tau|\tilde{\Psi}_n\rangle, \tag{2}$$

within the augmentation region Ω_R, then makes PS wave function $|\tilde{\Psi}_n\rangle$ computationally convenient.

Let us now choose a PS partial wave function $|\tilde{\phi}_i\rangle$, which is identical to the corresponding AE partial waves $|\psi_i\rangle$ outside the augmentation region and form a complete set within the augmentation region Ω_R. Within the augmentation region every PS wave function can be expanded into PS partial waves,

$$|\tilde{\Psi}\rangle = \sum_i c_i|\tilde{\phi}_i\rangle, \tag{3}$$

where the coefficients c_i are scalar products,

$$c_i = \langle\tilde{p}_i|\tilde{\Psi}_n\rangle, \tag{4}$$

with some fixed function $\langle\tilde{p}_i|$ of the PS wave function, which is called the projector function. By using the linear transformation[18],

$$\tau = 1 + \sum_i (|\phi_i\rangle - |\tilde{\phi}_i\rangle)\langle\tilde{p}_i|, \tag{5}$$

the corresponding AE wave function $|\Psi_n\rangle$ in the Eq. 1.2 is then,

$$|\Psi_n\rangle = |\tilde{\Psi}_n\rangle + \sum_i (|\psi_i\rangle - |\tilde{\psi}_i\rangle)\langle\tilde{p}_i|\tilde{\Psi}_n\rangle, \tag{6}$$

where i refers to the atomic site $\mathbf{R}$, the angular momentum quantum numbers $L = l, m$ and an additional index k for the reference energy ϵ_{kl} and, with $\tilde{p}_i$ being the projector functions, which within the augmentation region Ω_R satisfy the condition,

$$\langle\tilde{p}_i|\tilde{\phi}_j\rangle = \delta_{ij}. \tag{7}$$

The AE charge density at a given point $\mathbf{r}$ is the expectation value of the real-space projector operator $|r\rangle\langle r|$ and hence given by,

$$n(\mathbf{r}) = \tilde{n}(\mathbf{r}) + n^1(\mathbf{r}) - \tilde{n}^1(\mathbf{r}), \tag{8}$$

where $\tilde{n}(\mathbf{r})$ is the soft PS charge density calculated from the PS wave function

$$\tilde{n}(\mathbf{r}) = \sum_n f_n\langle\tilde{\Psi}_n|\mathbf{r}\rangle\langle\mathbf{r}|\tilde{\Psi}_n|\mathbf{r}\rangle, \tag{9}$$

and onsite charge densities are defined as,

$$n^1(\mathbf{r}) = \sum_{ij} \rho_{ij}\langle\phi_i|\mathbf{r}\rangle\langle\mathbf{r}|\phi_j\rangle, \tag{10}$$

and,

$$\tilde{n}^1(\mathbf{r}) = \sum_{ij} \rho_{ij} \langle \tilde{\phi}_i | \mathbf{r} \rangle \langle \mathbf{r} | \tilde{\phi}_j \rangle. \tag{11}$$

ρ_{ij} are the occupation of each augmentation channel (i, j) and are calculated from the PS wave functions applying the projector function,

$$\rho_{ij} = \sum_n f_n \langle \tilde{\Psi}_n | \tilde{p}_i \rangle \langle \tilde{p}_j | \tilde{\Psi}_n \rangle. \tag{12}$$

It should be pointed out here that for a complete set of projector functions the charge density $\tilde{n}^1$ is exactly same as $\tilde{n}$ within Ω_R.

2.2. The Total Energy Functional

The total energy can be written as a sum of three terms,

$$E = \tilde{E} + E^1 - \tilde{E}^1, \tag{13}$$

where the first term,

$$\begin{aligned}\tilde{E} &= \sum_n f_n \langle \tilde{\Psi}_n | -\frac{1}{2}\nabla^2 | \tilde{\Psi}_n \rangle + E_{xc}[\tilde{n} + \hat{n} + \tilde{n}_c] + E_H[\tilde{n} + \hat{n}] \\ &+ \int v_H[\tilde{n}_{Zc}][\tilde{n}(\mathbf{r}) + \hat{n}(\mathbf{r})]d\mathbf{r} + U(\mathbf{R}, Z_{ion}). \end{aligned} \tag{14}$$

$n_{Zc}(\tilde{n}_{Zc})$ is the sum of the point charge density of the nuclei $n_Z(\tilde{n}_z)$ and the frozen core charge density $n_c(\tilde{n}_c)$, i.e. $n_{Zc} = n_z + n_c$ and $\tilde{n}_{Zc} = \tilde{n}_z + \tilde{n}_c$ and $\hat{n}$ is the compensation charge, which is added to the soft charge densities $\tilde{n} + \tilde{n}_{Zc}$ and $n^1 + n_{Zc}$ to reproduce the correct multipole moments of the AE charge density $n^1 + n_{Zc}$ located in each augmentation region. As n_{Zc} and $\tilde{n}_{Zc}$ have same monopole $-Z_{ion}$ and vanishing multipoles, the compensation charge $\hat{n}$ is chosen so that $\tilde{n}^1 + \hat{n}$ has the same moments as AE valence charge density n^1 has, within each augmentation region. $U(\mathbf{R}, Z_{ion})$ is the electrostatic interaction potential between the cores.

The second term in the total energy is,

$$\begin{aligned}E^1 &= \sum_{ij} \rho_{ij} \langle \phi_i | -\frac{1}{2}\nabla^2 | \phi_j \rangle + \overline{E_{xc}[n^1 + n_c]} + \overline{E_H[n^1]} \\ &+ \int_{\Omega_r} v_H[n_{n_{Zc}}] n^1(\mathbf{r}) d\mathbf{r}. \end{aligned} \tag{15}$$

and the final term is,

$$\begin{aligned}\tilde{E}^1 &= \sum_{ij} \rho_{ij} \langle \tilde{\phi}_i | -\frac{1}{2}\nabla^2 | \tilde{\phi}_j \rangle + \overline{E_{xc}[\tilde{n}^1 + \hat{n} + \tilde{n}_c]} + \overline{E_H[\tilde{n}^1 + \hat{n}]} \\ &+ \int_{\Omega_r} v_H[\tilde{n}_{Zc}][\tilde{n}^1(\mathbf{r}) + \hat{n}(\mathbf{r})] d\mathbf{r}, \end{aligned} \tag{16}$$

In all these three energy terms, the electrostatic potential $v_H[n]$ and electrostatic energy $E_H[n]$ of charge density n is given by:

$$v_H[n](\mathbf{r}) = \int \frac{n(\mathbf{r}')}{|\mathbf{r}-\mathbf{r}'|} d\mathbf{r}', \tag{17}$$

$$E_H[n] = \frac{1}{2}\int d\mathbf{r} \int d\mathbf{r}' \frac{n(\mathbf{r})n(\mathbf{r}')}{|\mathbf{r}-\mathbf{r}'|}. \tag{18}$$

In the total energy functional the smooth part $\tilde{E}$ is evaluated on a regular grids in Fourier or real space, and the two one-center contributions $E^1$and $\tilde{E}^1$ are evaluated on radial grids for each sphere individually.

2.3. Compensation Charge Density

The compensation charge density $\hat{n}$ is defined such that $\tilde{n}^1 + \hat{n}$ has exactly the same moments as the AE charge density n^1 has, within the augmentation region, which then requires,

$$\int_{\Omega_r} (n^1 - \tilde{n}^1 - \hat{n})|\mathbf{r}-\mathbf{R}|^l Y_L^*(\widehat{\mathbf{r}-\mathbf{R}}) dr = 0. \tag{19}$$

The charge difference Q_{ij} between the AE and PS partial wave for each channel (i, j) within the augmentation region is defined by,

$$Q_{ij}(\mathbf{r}) = \phi_i^*(\mathbf{r})\phi_j(\mathbf{r}) - \tilde{\phi}_i^*(\mathbf{r})\tilde{\phi}_j(\mathbf{r}), \tag{20}$$

and their moments q_{ij}^L are,

$$q_{ij}^L = \int_{\Omega_r} Q_{ij}(\mathbf{r})|\mathbf{r}-\mathbf{R}|^l Y_L^*(\widehat{\mathbf{r}-\mathbf{R}}) d\mathbf{r}, \tag{21}$$

which has non zero values only for certain combinations of L, i and j. Then the compensation charge density can be rewritten as,

$$\hat{n} = \sum_{i,j,L} \rho_{ij} \hat{Q}_{ij}^L(\mathbf{r}), \tag{22}$$

where,

$$\hat{Q}_{ij}^L = q_{ij}^L g_l(|\mathbf{r}-\mathbf{R}|) Y_L(\widehat{\mathbf{r}-\mathbf{R}}), \tag{23}$$

where $g_l(r)$ are the functions for which the moment is equal to 1.

2.4. Operators

Let us consider some operator $\mathcal{O}$, so its expectation value $\langle \mathcal{O} \rangle = \sum_n f_n \langle \Psi_n | \mathcal{O} | \Psi_n \rangle$, where n is the band index and f_n is the occupation of the state. As in the PAW method we work with the PS wave function, we need to obtain observables as the expectation values of PS wave function. Applying the form of the transformation τ and using $\sum_i |\tilde{\phi}_i\rangle\langle\tilde{p}_i| = 1$ within the augmentation region Ω_R and $|\tilde{\phi}_i\rangle = |\phi_i\rangle$ outside the augmentation region, for quasilocal operators, the PS operator $\tilde{\mathcal{O}}$ has the following form[1]:

$$\begin{aligned}\tilde{\mathcal{O}} &= \tau^\dagger \mathcal{O} \tau \\ &= \mathcal{O} + \sum_{ij} |\tilde{p}_i\rangle(\langle\phi_i|\mathcal{O}|\phi_j\rangle - \langle\tilde{\phi}_i|\mathcal{O}|\tilde{\phi}_j\rangle)\langle\tilde{p}_j|. \end{aligned} \tag{24}$$

Overlap operator: The PS wave function obey orthogonality condition,

$$\langle\tilde{\Psi}_n|\mathcal{S}|\tilde{\Psi}_m\rangle = \delta_{nm}, \tag{25}$$

where $\mathcal{S}$ is the overlap operator in the PAW approach. The overlap matrix in the AE representation is given by the matrix elements of unitary operator. Therefore, $\mathcal{S}$ has the form given by,

$$\mathcal{S}[\{\mathbf{R}\}] = 1 + \sum_i |\tilde{p}_i\rangle[\langle\phi_i|\phi_j\rangle - \langle\tilde{\phi}_i|\tilde{\phi}_j\rangle]\langle\tilde{p}_j|. \tag{26}$$

Hamiltonian operator: The Hamiltonian operator is defined as the first derivative of the total energy functional with respect to the density operator, $\tilde{\rho} = \sum_n f_n|\tilde{\Psi}_n\rangle\langle\tilde{\Psi}_n|$,

$$H = \frac{dE}{d\tilde{\rho}}, \tag{27}$$

where the PS density operator $\tilde{\rho}$ depends on the PS charge density $\tilde{n}$ and on the occupancies of each augmentation channel ρ_{ij}. So the variation of the total energy functional can be rewritten as,

$$\frac{dE}{d\tilde{\rho}} = \frac{\partial E}{\partial\tilde{\rho}} + \int \frac{\delta E}{\delta\tilde{n}(\mathbf{r})}\frac{\partial\tilde{n}(\mathbf{r})}{\partial\tilde{\rho}}d\mathbf{r} + \sum_{i,j}\frac{\partial E}{\partial\rho_{ij}}\frac{\partial\rho_{ij}}{\partial\tilde{\rho}}. \tag{28}$$

The partial derivative of $\tilde{E}$ with respect to the PS density operator is the kinetic energy operator and the variation with respect to $\tilde{n}(\mathbf{r})$ leads to the usual one-electron potential $\tilde{v}_{eff}$,

$$\frac{\partial\tilde{E}}{\partial\tilde{\rho}} = -\frac{1}{2}\nabla^2, \tag{29}$$

and,

$$\frac{\delta\tilde{E}}{\delta\tilde{n}(\mathbf{r})} = \tilde{v}_{eff} = v_H[\tilde{n} + \hat{n} + \tilde{n}_{Zc}] + v_{xc}[\tilde{n} + \hat{n} + \tilde{n}_c]. \tag{30}$$

As in the smooth part of the total energy functional, $\tilde{E}$, the occupancies ρ_{ij} enter only through the compensation charge density $\hat{n}$, the variation of $\tilde{E}$ with respect to ρ_{ij} is given by,

$$\begin{aligned}\hat{D}_{ij} = \frac{\partial\tilde{E}}{\partial\rho_{ij}} &= \int \frac{\delta\tilde{E}}{\delta\hat{n}(\mathbf{r})}\frac{\partial\hat{n}(\mathbf{r})}{\partial\rho_{ij}}d\mathbf{r} \\ &= \sum_L \int \tilde{v}_{eff}(\mathbf{r})\hat{Q}_{ij}^L(\mathbf{r})d\mathbf{r}. \end{aligned} \tag{31}$$

[1]The kinetic energy operator $-1/2\nabla^2$ and the real space projector operators $|r\rangle\langle r|$ are quasi local operators. For truely nonlocal operators an extra term must be added to the $\tilde{\mathcal{O}}$ expression.

In the remaining two terms E^1 and $\tilde{E}^1$ in the energy functional, ρ_{ij} enters directly via kinetic energy or through n^1, $\tilde{n}^1$ and $\hat{n}$. Now the variation of E^1 with the occupancies ρ_{ij} is given by,

$$D^1_{ij} = \frac{\partial E^1}{\partial \rho_{ij}} = \langle \phi_i | -\frac{1}{2}\nabla^2 + v^1_{eff} | \phi_j \rangle, \tag{32}$$

where,

$$v^1_{eff}[n^1] = v_H[n^1 + n_{Zc}] + v_{xc}[n^1 + n_c],$$

and the variation of $\tilde{E}^1$ is given by,

$$\tilde{D}^1_{ij} = \frac{\partial \tilde{E}^1}{\partial \rho_{ij}} = \langle \tilde{\phi}_i | -\frac{1}{2}\nabla^2 + \tilde{v}^1_{eff} | \tilde{\phi}_j \rangle + \sum_L \int_{\Omega_r} d\mathbf{r} \tilde{v}^1_{eff}(\mathbf{r}) \hat{Q}^L_{ij}(\mathbf{r}), \tag{33}$$

where,

$$\tilde{v}^1_{eff}[\tilde{n}^1] = v_H[\tilde{n}^1 + \hat{n} + \tilde{n}_{Zc}] + v_{xc}[\tilde{n}^1 + \hat{n} + \tilde{n}_c].$$

The onsite terms D^1_{ij} and $\tilde{D}^1_{ij}$ are evaluated on radial grid within each augmentation region and they restore the correct shape of the AE wave function within the sphere. The final form of the Hamiltonian operator:

$$H[\rho, \{\mathbf{R}\}] = -\frac{1}{2}\nabla^2 + \tilde{v}_{eff} + \sum_{i,j} |\tilde{p}_i\rangle (\hat{D}_{ij} + D^1_{ij} - \tilde{D}^1_{ij}) \langle \tilde{p}_j |. \tag{34}$$

2.5. Forces in the PAW Method

Forces are usually defined as the total derivative of the energy with respect to the ionic position $\mathbf{R}$,

$$\mathbf{F} = -\frac{dE}{d\mathbf{R}}. \tag{35}$$

The total derivative consists of two terms, first one is the forces on the ionic core and the second term is the correction due to the change of AE wave functions for fixed PS wave functions when ions are moved. This correction term comes because augmentation depends on the ionic positions and are known as Pulay force [20]. When calculating this Pulay correction one must consider the overlap between the wave functions due to the change in ionic position. According to Goedecker and Maschke[21], the total derivative can be written as,

$$\frac{dE}{d\mathbf{R}} = \frac{\partial U(\mathbf{R}, Z_{ion})}{\partial \mathbf{R}} + \sum_n f_n \left\langle \tilde{\Psi}_n \left| \frac{\partial (H[\rho, \{\mathbf{r}\}] - \epsilon_n S[\{\mathbf{R}\}])}{\partial \mathbf{R}} \right| \tilde{\Psi}_n \right\rangle, \tag{36}$$

where the first term is the forces between the ionic cores and ϵ_n are the Kohn-Sham eigenvalues, and the PS wave functions assumed to satisfy the orthogonality condition $\langle \tilde{\Psi}_n | S | \tilde{\Psi}_m \rangle = \delta_{nm}$ and the corresponding Kohn-Sham equation, reads $H|\tilde{\Psi}_n\rangle = \epsilon_n S |\tilde{\Psi}_n\rangle$.

The first contribution, $\mathbf{F}^1$ to the second term of the total derivative comes from the change in the effective local potential $\tilde{v}_{eff}$ when the ions are moved and $\tilde{v}_{eff}$ depends explicitly on the ionic positions via $\hat{n}_{Zc}$,

$$\mathbf{F}^1 = -\int \left[\frac{\delta Tr[H\tilde{\rho}]}{\delta v_H[\tilde{n}_{Zc}](\mathbf{R})} \frac{\partial v_H[\tilde{n}_{Zc}](\mathbf{R})}{\partial \mathbf{R}} \right] d\mathbf{R}. \tag{37}$$

This equation can be further simplified to,

$$\begin{aligned}\mathbf{F}^1 &= -\int \left(\tilde{n}(\mathbf{r}) + \sum_{i,j,L} \hat{Q}_{ij}^L(\mathbf{r})\rho_{ij} \right) \frac{\partial v_H[\tilde{n}_{Zc}](\mathbf{r})}{\partial \mathbf{R}} d\mathbf{r} \\ &= -\int [\tilde{n}(\mathbf{r}) + \hat{n}(\mathbf{r})] \frac{\partial v_H[\tilde{n}_{Zc}](\mathbf{r})}{\partial \mathbf{R}} d\mathbf{r}. \end{aligned} \tag{38}$$

The second contribution arise from $\hat{D}_{ij}$ due to the changes in the compensation charge density $\hat{n}$, when ions are moved,

$$\mathbf{F}^2 = -\sum_{i,j,L} \int \tilde{v}_{eff}(\mathbf{r})\rho_{ij} q_{ij}^L \frac{\partial g_l(|\mathbf{r}-\mathbf{R}|) Y_L(\widehat{\mathbf{r}-\mathbf{R}})}{\partial \mathbf{R}} d\mathbf{r}. \tag{39}$$

These two terms, $\mathbf{F}^1$ and $\mathbf{F}^2$, together describe the electrostatic contributions to the force. The third term comes due to the change in the projector function $\tilde{p}_i$ as ions are moved,

$$\mathbf{F}^3 = -\sum_{n,i,j} (\hat{D}_{ij} + D_{ij}^1 - \tilde{D}_{ij}^1 - \epsilon_n q_{ij}) f_n \left\langle \tilde{\Psi}_n \left| \frac{\partial |\tilde{p}_i\rangle\langle\tilde{p}_j|}{\partial \mathbf{R}} \right| \tilde{\Psi}_n \right\rangle, \tag{40}$$

where $q_{ij} = \langle\phi_i|\phi_j\rangle - \langle\tilde{\phi}_i|\tilde{\phi}_j\rangle$.

As the exchange-correlation potential depends on the nonlinear core corrections $\tilde{n}_c$ gives an additional contribution, which can be obtained from the total energy functional,

$$\mathbf{F}^{\text{nlcc}} = -\int v_{xc}[\tilde{n} + \hat{n} + \tilde{n}_c] \frac{\partial \tilde{n}_c(\mathbf{r})}{\partial \mathbf{R}} d\mathbf{r}. \tag{41}$$

All the differences between the PAW method and ultra-soft pseudopotential are automatically absorbed in the in the definition of $(\hat{D}_{ij} + D_{ij}^1 - \tilde{D}_{ij}^1)$. $D_{ij}^1 - \tilde{D}_{ij}^1$ are constant for US-PP, where they are calculated once and forever whereas in PAW method they vary during the calculation of the electronic ground state.

3. Computational Method

The electronic and magnetic calculations of the pure and doped- manganese clusters were carried out using density functional theory (DFT), within the pseudopotential plane wave method [16, 17], using projected augmented wave (PAW) formalism [18, 19] and Perdew-Burke-Ernzerhof (PBE) exchange-correlation functional [22] for the spin-polarized generalized gradient approximation (GGA), as it is implanted in VASP code [23]. The wave functions are expanded in a plane wave basis set with the kinetic energy cutoff equal to 337.3 eV and reciprocal space integrations were carried out at the Γ point. The $3d$, $4s$ for Mn-atom and $4s$, $4p$ orbitals for As-atom were treated as valence states. Symmetry unrestricted geometry optimizations were performed using quasi Newtonian and conjugate gradient methods until all the force components are less than 0.005 eV/Å. Simple cubic supercells are used with neighbouring clusters separated by at least 12Å vacuum regions. Several initial structures were studied to ensure that the globally optimized geometry does not correspond to the local minima, as well as, for all clusters, we have explicitly considered *all possible* spin multiplicities to determine the magnetic ground state.

4. Results

4.1. Magnetic Transition: Pure Mn_n ($n \leq 20$) Clusters

We start with the magnetism in the pure manganese clusters. Theory and experimental reports are in contradiction for the Mn_2 dimer. More than 30 years ago, Nesbet [24] calculated binding energy, bond length and magnetic moment of the dimer at the restricted Hartree-Fock (RHF) level and predicted an antiferromagnetic (AFM) ground state with bond length 2.88 Å. Later on, unrestricted Hartree-Fock calculation was done by Shillady *et al.* [25] and found a ferromagnetic (FM) ground state with total spin 10 μ_B and bond length 3.50 Å. The experiments based on resonant Raman spectroscopy [26] and Electron Spin Resonance (ESR)[27] predicted an AFM ground state with a bond length 3.4 Å. However, all DFT calculations [28, 29, 8, 30] within different levels of approximation and using different exchange-correlation functionals, predict a FM ground state with much smaller bond length, $\sim$ 2.60 Å than the experimental bond length. A comparison among different levels of theory is given in the Table1.

Table 1. Summary of binding energy (E_b), bond length (R_e) and magnetic moment (μ) of Mn_2 reported by various authors.

Authors	Method	E_b (eV)	R_e (Å)	μ (μ_B)
Nesbet [24]	RHF + Heisenberg exchange	0.79	2.88	0
Wolf and Schmidtke [31]	RHF		1.52	0
Shillady *et al.* [25]	UHF	0.08	3.50	10
Fujima and Yamaguchi [32]	LSDA	0.70	3.40	0
Harris and Jones [33]	LSDA	1.25	2.70	10
Salahub and Baykara [34]	LSDA	0.86	2.52	0
Nayak *et al.* [28]	All-electron + GGA (BPW91)	0.91	2.60	10
Parvanova *et al.*[29]	All-electron + GGA (PBE)	0.98	2.60	10
Kabir *et al.* [8]	Pseudopotential + GGA (PBE)	1.06	2.58	10
Pederson *et al.* [30]	All-electron + GGA	0.99	2.61	10
Experiment	ESR in rare-gas matrix [27]	0.1 ± 0.1	3.4	0

Due to the filled $4s$ and half-filled $3d$ electronic configuration, as well as high (2.14 eV) $4s^2\,3d^5 \rightarrow 4s\,3d^6$ promotion energy Mn atoms do not bind strongly as two Mn atoms come closer to form Mn_2 dimer, and as a result Mn_2 is very weakly bond van der Walls dimer, which is also evident from the low experimental value, 0.01 ± 0.1 — 0.56 ± 0.26 eV/atom [6]. However, no experimental results available in the gas-phase and all the experiments are done in the gas matrix, and therefore, it is quite possible that the Mn atoms interact with the matrix, which could stretch R_e and could lead to the AFM ground state.

From the Table2, we see that R_e decreases monotonically as the net moment decreases. It is simply because the reduction of the atomic spacing leads to comparatively stronger overlap of the atomic orbitals which reduces the magnetic moment. The manganese trimer and tetramer is found to have FM ground state with 5 μ_B/atom magnetic moment. Resonance Raman spectroscopy study by Bier *et al.*[26] suggested a distorted D_{3h} structure with

Table 2. The binding energy and equilibrium bond length of Mn_2 dimer for all possible spins.

spin	Binding energy (eV)	Bond length
0 (AFM)	0.51	2.57
2	0.44	1.53
4	unbound	1.73
6	unbound	1.94
8	0.47	2.24
10	1.06	2.58

odd-integer magnetic moment for trimer. Manganese tetramer in solid silicon was studied by Ludwig *et al.* [35] and observed a 21-line hyperfine pattern which confirmed that the all four atoms are equivalent with a total magnetic moment 20 μ_B.

Calculated binding energy per atom, relative energies of the isomers, magnetic moments per atom with the experimental comparison are given in the Table3 (n = 2-10) and Table4 (n = 2-20) and the calculated magnetic moment per atom corresponding to the ground state are plotted in the Figure1. We see a FM $\rightarrow$ ferrimagnetic transition takes place at n=5. and remains the same for larger size clusters [8]. Figure1 shows very good agreement with the recent Stern-Gerlach experiments[4, 5]. However, we find, for all sizes there exists several

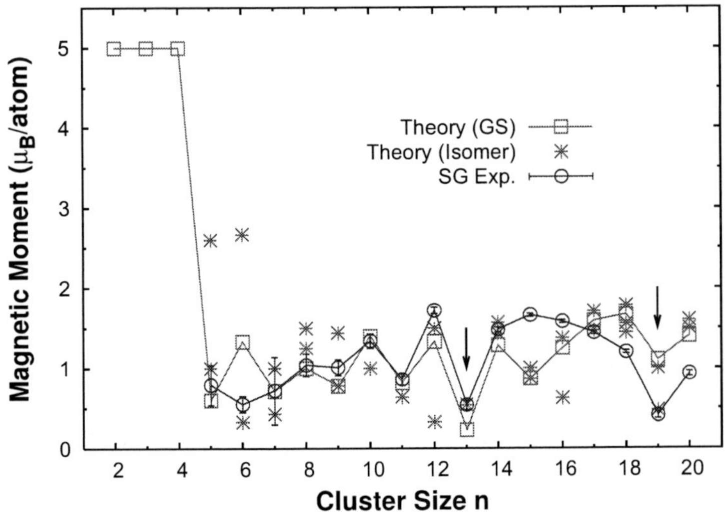

Figure 1. Variation of magnetic moment with cluster size for the size range $n = 2-20$. A comparison is made with the experimentally measured value [5]. For all sizes, the magnetic moments of closely lying isomers are also plotted.

Table 3. Binding energy per atom, relative energy to the ground state ($\triangle E = E - E_{GS}$) and per atom magnetic moment for Mn_n ($n \leq 10$) clusters. Predicted magnetic moments are compared with the Stern-Gerlach experiment.[4, 5]

Cluster	BE (eV/atom)	$\triangle E$ (eV)	Magnetic Moment (μ_B/atom)	
			Theory[8]	Experiment[4, 5]
Mn_2	0.530	0.000	5.000	—
	0.255	0.531	0.000	
Mn_3	0.823	0.000	5.000	—
	0.808	0.046	1.667	
Mn_4	1.179	0.000	5.000	—
	1.160	0.078	2.500	
	1.130	0.196	0.000	
	1.130	0.195	2.000	
Mn_5	1.413	0.000	0.600	0.79 ± 0.25
	1.401	0.059	2.600	
	1.399	0.069	1.000	
	1.374	0.193	4.600	
Mn_6	1.567	0.000	1.333	0.55 ± 0.10
	1.564	0.017	0.333	
	1.559	0.045	2.667	
	1.543	0.142	4.333	
Mn_7	1.726	0.000	0.714	0.72 ± 0.42
	1.713	0.091	1.000	
	1.699	0.193	0.429	
Mn_8	1.770	0.000	1.000	1.04 ± 0.14
	1.770	0.000	1.500	
	1.765	0.039	1.250	
Mn_9	1.867	0.000	0.778	1.01 ± 0.10
	1.856	0.099	1.444	
	1.850	0.151	0.778	
	1.844	0.211	1.000	
Mn_{10}	1.936	0.000	1.400	1.34 ± 0.09
	1.934	0.015	1.000	
	1.935	0.007	0.400	
	1.935	0.009	0.400	

isomers with different magnetic structure[8].

In the SG experiment [5] it has been seen that the experimental uncertainty in measurement of magnetic moment generally decreases with the cluster size as the cluster production efficiency increases with the size. However, this uncertainty for Mn_7 is quite high, 0.72 ± 0.42 μ_B/atom[5], which is 58% of the measured value. Now we discuss what might be the possible reason for this. We predict [8] the Mn_7 to be a pentagonal bi-pyramidal (PBP) structure in its ground state with magnetic moment 5 μ_B, which is exactly the experimental value. In addition, we found another two PBP isomers, with magnetic moment 7 μ_B and 3 μ_B, to be close in energy to the ground state, which are shown in the Figure2. They lie 0.09

Table 4. Same as Table3 for Mn_n ($n = 11$— 20) clusters.

Cluster	BE (eV/atom)	$\triangle E$ (eV)	Magnetic Moment (μ_B/atom) Theory[8]	Experiment[4, 5]
Mn_{11}	1.993	0.000	0.818	0.86 ± 0.07
	1.984	0.107	0.455	
	1.980	0.153	0.636	
Mn_{12}	2.081	0.000	1.333	1.72 ± 0.04
	2.077	0.052	0.333	
	2.071	0.114	1.500	
Mn_{13}	2.171	0.000	0.231	0.54 ± 0.06
	2.165	0.076	0.538	
Mn_{14}	2.171	0.000	1.286	1.48 ± 0.03
	2.170	0.021	1.429	
	2.167	0.055	1.571	
Mn_{15}	2.231	0.000	0.867	1.66 ± 0.02
	2.229	0.033	0.333	
	2.228	0.057	0.467	
	2.228	0.049	0.867	
	2.227	0.064	1.000	
	2.213	0.281	0.467	
Mn_{16}	2.274	0.000	1.250	1.58 ± 0.02
	2.273	0.017	1.375	
	2.271	0.055	0.625	
	2.268	0.097	0.500	
Mn_{17}	2.327	0.000	1.588	1.44 ± 0.02
	2.323	0.069	1.471	
	2.322	0.087	1.706	
Mn_{18}	2.351	0.000	1.667	1.20 ± 0.02
	2.350	0.018	1.556	
	2.350	0.020	1.444	
	2.348	0.064	1.778	
Mn_{19}	2.373	0.000	1.105	0.41 ± 0.04
	2.373	0.009	1.000	
	2.369	0.076	0.474	
Mn_{20}	2.370	0.000	1.400	0.93 ± 0.03
	2.370	0.003	1.500	
	2.368	0.055	1.600	
	2.367	0.067	0.800	

eV and 0.20 eV higher in energy. Therefore, the presence of these two isomers along with the ground state in the SG molecular beam could be a possible explanation to the observed large uncertainty. But it should be pointed out here that this might be a possible argument but never the certain reason, because for all sizes there exist more than one isomer with different magnetic structure, but the corresponding experimental uncertainty is not large.

Now we move to the sudden deep observed in the measured magnetic moment at n=13 and 19 compared to their neighbours[4, 5]. Through *ab initio* calculation, we reproduced

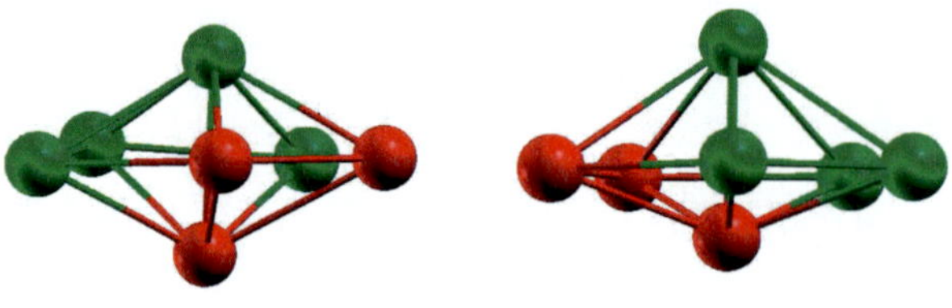

$5\ \mu_B$ $7\ \mu_B$ $3\ \mu_B$

Figure 2. Ground state and isomeric configuration for Mn_7. Note that for all configurations number of spin up atoms and spin down atoms are the same ($N_\uparrow = 4$, $N_\downarrow = 3$). Green(Red) ball represents up(down) atom and we follow the same convention throughout.

this behaviour[8]. For Mn_{13}, we studied icosahedral, cubooctahedral and hexagonal structures for all possible multiplicities and we found the icosahedral structure with magnetic moment 3 μ_B to be the ground state. The optimal cuboctahedral and hexagonal structure have 9 and 11 μ_B magnetic moment, which is much higher than the SG experimental value,$0.54 \pm 0.06\ \mu_B$[4, 5], and they lie much higher in energy than the icosahedral ground state. They are shown in the Figure3. Similarly, the ground state of Mn_{19} is found to be double icosahedral structure with magnetic moment 21 μ_B (Figure4). The optimal FCC structure (Figure4) lie 1.53 eV higher in energy. So in our recent paper we concluded that the "closed" icosahedral geometric structure is responsible for the observed sudden deep.

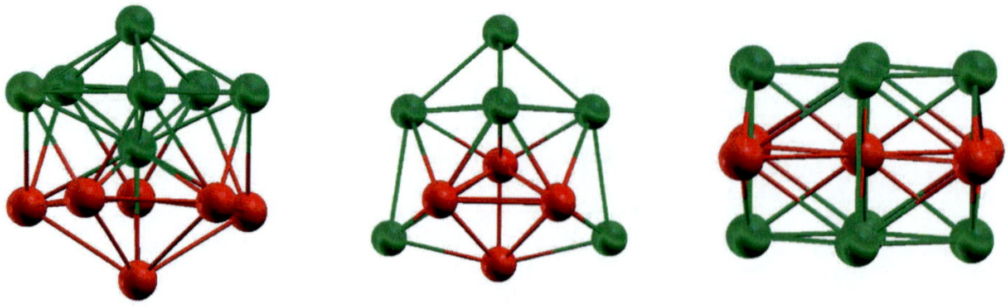

Icosahedral	Hexagonal	Cubooctahedral
$\Delta E = 0.00$ eV	$\Delta E = 0.89$ eV	$\Delta E = 1.12$ eV
$3\ \mu_B$	$9\ \mu_B$	$11\ \mu_B$

Figure 3. Spin configuration for the icosahedral ground state of Mn_{13}. Lowest energy spin configurations for hexagonal and cubooctahedral structure have also been shown.

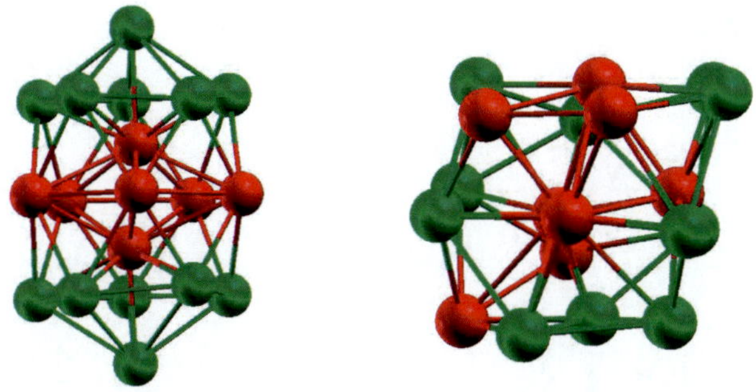

Double-icosahedral
$\triangle E = 0.00$ eV
21 μ_B

FCC
$\triangle E = 1.53$ eV
17 μ_B

Figure 4. Ground state configuration for Mn_{19} is a double icosahedron. Optimal FCC structure lies much higher in energy.

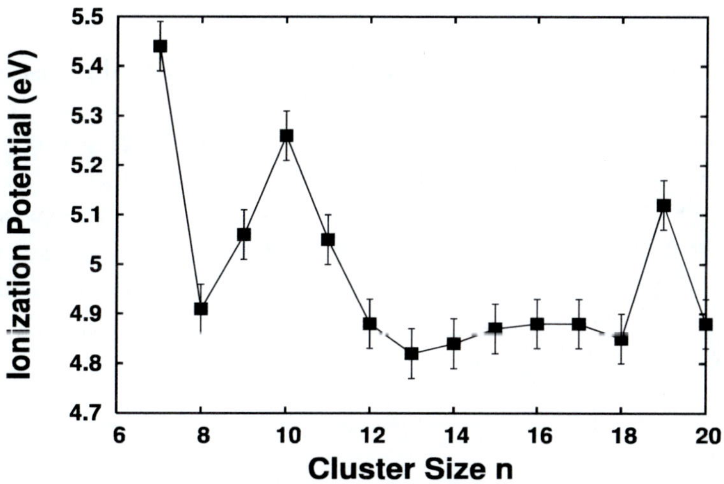

Figure 5. Experimentally measured ionization potential for Mn_n clusters in the size range n = 2-20.(From Ref.[37].)

4.2. Non-metal to Metal Transition?

Parks *et al.* produced Mn_n clusters containing up to 70 atoms by cooling the inert gas condensation source to -160^{0}C to study their reaction rate with the molecular hydrogen[9]. They found that the clusters with $n \leq 15$ are nonreactive toward H_2, whereas they form stable hydride above $n = 15$ and the reaction rate varies considerably with cluster size. It was thought that this might arise due to the possible abrupt change in the bonding character at $n = 16$: For small clusters $n \leq 15$, the bonding is weak and of van der Walls kind, which perhaps become metallic at $n = 16$ and remains the same for larger clusters. If it is so then once the cluster is metallic, it is energetically possible to transfer charge from metallic

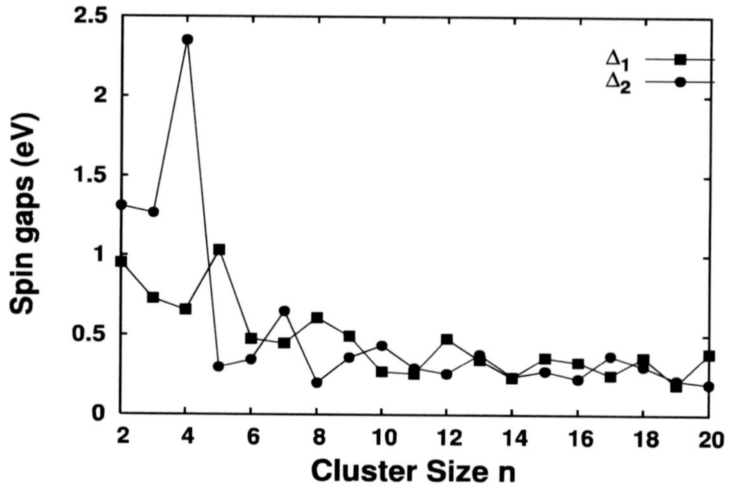

Figure 6. Variation of spin gaps with cluster size in the size range $n = 2$—20. (From Ref.[8].)

cluster to the antibonding state of the H_2 molecule and H-H bond would consequently break and H atom would attach to the metal cluster. If this is indeed the case, it is likely that a significant downward change in the ionization potential would be observed at Mn_{16}. The similar effect has observed in the free mercury clusters[36], where a steady decrease in the ionization potentials attribute to the nonmetal-metal transition. We also expect the closing up of the two spin gaps for this kind of transition. Koretsky *et al.* measured ionization potentials of Mn_n clusters in the size range $n = 7 — 64$[37]. However, no sudden decrease is observed at $n = 16$ (Figure5).

Spin gaps[2] are defined as follows:

$$\begin{aligned} \Delta_1 &= -\left[\epsilon_{\text{HOMO}}^{\text{majority}} - \epsilon_{\text{LUMO}}^{\text{minority}}\right] \\ \Delta_2 &= -\left[\epsilon_{\text{HOMO}}^{\text{minority}} - \epsilon_{\text{LUMO}}^{\text{majority}}\right], \end{aligned} \tag{42}$$

and we plot them in the Figure6. We did not find any closing up of these spin gaps. Therefore, there is no evidence of nonmetal-metal transition and the observed sudden change in the reactivity is not due to any kind of structural transition at $n = 16$ either, as all the medium sized clusters adopt icosahedral growth pattern. So the reason for the abrupt change in the reaction rate of Mn_n clusters with H_2 at $n = 16$ is yet unknown.

4.3. Localization of $d-$electrons

In finite size clusters coordination number may very drastically for different sites and, therefore, they are very good system to study the fact that how the $d-$electrons get localized with

[2] Any spin arrangement for these Mn_n clusters is magnetically stable only if both the spin gaps are positive. This means that the lowest unoccupied molecular orbital (LUMO) of the minority(majority) spin lies above the highest occupied molecular orbital(HOMO) of the majority(minority) spin.

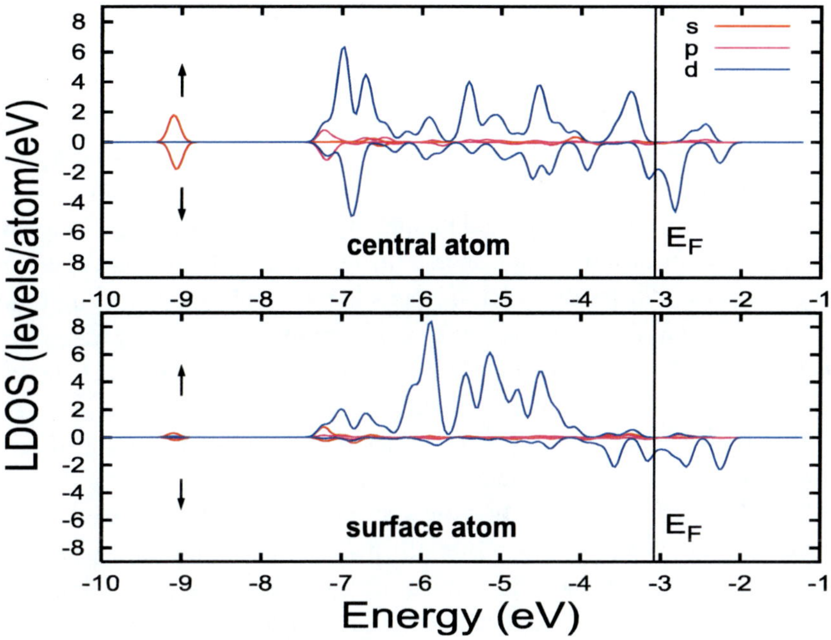

Figure 7. The local density of states (LDOS) for the central and a surface atom for Mn_{13} cluster.

the coordination. Take an example of icosahedral Mn_{13} cluster, where the bonding character of the 12 coordinated central atom is very much different from the the 6-coordinated surface atoms. Here to study the effect of localization, mainly of $d-$ electrons, on the coordination, we calculate the angular momentum projected local density of states, which is plotted in the Figure7 for Mn_{13} cluster. We see that the $d-$ projected local density of states (LDOS) of the central atom are broad for both the majority and minority spin states, whereas the same for the surface atoms are rather localized and the majority spin states are nearly fully occupied. This is also reflected from the local magnetic moments of the central and surface atoms, which is defined as follows:

$$\mathcal{M}_i = \int_0^R \left[\rho^{\uparrow}(\mathbf{r}) - \rho^{\downarrow}(\mathbf{r})\right] d\mathbf{r}, \tag{43}$$

where $\rho^{\uparrow}(\mathbf{r})$ and $\rho^{\downarrow}(\mathbf{r})$ are spin up and spin down densities, respectively and R is the radius of the integrating sphere centering each atom. The local magnetic moment of the central atom is small, 1.42 μ_B, compared to the surface atoms $\sim$ 3.60 μ_B due to the above mentioned reason.

5. Single As-doping in Mn_n Clusters: Mn_nAs ($n \leq 10$)

We already have discussed the pure Mn_n clusters and have seen that very small clusters containing up to 4 atoms are ferromagnetic and they undergo a magnetic (ferromagnetic to ferrimagnetic) transition at n=5. Now it will be interesting to see that what happens to the magnetic structure of those if we dope a single arsenic into it to form Mn_nAs species. In the previous section, we have also noticed that the binding energy of pure Mn_n clusters are small compared to other $3d$−transition metal clusters due to its $3d^5\ 4s^2$ electronic structure. Here we will also discuss that what happens to the bonding due to single As-doping. Particularly magnetism in these Mn_nAs clusters are more interesting, as manganese-doped semiconductors, such as (GaMn)N, (GaMn)As and (InMn)As, have attracted considerable attention because of their carrier induced ferromagnetism[38, 39, 40, 41, 42]. The Mn dopants in these III-V dilute magnetic semiconductors serve the dual roles of provision of magnetic moments and acceptor production. A wide range of T_c, 10-940 K, has been reported for $Ga_{1-x}Mn_xN$ by varying Mn concentration (x ∼7-14%) as well as by varying the growth conditions[43, 44, 45, 46]. But still it is unclear at present, whether all these reports of ferromagnetism are indeed intrinsic or arise due to the some kind of 'defects' originated during the growth, as magnetic atoms are not thermodynamically stable in the semiconductor host and tend to form 'defects'. Recent investigation[51] found that ferromagnetism persists up to ∼ 980 K for metastable Mn-doped ZnO and further heating transforms the metastable phase and kills the ferromagnetism. Earlier experimental results on $Ga_{1-x}Mn_xAs$ indicate a nonmonotonic behaviour of $T_c(x)$, first increases with the Mn concentration x, reaching a maximum of 110 K for x ∼5% and then decreases with the further increase of x[47, 48, 49, 50]. However, recent experimental studies [52, 48, 53], under carefully controlled growth and annealing conditions, suggest that the 'metastable' nature and high 'defect' content (such as clustering of Mn) of low temperature of MBE grown $Ga_{1-x}Mn_xAs$ may be playing an important role in the magnetic properties and could enhance T_c with Mn content. Chiba *et al.* [52] have reported T_c as high as 160 K for 7.4 % Mn concentration, in a layered structure. These studies rule out the earlier suggestion that the T_c in $Ga_{1-x}Mn_xAs$ is fundamentally limited to 110 K and certainly opened up a hope for further possible increase in T_c for such metastable phase under different suitable growth and annealing conditions. Towards the end of the chapter, we will discuss this recently debated clustering phenomena, which at least puts some light on the issue.

5.1. Geometry and Bonding

We start with the MnAs dimer, which has much higher binding energy of 1.12 eV/atom and much shorter bond length, 2.21 Å compared with the pure Mn_2 cluster, which has the values 0.53 eV/atom and 2.58 Å, respectively[10]. The ground state of the Mn_2As is an isosceles triangle and the binding energy gets substantially increased to 1.63 eV/atom. The Mn-Mn bond, 2.59 Å, in this Mn_2As cluster is nearly equal to that of the Mn_2 dimer. As more Mn-atoms are added to the Mn_2As cluster, the resultant clusters adopt three dimensional shape and, moreover, the determination of the ground state is far more difficult as it is known that the number of local minima in the potential energy surface is an exponential function of the number of atoms present in that cluster.

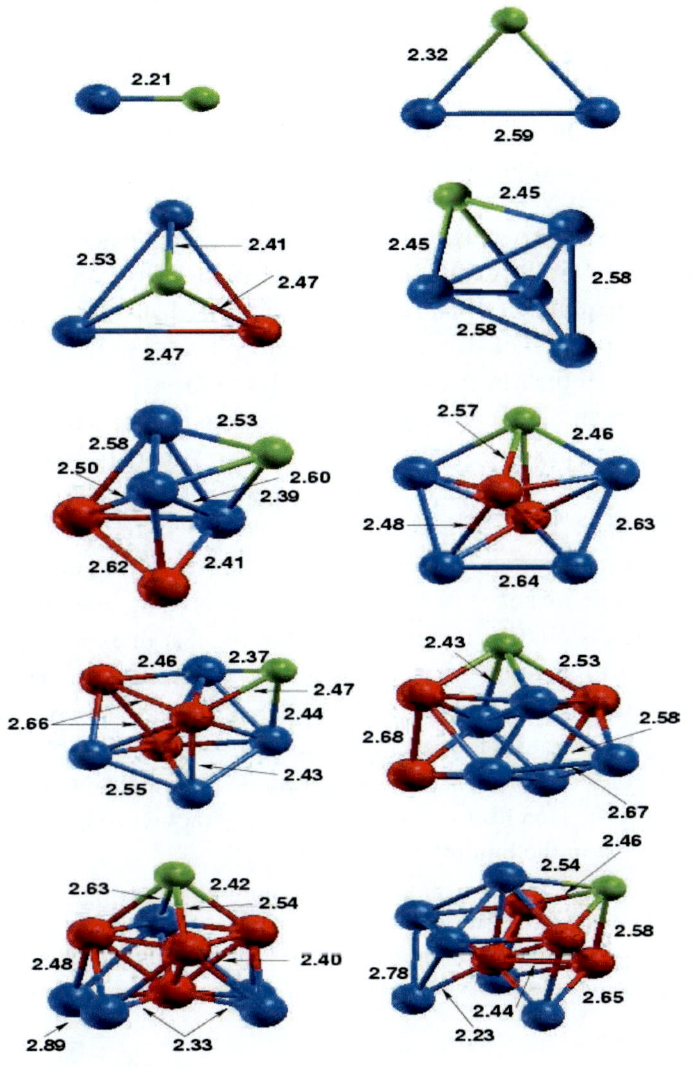

Figure 8. Ground state geometries and spin ordering of Mn_xAs clusters, x=1-10. Blue and red balls represent $Mn_\uparrow$ and $Mn_\downarrow$ atoms, respectively. Green ball represents the As atom. The bond lengths are given in Å. Magnetic polarization of As, is negative for MnAs - Mn_8As and, whereas, positive for Mn_9As and $Mn_{10}As$. Note, Mn atoms are coupled ferromagnetically only in Mn_2As and Mn_4As clusters.

Calculated binding energy, total magnetic moment, shortest Mn-Mn and Mn-As bond lengths and two spin gaps are given in the Table5 for the ground state along with the isomers for Mn_nAs cluster. The presence of an As-atom makes Mn_3As cluster tetrahedral, however, Mn_4As is a Mn_4 tetrahedra with As at a face cap. Generally, the ground state structures of Mn_nAs clusters can be seen as that of the corresponding pure Mn_n cluster with As capped at some face, i.e. As-doping does not give rise to any considerable structural change, but a moderate perturbation to the corresponding pure cluster structure. However, the Mn_6As

cluster is the only cluster whose geometry differs significantly from the pure Mn_6 cluster. Mn_6 is octahedral, whereas Mn_6As is a pentagonal bipyramid, where the As atom is trapped in the pentagonal ring. As n in the Mn_nAs increases, binding energy increases monotonically due to the increase in the coordination number. However, the increase rate is very slow after $n = 2$: Increases slowly from 1.73 eV/atom for Mn_3As to 2.22 eV/atom for $Mn_{10}As$. The shortest Mn-As bond length increases from 2.21 Å for MnAs dimer to 2.46 Å for $Mn_{10}As$ cluster, whereas the shortest Mn-Mn bond length in these clusters decreases with cluster size (2.59 Å for Mn_2As to 2.23 Å for $Mn_{10}As$). Generally, we find, the bonds between Mn atoms of opposite spin to be somewhat shorter (2.20-2.60 Å) than the bonds between Mn atoms of like spin (2.50-2.90 Å), whereas Mn-As distance varies between 2.20-2.60 Å. All the Mn-As-Mn bond angles in these clusters vary in between $\sim$ 60-70^0. All the clusters in the Figure8 and their respective isomers are magnetically stable i.e. both the spin gaps are positive. These two spin gaps, $\triangle_1$ and $\triangle_2$, for Mn_2As (0.83 and 1.34 eV) and Mn_4As (0.89 and 1.14 eV) are the highest among all clusters. As x increases, $\triangle_1$ and $\triangle_2$ decrease to a value 0.47 and 0.35 eV, respectively, for $Mn_{10}As$ (see Table5).

Now we discuss the important issue whether the Mn-clustering around the single As-atom are at all fafourable or not. First we look into the binding energy of pure Mn_n and Mn_nAs clusters, which is plotted in the Figure9. We see that due to single As-doping the binding energy of the Mn_nAs clusters are substantially enhanced from their respective pure counterpart. As it is discussed earlier that due to the half-filled $3d$ and filled $4s$ shell and high $3d^5\,4s^2 \rightarrow 3d^6\,4s^1$ promotion energy (2.12 eV), Mn atoms do not bind strongly when they are brought together to form clusters. However, as a single As-atom is attached to the pure Mn_n clusters, the $4s^2$ electrons of the Mn-atom interact with the $4p^3$ electrons of As and results enhancement in the binding energy.

To understand this better, one can calculate the energy gains which are of two kind[10]:

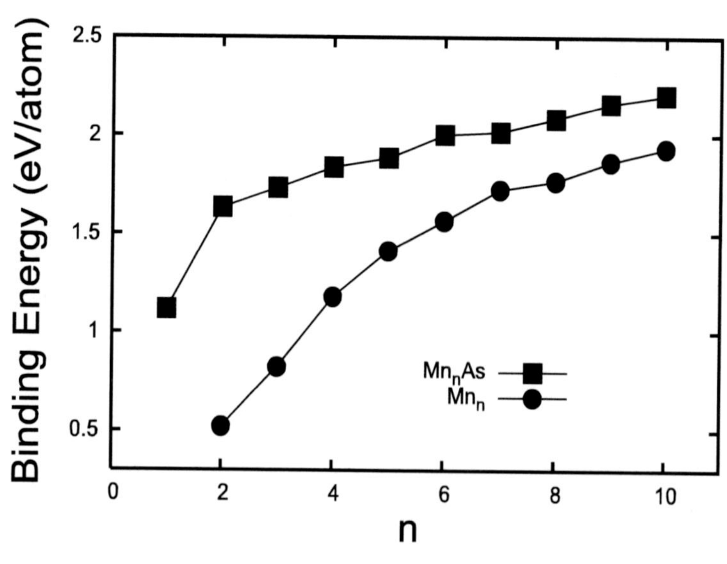

Figure 9. Plot of binding energy per atom for pure and doped manganese clusters, Mn_n and Mn_nAs, in the size range $n \leq 10$.

- The energy gain due to adding an As-atom to a pure Mn_n cluster to form $\mathrm{Mn}_n\mathrm{As}$ cluster,

$$\triangle^1 = -[E(\mathrm{Mn_nAs}) - \mathrm{E}(\mathrm{Mn_n}) - \mathrm{E}(\mathrm{As})], \tag{44}$$

and,

- The energy gain due to adding a Mn-atom to a existing $\mathrm{Mn}_{n-1}\mathrm{As}$ cluster to from $\mathrm{Mn}_n\mathrm{As}$ cluster,

$$\triangle^2 = -[E(\mathrm{Mn_nAs}) - \mathrm{E}(\mathrm{Mn_{n-1}As}) - \mathrm{E}(\mathrm{Mn})]. \tag{45}$$

These two energy gains are plotted as a function of Mn concentration in the Figure10. The $\triangle^1$ increases with n: The increment from $n = 1$ to $n = 3$ is monotonous and much stiff but afterwards it is not monotonic and tends to saturate. The other energy gain, $\triangle^2$, gives the number that how many Mn-atoms can be bonded to a single As-atom, which is significant, 2.65 eV, even for $\mathrm{Mn}_{10}\mathrm{As}$ cluster. Both the energy gains are positive and the energy gain in adding an As-atom is much greater than that of adding a Mn-atom ($\triangle^1 >> \triangle^2$) and, therefore, we can conclude that the clustering of Mn-atom around a single As is favourable[10]. So, the clustering of Mn is possible around As during the growth in Mn-doped GaAs/InAs. However, the other factors like lattice distortions and available space will play a role in determining the size of Mn clustering in these semiconductors.

5.2. Magnetic Order

Now we ask the next important question that what is the Mn-Mn magnetic coupling in these $\mathrm{Mn}_n\mathrm{As}$ clusters? The total magnetic moments of $\mathrm{Mn}_n\mathrm{As}$ clusters corresponding to

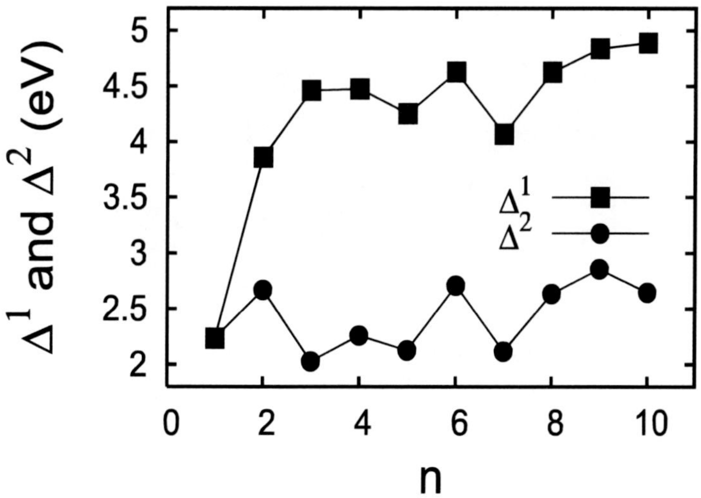

Figure 10. Plot of two different energy gains, $\triangle_1$ and $\triangle_2$ as a function of size.

Table 5. Binding energy, total moment, lowest Mn-Mn ($d_{\mathrm{Mn-Mn}}$) and Mn-As ($d_{\mathrm{Mn-As}}$) bond length and two spin gaps ($\triangle_1$ and $\triangle_2$) for Mn$_n$As (n = 1-10) clusters for their respective ground and isomeric state.

Cluster	B.E. (eV/atom)	Total moment μ_B	$d_{\mathrm{Mn-Mn}}$ Å	$d_{\mathrm{Mn-As}}$ Å	$\triangle_1$ eV	$\triangle_2$ eV
MnAs	1.116	4	—	2.208	0.709	1.016
Mn_2As	1.634	9	2.591	2.316	0.826	1.342
	1.597	1	2.428	2.246	0.582	1.345
Mn_3As	1.733	4	2.473	2.413	0.651	0.504
	1.702	12	2.481	2.372	1.295	0.523
Mn_4As	1.838	17	2.573	2.453	0.891	1.144
Mn_5As	1.886	2	2.417	2.389	0.856	0.381
	1.871	12	2.401	2.464	0.620	0.665
Mn_6As	2.004	9	2.469	2.461	0.540	0.669
	1.974	1	2.362	2.435	0.014	0.623
Mn_7As	2.018	6	2.430	2.375	0.409	0.593
	2.014	14	2.405	2.436	0.470	0.611
	2.012	6	2.323	2.472	0.456	0.402
Mn_8As	2.086	7	2.307	2.424	0.405	0.347
	2.075	3	2.305	2.478	0.072	0.552
	2.074	5	2.227	2.468	0.468	0.378
Mn_9As	2.163	10	2.306	2.416	0.418	0.250
	2.158	12	2.308	2.449	0.159	0.826
$Mn_{10}As$	2.207	13	2.225	2.455	0.467	0.347
	2.204	3	2.200	2.501	0.484	0.372

Figure 11. Constant spin density surfaces for Mn_2As and Mn_4As corresponding to 0.04 e/Å^3. Red and blue surfaces represent positive and negative spin densities, respectively. Green ball is the As atom, which has negative polarization in both the structures. Note ferromagnetic coupling among Mn atoms in the both Mn_2As and Mn_4As clusters.

the ground state geometries are 4, 9, 4, 17, 2, 9, 6, 7, 10 and 13 μ_B, respectively, for n=1-10 (Table5). These large magnetic moments generally arise form the ferrimagnetic coupling between the moments at Mn sites with exceptions for Mn_2As and Mn_4As, where the magnetic coupling is found to be ferromagnetic. We note, generally, no change in the magnetic coupling between the Mn sites in Mn_nAs clusters from their respective Mn_n, however, for Mn_3As, the Mn-Mn coupling behaviour changes form ferromagnetic to ferrimagnetic, due to As doping in Mn_3. Another tetrahedral isomer with magnetic moment 12 μ_B is 0.12 eV higher, where all the Mn atoms are ferromagnetically ordered. Here we should point out, as mentioned in the Table3, a frustrated antiferromagnetic structure with total magnetic moment 5 μ_B is only 0.05 eV higher in energy than the ferromagnetic Mn_3 cluster. The ground state magnetic moments of Mn_nAs, for n=1, 2, 3, 5, 7, and 10 can be represented as $(\mu_n - 1)$ μ_B, whereas for n=4, 6, 8, 9, it can be expressed as $(\mu_n - 3)$ μ_B, where μ_n is the total magnetic moment of the Mn_n cluster corresponding to the ground state or the 'first' isomer. It should be pointed out here that, for n=6 and 7, the ground state magnetic moments of Mn_nAs clusters are higher than that of Mn_n, as due to As doping, Mn_nAs can favour the isomeric magnetic structure of the corresponding Mn_n cluster without changing the overall magnetic behaviour between the Mn sites. For MnAs dimer, the magnetic moment at Mn site, $\mathcal{M}_{\rm Mn}$, is 3.72 μ_B and $\mathcal{M}_{\rm As}$ is -0.26 μ_B for As. This large negative polarization of the anion, As, is due to the strong $p-d$ interaction. Polarized neutron diffraction study found a local magnetic moment -0.23 ± 0.05 μ_B at the As sites for NiAs-type MnAs [54], which is very close to the present value for MnAs dimer. In the Mn_2As cluster, Mn atoms are ferromagnetically coupled with $\mathcal{M}_{\rm Mn}$=3.79 μ_B each, whereas the magnetic polarization is negative for As atom, -0.14 μ_B. Mn atoms show ferrimagnetic ordering with $\mathcal{M}_{\rm Mn}$ 3.1, 3.1 and -3.9 μ_B for Mn_3As, whereas, As has $\mathcal{M}_{\rm As}$=-0.21 μ_B, which gives a total magnetic moment 4 μ_B. For Mn_4As, Mn atoms are ferromagnetically arranged with average M_{Mn}=3.66 μ_B and are coupled antiferromagnetically with As atom, $\mathcal{M}_{\rm As}$=-0.22 μ_B. Nature of magnetic coupling can be visualize through their respective spin iso-density plots, which is shown in the Figure11 for Mn_2As and Mn_4As clusters. For Mn_5As, $\mathcal{M}_{\rm Mn}$ varies between 3.04-3.72 μ_B with a local magnetic moment -0.23 μ_B at As site. We observe, the negative polarization of As, $\mathcal{M}_{\rm As}$, decreases sharply to -0.08 μ_B for Mn_6As and further decreases monotonically to $-$ 0.02 μ_B for Mn_8As, however it become positive, 0.04 and 0.02, for Mn_9As and $Mn_{10}As$, respectively. For all Mn_xAs clusters, x =6-10, Mn atoms are coupled ferrimagnetically, where $\mathcal{M}_{\rm Mn}$ varies between 0.8 - 3.7 μ_B. These are unlike the study of Rao and Jena [55], where Mn atoms were found to be coupled ferromagnetically for all sizes, n=1-5, in nitrogen doped Mn_n clusters.

6. Exchange Coupling

To determine the exchange interactions J_{ij}'s between the atoms i and j, we map the magnetic energy onto a classical Heisenberg Hamiltonian:

$$\mathcal{H} = - \sum_{ij} J_{ij}(r)\, \mathbf{S}_i \cdot \mathbf{S}_j, \tag{46}$$

where the sum is over all distinct magnetically coupled atoms, $\mathbf{S}_i(\mathbf{S}_j)$ denotes the localized magnetic moments at $i(j)$-th Mn-site. $r = |\mathbf{R}_i - \mathbf{R}_j|$ is the Mn-Mn spatial separation. Here, rather than the conventional $|\mathbf{S}_i| = 1$ consideration, it is taken as $|\mathcal{M}_i|/2$, where

$$\mathcal{M}_i = \int_0^{R'} [\rho_i^{\uparrow}(\mathbf{r}') - \rho_i^{\downarrow}(\mathbf{r}')] \, d\mathbf{r}'. \tag{47}$$

The reason for this kind of consideration is that, for Mn_nAs clusters Mn atoms have different environment and so as the bonding, and therefore, it is very much likely that $|\mathbf{S}_i|$ have (slightly or much) different value at different site i. In the Eqn 1.47, R' is the radius of the integrating sphere centering each atom and taken as the half of the lowest Mn-Mn separation in the cluster to make sure that no two spheres overlap. $\rho^{\uparrow}(\mathbf{r}')$ and $\rho^{\downarrow}(\mathbf{r}')$ are the up and down spin densities, respectively, at a point $\mathbf{r}'$. It should be noted that in the construction of the Hamiltonian $\mathcal{H}$ in the Eqn.1.46, the signs (but not the magnitudes) are already absorbed in the definition of J_{ij}: positive for ferromagnetic and negative for antiferromagnetic coupling. We can calculate J_{ij}'s from Eqn.1.46, by computing the total energy for a judicious choice of spin configurations with inequivalent combinations of pair correlation functions $\mathbf{S}_i \cdot \mathbf{S}_j$, which results in a set of linear equations for the J_{ij}'s.

We can then compare the calculated J_{ij}'s with the Ruderman-Kittel-Kasuya-Yosida (RKKY)-like theory. This kind of theory describes the magnetic interaction between the localized Mn moments induced by the free carrier spin polarization. This indirect Mn-Mn exchange interaction arise from the local Zener coupling (or the $p-d$ hybridization) between the holes and the Mn d-levels, which then leads to the effective Mn-Mn RKKY interaction:

$$J_{ij}^{\mathrm{RKKY}}(r) \propto r^{-4} \left[\sin(2k_F r) - 2k_F r \cos(2k_F r)\right], \tag{48}$$

or, in the simplest approximation it reads,

$$J_{ij}^{\mathrm{RKKY}}(r) \propto r^{-3} \cos(2k_F r), \tag{49}$$

k_F is the Fermi wave vector. In dilute magnetic semiconductors, RKKY-like models predict that J_{ij} increases with concentration as $n^{1/3}$ at 0K and for fixed n, are independent of environment. We will check the case for As-doped Manganese (Mn_nAs) clusters.

For the smallest possible cluster Mn_2As, the Mn-Mn exchange coupling J can be calculated from the energy difference between the two different spin orientations: parallel ($\uparrow\uparrow$) and antiparallel ($\uparrow\downarrow$) as,

$$E_{\uparrow\downarrow}(r) - E_{\uparrow\uparrow}(r) = J(r) \left[|S_1^{\uparrow}| \, |S_2^{\downarrow}| + |S_1^{\uparrow}| \, |S_2^{\uparrow}| \right], \tag{50}$$

which is the energy measure to flip either one of the moments. Spatial ($r_{\mathrm{Mn-Mn}}$) dependence of $E_{\uparrow\downarrow}$, $E_{\uparrow\uparrow}$ and exchange coupling J is plotted in the Figure12. Calculated J is compared with the simplest RKKY analytic form $J^{\mathrm{RKKY}}(r) \propto r^{-3} \cos(2k_F r)$. Exchange coupling J oscillates between positive and negative with r favouring FM and AFM solutions, respectively and dies down as $1/r^3$ — a typical RKKY type behaviour.

The case of Mn_3As is very interesting because of the possible magnetic frustration. The Mn moment orientation in the ground state of Mn_3As nanostructure is $\uparrow\uparrow\downarrow$ with a total moment 4 μ_B (Figure13). Energies of different Mn-spin orientations are given in the Table6.

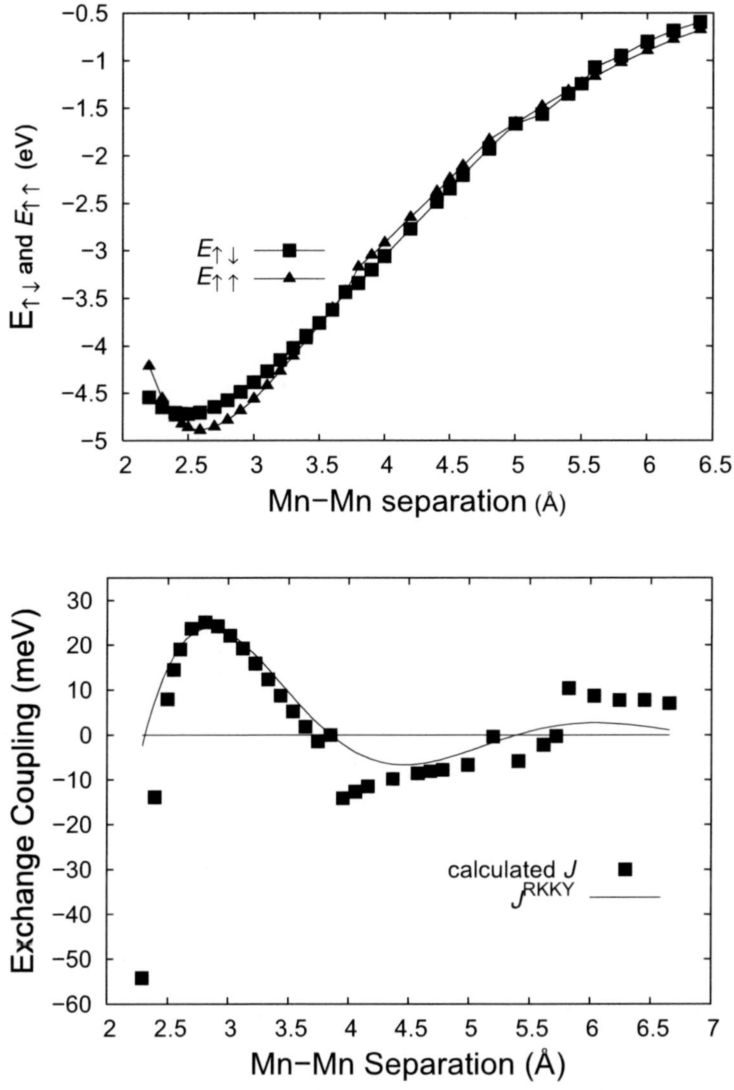

Figure 12. (upper) Spatial dependence of total energy for ↑↑ and ↑↓ configuration of Mn moment in Mn_2As cluster. (lower) Plot of Mn-Mn exchange coupling J with $r_{\text{Mn}-\text{Mn}}$. and compared with the simplistic RKKY-like form fitted with $k_F = 1.02$ Å^{-1}.

However, instead of two negative and one positive values, all the computed exchange couplings J_{ij}'s turn out to be negative ($J_{12} = -11$ meV and $J_{23}, J_{13} = -14$ meV) indicating the magnetic frustration of Mn-spins.

Now, it is interesting to see, how exchange couplings behave with the increasing cluster size n in Mn_nAs clusters. We plot averaged exchange coupling $\bar{J}_{ij}$ in the Figure 14, which behaves quite differently form RKKY-like theory: $\bar{J}_{ij}$ decreases as n increases with an exceptional increase for ferromagnetic Mn_4As cluster and it has a strong environment dependency.

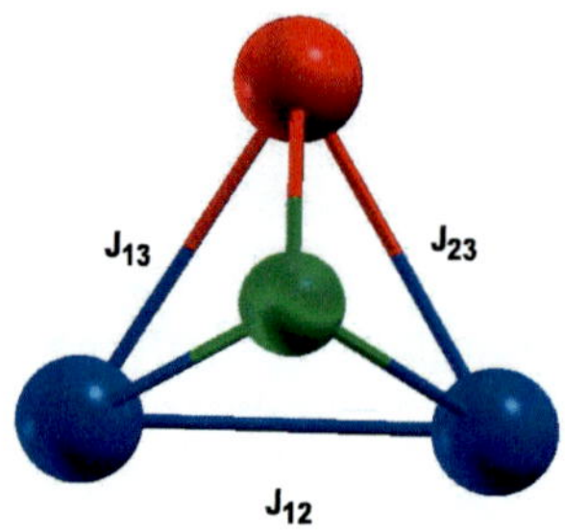

Figure 13. Ground state spin configuration for Mn_3As. Red(Blue) ball(s) represent Mn(↓) and Mn(↑) spin. The structure has a total magnetic moment $4\mu_B$.

7. Summary

In this chapter we discussed the electronic and magnetic properties of pure and doped manganese , Mn_n (n = 2 - 20) and Mn_nAs (n = 1 - 10), clusters from the density functional theory within the pseudopotential scheme. Here we use projector augmented wave method [19] which has been discussed briefly in the Section2.. The spin polarized generalized gradient approximation has been used to treat the exchange-correlation energy. To determine the ground state structure (both geometrical and magnetic), many different initial geometrical structures have been considered and all the possible spin multiplicities have also been investigated for each geometrical conformation. For both pure and doped clusters, we found many isomers with different magnetic structures are possible for a particular size.

Very small pure clusters, Mn_n, up to 4 atoms show Mn-Mn ferromagnetic coupling with 5 μ_B/atom magnetic moment. An addition of a single Mn atom then makes the magnetic coupling ferrimagnetic for Mn_5 and remains the same for the entire size range n = 5–20. These results show very good agreement with the recent experiments[4, 5]. The extraordinarily large experimental uncertainty for Mn_7 cluster[5] is explained due to the possible presence of the isomers in the experimental SG beam. However, the present density functional theory calculation predicts that for almost all the sizes there exist more than one isomer with different magnetic structure, but we do not observe any sufficiently large uncertainty corresponding to the size. The sudden deep in the experimentally measured

Table 6. Total energies of the Mn_3As clusters for different spin orientations.

Mn-spin configuration	Total energy (eV)
↑↑↑	-6.734
↑↑↓	-6.926
↑↓↑	-6.904
↓↑↑	-6.904

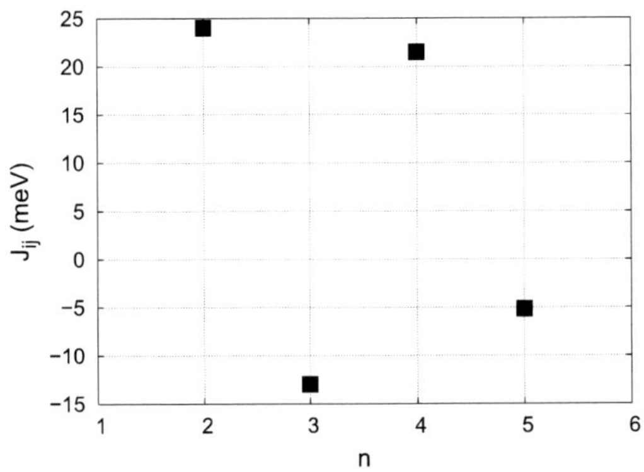

Figure 14. Plot of average exchange coupling $\bar{J}_{ij}$ with the number of Mn-atoms in the cluster n in Mn_nAs.

magnetic moment at Mn_{13} and Mn_{19} [4, 5] is well reproduced within the density functional calculation. This is attributed to their closed icosahedral structure[8].

Previously it was argued that the the sudden increase in the reaction rate of the Mn_n clusters with molecular hydrogen [9] is attributed to the nonmetal-to-metal transition at $n = 16$. If this is indeed the case then there must be a large decrease in the ionization potential and a closing-up of spin gaps at that particular size $n = 16$. However, such kind of downward discontinuity in the ionization potential [37] or any closing-up of spin gaps [8] are not observed at $n = 16$ and the reason for this abrupt change in the reaction rate remains unknown.

Finite systems like cluster are very good system to study the effect of coordination on the electron localization. For an example we discussed Mn_{13} cluster: The $d-$ electrons of the 12-coordinated central atom are more delocalized than those of the 6-coordinated surface atoms, i.e. the $d-$ electron localization decreases with the increase in coordination[8].

Next we discussed the issue how electronic and magnetic structure of these Mn_n clusters do change due to single As-doping. The geometrical structures do not change substantially due to As-doping, but only a moderate perturbation to the corresponding pure Mn_n cluster, i.e. a Mn_nAs cluster, more or less, could be viewed as pure Mn_n cluster with a As-atom caped somewhere. However, the magnetic structure of the Mn_nAs cluster change substantially from their pure counterpart. Only in Mn_2As and Mn_4As clusters, the Mn-Mn coupling is ferromagnetic, and for all other size range this coupling is ferrimagnetic. It is seen that in Mn_nAs clusters the exchange couplings are anomalous and behave quite differently from the RKKY-type predictions[10]. It is known that due to $4s^2$, $3d^5$ electronic structure as well as for high enough $4s^2$, $3d^5 \rightarrow 4s$, $3d^6$ promotion energy of the isolated Mn-atom , they do not bind strongly and as a result Mn_n clusters have lowest binding energies among all other $3d-$transition metal clusters. This scenario has been changed when a single As-atom is attached to them to form a resultant Mn_nAs cluster. The binding energy of the resultant Mn_nAs clusters are enhanced by the hybridization of the $3d-$electrons of the Mn-atom with the $4p-$electrons of the As-atom. The corresponding energy gain in adding an As-atom to a existing Mn_n cluster ($\triangle^1$) is much larger than that of adding an Mn-atom to an existing $Mn_{1-n}As$ cluster ($\triangle^2$), which finally points out that the clustering of the Mn-atoms around an As-atom is energetically favourable.

This point is very important and should be discussed in the context of the observed ferromagnetism in the Mn-doped GaAs/InAs semiconductors. From the free cluster calculations, one come to the conclusion that As-atom act as a nucleation centre for Mn-atoms and it is very much likely that Mn-clusters might be present in those low temperature molecular beam epitaxy grown (Ga, Mn)As/(In, Mn)As samples. Now how the presence of these clusters influence the observed ferromagnetism and Curie temperature is a debatable question for a long time. The ferromagnetic ordering of Mn-atoms are intrinsic for Mn_2As and Mn_4As clusters and are the source of effective internal magnetic field, which influences the energy structure and transport properties. They provide (high temperature) ferromagnetic contribution to the total magnetization of the GaMnAs samples. Although the Mn_2As and/or Mn_4As phases can be mostly responsible for the observed ferromagnetic behaviour of GaMnAs, it is not excluded that predicted carrier-induced ferromagnetic or other mechanism leading to ferromagnetic behaviour can be effective. The presence of ferromagnetic Mn_2As and/or Mn_4As phases in the GaMnAs samples, which have large magnetic moment and very high exchange coupling would consequently enhance the Curie temperature. On the other hand, the presence of the other sized clusters would eventually lead to the low Curie temperature due to their low magnetic moment and low exchange coupling value.

Finally, it should be pointed out that in this chapter we talked about collinear spin arrangements. However, it will be interesting so see whether these clusters show noncollinear magnetic structure or not. Indeed, for larger clusters both the pure and As-doped manganese clusters with more that 5 manganese atoms into it might have noncollinear magnetic structure with substantially different total magnetic moment [8, 10].

References

[1] A. J. Cox, J. G. Louderback and L. A. Bloomfield, *Phys. Rev. Lett.* **71**, 923 (1993); A. J. Cox, J. G. Louderback, S. E. Apsel and L. A. Bloomfield, *Phys. Rev. B* **49**, 12295 (1994).

[2] D. M. Cox, D. J. Trevor, R. L. Wheetten, E. A. Rohlfing and A. Kaldor, *Phys. Rev. B* **32**, 7290 (1985).

[3] L. A. Bloomfield, J. Deng, H. Zhang and J. W. Emmert, in *Proceedings of the International Symposium on Cluster and Nanostructure Interfaces*, edited by P. Jena, S. N. Khanna and B. K. Rao (World Publishers, Singapore, 2000), p.131

[4] M. B. Knickelbein, *Phys. Rev. Lett.* **86**, 5255 (2001).

[5] M. B. Knickelbein, *Phys. Rev. B* **70**, 14424 (2004).

[6] M. D. Morse, *Chem. Rev.* **86**, 1049 (1986).

[7] D. A. Young, *Phase Diagrams of the Elements* (Berkeley, Los Angeles, CA: University of California Press).

[8] M. Kabir, A. Mookerjee and D. G. Kanhere, (To be published.)

[9] E. K. Parks, G. C. Nieman and S. J. Riley, *J. Chem. Phys.* **104**, 3531 (1996).

[10] M. Kabir, D. G. Kanhere and A. Mookerjee, physics/0503009, (2005).

[11] O.K. Andersen, *Phys. Rev. B* **12**, 3060 (1975).

[12] J.C. Slater, *Phys. Rev.* **51**, 846 (1937). .

[13] P.M. Marcus, *Int. J. Quantum. Chem.* **1S**, 567 (1967).

[14] J. Korringa, *Physica* **13**, 392 (1947).

[15] W. Kohn and N. Rostocker, *Phys. Rev.* **94**, 111 (1954).

[16] D.R. Hamann, M. Schlüter and C. Chiang, *Phys. Rev. Lett.* **43**, 1494 (1979).

[17] D. Vanderbilt, *Phys. Rev. B* **41**, 7892 (1985) .

[18] P. E. Blöchl, *Phys. Rev. B* **50**, 17953 (1994).

[19] G. Kresse and D. Joubert, *Phys. Rev. B* **59**, 1758 (1999).

[20] P. Pulay, in *Modern Theoretical Chemistry*, edited by H. F. Schaefer (Plenum, New York, 1977); *Mol Phys.* **17**, 197 (1969).

[21] S. Goedecker and K. Maschke, *Phys. Rev. B* **45**, 1597 (1992).

[22] J. P. Perdew, K. Burke and M. Ernzerhof, *Phys. Rev. Lett.* **77**, 3865 (1996).

[23] G. Kresse and J. Furthmuller, *Phys. Rev. B 54*, 11169 (1996).

[24] R. K. Nesbet, *Phys. Rev.* **135**, A460 (1964).

[25] D. D. Shillady, P. Jena, B. K. Rao, and M. R. Press, *Int. J. Quantum Chem.* **22**, 231 (1998).

[26] K. D. Bier, T. L. Haslett, A. D. Krikwood and M. Moskovits, *J. Chem. Phys.* **89**, 6 (1988).

[27] R. J. Van Zee and W. Weltner,Jr., *J. Chem. Phys.* **89**, 4444 (1988); C. A. Bauman, R. J. Van Zee, S. Bhat, and W. Weltner,Jr., *J. Chem. Phys.* **78**, 190 (1983); R. J. Van Zee, C. A. Baumann and W. Weltner, *J. Chem. Phys.* **74**, 6977 (1981).

[28] S. K. Nayak, B. K. Rao and P. Jena, *J. Phys.: Condens Matter* **10**, 10863 (1998).

[29] P. Bobadova-Parvanova, K. A. Jackson, S. Srinivas and M. Horoi, *Phys. Rev. A* **67**, 61202 (2003); *J. Chem. Phys* **122**, 14310 (2005).

[30] M. R. Pederson, F. Ruse and S. N. Khanna, *Phys. Rev. B* **58**, 5632 (1998).

[31] A. Wolf and H-H. Schmidtke, *Int. J. Quantum Chem.* **18**, 1187 (1980).

[32] N. Fujima and T. Yamaguchi, *J. Phys. Sols Japan* **64**, 1251 (1995).

[33] J. Harris and R. O. Jones, *J. Chem. Phys.* **70**, 830 (1979).

[34] D. R. Salahub and N. A. Baykara, *Surf. Sci.* **156**, 605 (1985).

[35] G. W. Ludwig, H. H. Woodbury and R. O. Carlson, *J. Phys. Chem. Solids* **8**, 490 (1959).

[36] K. Rademann, B. Kaiser, U. Even, and F. Hensel, *Phys. Rev. Lett.* **59**, 2319 (1987).

[37] G. M. Koretsky and M. B. Knickelbein, *J. Chem. Phys.* **106**, 9810 (1997).

[38] H. Ohno, *J. Magn. Magn. Mater.* **200**, 110 (1999).

[39] H. Ohno and F. Matsukura, *Solid State Commun.* **117**, 179 (2001).

[40] T. Dietl, H. Ohno, F. Matsukura, J. Cibert, and D. Ferrand, *Science* **287**, 1019 (2000).

[41] H. Ohno, H. Munekata, T. Penney, S. Von Molnar, and L. L. Chang, *Phys. Rev. Lett.* **68**, 2664 (1992).

[42] H. Ohno, A. Shen, F. Matsukura, A. Oiwa, A. Endo, S. Katsumoto, and Y. Iye, *Appl. Phys. Lett.* **69**, 363 (1996).

[43] M. E. Overberg, C. R. Abernathy, and S. J. Pearton, *Appl. Phys. Lett.* **79**, 1312 (2001).

[44] M. L. Reed, N. A. El-Masry, H. H. Stadelmaier, M. K. Ritums, M. J. Reed, C. A. Parker, J. C. Roberts, and S. M. Bedair, *Appl. Phys. Lett.* **79**, 3473 (2001).

[45] S. Sonoda, S. Shimizu, T. Sasaki, Y. Yamamoto, and H. Hori, cond-mat/0108159, (2001).

[46] A. F. Guillermet and G. Grimvall, *Phys. Rev. B* **40**, 10582 (1989).

[47] H. Shimizu, T. Hayashi, T. Nishinaga, and M. Tanaka, *Appl. Phys. Lett.* **74**, 398 (1999).

[48] S. J. Potashnik, K. C. Ku, R. Mahendiran, S. H. Chun, R. F. Wang, N. Samarth, and P. Schier, *Phys. Rev. B* **66**, 012408 (2002).

[49] S. J. Potashnik, K. C. Ku, S. H. Chun, J. J. Berry, N. Samarth, and P. Schier, *Appl. Phys. Lett.* **79**, 1495 (2001).

[50] T. Hayashi, Y. Hashimoto, S. Katsumoto, and Y. Iye, *Appl. Phys. Lett.* **78**, 1691 (2001).

[51] Darshan C. Kundaliya, S. B. Ogale, S. E. Lofland, S. Dhar, C. J. Metting, S. R. Shinde, Z. Ma, B. Varughese, K.V. Ramanujachary, L. Salamanca-Riba, T. Venkatesan, *Nature Materials* **3**, 709 (2004).

[52] D. Chiba, K. Tankamura, F. Matsukura and H. Ohno, *Appl. Phys. Lett.* **82**, 3020 (2003).

[53] K. W. Edmonds, K. Y. Wang, R. P. Campion, A. C. Neumann, N. R. S. Farley, B. L. Gallagher and C. T. Foxon, *Appl. Phys. Lett.* **81**, 4991 (2002); K. C. Ku, S. J. Potashnik, R. F. Wang, S. H. Chun, P. Schiffer, N. Samarth, M. J. Seong, A. Mascarenhas, E. Johnston-Halperin, R. C. Myers, A. C. Gossard, and D. D. Awschalom, ibid **82**, 2302 (2003).

[54] Y. Yamaguchi and H. Watanabe, *J. Magn. Magn. Mater.* **31-34**, 619 (1983).

[55] B. K. Rao and P. Jena, *Phys. Rev. Lett.* **89**, 185504 (2002).

In: Horizons in World Physics, Volume 268
Editors: M. Everett and L. Pedroza, pp. 97-136
ISBN 978-1-60692-861-5

Chapter 3

Resonant Ultrasound Spectroscopy Close to Its Applicability Limits

Michal Landa[a], Hanuš Seiner[a,b], Petr Sedlák[a,b], Lucie Bicanová[a,b], Jan Zídek[a] and Luděk Heller[a]
[a]Institute of Thermomechanics v.v.i.,
Academy of Sciences of Czech Republic, Dolejškova 5,
18200 Prague 8, Czech Republic
[b]Faculty of Nuclear Sciences and Physical Engineering,
Czech Technical University in Prague, Trojanova 13,
12000 Prague 2, Czech Republic

Abstract

This chapter brings a critical review of the applicability of the resonant ultrasound spectroscopy (RUS) for determination of all independent elastic coefficients of anisotropic solids. Such applicability limits are sought which follow from the properties of the examined materials, i.e. from the strength and class of the anisotropy, etc.. After introducing the general theoretical background of RUS, particular limiting factors are illustrated on experimental results, namely on the investigation of extremely strongly anisotropic single crystals, of weakly anisotropic polycrystals (where neither the class nor the orientation of the anisotropy are known) and of single crystals with strong temperature-dependent magneto-elastic attenuation. In all these cases, a sensitivity analysis is carried out to show which elastic coefficients (their combinations) can be accurately determined form RUS measurements and which cannot, whereto the complementarity of the RUS and pulse-echo methods is shown and utilized. The general findings of both the theoretical introduction and the experimental part are summarized in a concluding section, which tries to formulate the most essential open questions of the RUS method.

1. Introduction

Although the fundamentals of resonant ultrasound spectroscopy (RUS) are known to the physics community for more than fifteen years [1, 2], this method cannot still be counted among well established or routinely used experimental techniques for evaluation of elastic properties of anisotropic solids. In comparison with methods based on acoustic wave velocity measurements, as are the family of pulse-echo techniques (either contact or with immersion in a liquid), point-source/point-receiver (PS/PR) or surface acoustic waves (SAW) methods, the employment and the scientific impact of RUS are undoubtedly minor, except of some cases where e.g. measurements in wide ranges of temperatures are required [3–5] which disqualifies the above more conventional methods because of technical difficulties.

The reason why such reliable and accurate method finds only hardly its way into a broader awareness could be sought neither in requirements on the tested specimens (RUS has been already successfully applied to prisms [1,2], plates [6], membranes [7], spherical balls [8], cylindrical nanotubes [9], rods [10], or samples of even more general shapes [11]) nor in requirements on the experiment instrumentation. The problem lies in the *inverse procedure*, i.e. in the procedure necessary to obtain the desired information (the elastic coefficients) from the experimental data. Whereas the pulse-echo measurements result in sets of ultrasound velocities in various directions, indicating clearly and understandably the anisotropy of the material, the outputs of RUS (resonant spectra of mechanical vibrations of a chosen specimen) require a sophisticated postprocessing to reveal the information on the material encrypted in it. This inverse procedure cannot be constructed universally, once for ever – each particular application of RUS requires slight modification of the procedure, taking the geometry of the specimen, strength of the anisotropy or the attenuation in the material into account. That is the reason why the RUS techniques appear to be unsuitable for automation, and thus, for massive use in commercial devices, industry or applied research.

This chapter aims to bring an analysis of what the applicability limits of RUS can be, searching for the novel ways how the inverse procedures can be constructed to push these limits at least a tiny bit further. It focuses on the limits given by the nature of the method itself, rather than these resulting from the experimental setup, although some such problems are mentioned as well. In the first half of this chapter, general ideas of the RUS method is overviewed with special emphasis laid on the relation between the properties of the examined material and the information obtainable on them by RUS measurements. The essentiality of the knowledge of such relations is revealed in the second half of the chapter, where the RUS method is applied to particular issues from solid state physics and materials science.

2. Resonant Ultrasound Spectroscopy - A General Background

2.1. Historic Development of RUS

The fact that the resonant spectra of free vibrations of a homogeneous, elastically anisotropic, rectangular parallelepiped contain a sufficient information on the elastic anisotropy of the material became fully understood in the first half of 1970s by Demarest et al. [13]. After significant extensions by Ohno [14], this finding has found a broad applicability in geophysics, which, according to [15], motivated Migliori et al. [2] to develop a similar method for investigation of elastic properties of small crystalline samples, and introduce, thus, these approaches to the general physics community. As the resonant frequencies of such small specimens were in the ultrasonic domain, Migliori et al. decided to refer this new method to as the *resonant ultrasound spectroscopy* (abbreviated as RUS), which was an equivalent of the term *rectangular parallelepiped resonance method* (RPR) used in geophysics.

The main idea of RUS was following: Let us consider that the first n resonant frequencies ($f_{p=1...n}$) of free elastic vibrations of a small rectangular parallelepiped of the examined material are obtained experimentally. Then, let us construct a numerical procedure which for every guess of elastic coefficients c_{ij} calculates an estimate of the first n resonant frequencies of such rectangular parallelepiped ($f^{\text{calc.}}_{p=1...n}(c_{ij})$). By matching the experimental and calculated frequencies, i.e. by minimizing the difference

$$\Delta(c_{ij}) = \sum_{p=1}^{n} \left(f_p - f_p^{\text{calc.}}(c_{ij})\right)^2 \tag{1}$$

over all c_{ij}, one can easily reach the coefficients which describe the vibrational properties (and, consequently, the elasticity) in some 'optimal' way.

For the numerical calculation of the vibrational modes for given c_{ij}, Migliori et al. overtook the original RPR algorithm and used a variational (Rayleigh-Ritz) method. For the consequent minimization of (1), a gradient search routine was adopted. During the next fifteen years, this basic scheme of the inverse method remained nearly unchanged. The only significant improvement came from Ogi et al. [17], who proposed to identify particular modes of vibrations by scanning the surface of the specimen by a laser interferometer during the measurements. This mode identification enabled a correct association of the pairs of resonant frequencies (the measured and the calculated) appearing in (1), which stabilized significantly the minimization procedure. The RUS methods with such mode identification became later called *modal resonant ultrasound spectroscopy* (MRUS) to emphasize the role of the shapes of the vibrational modes played in the inverse procedure.

The RUS method was successfully tested on known materials (namely $SrTiO_3$ in [2]) and immediately applied for determination of the elastic properties of advanced materials, such as high-temperature superconductors [2, 18, 19], or quasicrystals [20, 21], where both the need of measurements in extremely low temperatures and the small dimensions of the obtainable specimens precluded the use of the pulse-echo methods. Along with the application of RUS for particular materials, the method itself became more and more general. Whereas the use of the original algorithm described in [2] was restricted to a rectangular parallelepiped cut exactly along the principal axes of an orthorhombic (or higher) symmetry, Sarrao et al. [16] have shown soon that the method can be also used for identification of the crystallographic orientation of the material anisotropy inside the specimen. In the widely cited paper [15], Maynard writes about the applicability of the RUS method to 'prisms, spheroids, ellipsoids, shells, bells, eggs, potatoes, sandwiches and other shapes', meaning that the inverse procedure can be easily modified for any geometrically well-defined shape. However, the first successful attempts to use RUS for investigation of thin films and coatings came about ten years later [22–24].

In the most recent applications, the outputs of RUS are often not restricted to the elastic coefficients only. The results of the RUS measurements could be also the piezoelectric constants [25] or the internal friction parameters [26], or both [27]. Moreover, the changes in the resonant spectra can simply serve as reliable indicators of damage in the material [28–30]. The main idea of the method, however, remains nearly the same, always consisting

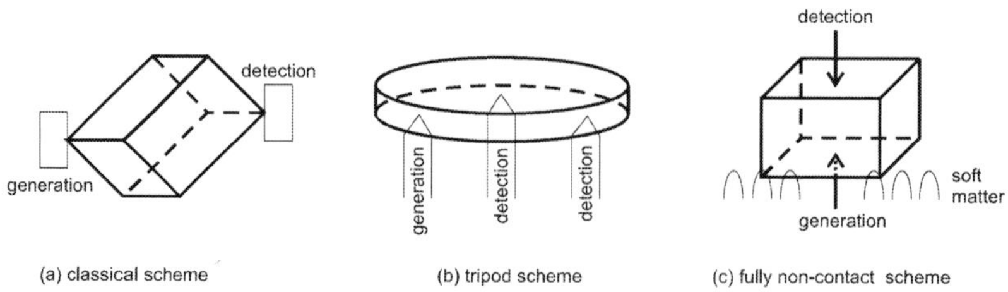

Figure 1. The main experimental schemes of RUS: The classical scheme, the tripod scheme and the fully non-contact scheme.

of measurements of the resonant spectra and, consequently, of the analysis of these spectra by an inverse algorithm.

For completeness, let us here also summarize the recent development in the experimental part of RUS, i.e. the methodology how the resonant spectra are experimentally obtained. Three principally different experimental arrangements of RUS measurements can be found in the available literature (Fig. 1):

1. The **classical scheme** was adopted by the pioneering works in RUS and remained as a most widely used RUS experimental methodology till nowadays. This scheme follows the setups used in the RPR measurements in geophysics. In Fig.1(a), the main idea of this scheme is outlined: The specimen (parallelepiped, sphere, or any other bulk shape) is placed between two transducers such that the contact area between the specimen and the transducers is minimal (e.g. a cube is placed such that it touches the transducers by two opposite corners only) to ensure the best possible approximation of fully free vibrations. Then, one of the transducers is used as a generator of ultrasonic waves (either scanning slowly the frequencies within a chosen range, or generating a broadband pulse), whereas the second as a detector. Obviously, the main disadvantage of such method lies in the fact that the vibrations are not purely free, as the specimen is restricted by the contact forces from the transducers. This effect was repeatedly shown to be negligible [33, 34], but for the bulk specimens only. For thin plates or shells, which have bending stiffness in some directions comparable to the forces applied by the contact of the transducers, the classical scheme becomes unsuitable.

   As mentioned above, the classical scheme can be significantly improved by scanning the specimen by a laser vibrometer during the measurements in order to obtain the shapes of particular eigenmodes.

2. The **tripod scheme** solves the problem of contact forces for thin plates. First adopted by Ogi [17], this scheme uses the arrangement outlined in Fig.1(b). The contact forces are minimized to the gravitation of the specimen lying of a tripod of rod-like transducers. Alternatively, either all the rods can be transducers, one used as a generator, the other two detecting the vibrations, or one of the rods can be just supporting the specimen and the remaining two be the generator and detector. Another crucial advantage of the tripod scheme is that the rods can be just waveguides, transmitting the ultrasonic signal to/from the specimen from/to the piezoelectric transducers

situated relatively far away from the specimen. This enables the tripod scheme to be used for measurements at extreme temperatures with the transducers (which are temperature-sensitive as well) situated safely outside the furnace or the cryostat [35].

Again, the tripod scheme can be improved by detecting the modes of vibrations.

3. The **fully non-contact scheme** (Fig.1(c)) is a logical extension of the tripod. In this arrangement, the specimen is not laid on a firm tripod but either hung on thin wires (glued to two 250μm wires using a minimal amount of Torr-seal epoxy in [36]) or laid on a soft underlay made of a material with extremely low acoustic impedance (cork wood used in most of the results presented by the authors within this chapter). The vibrations here are both generated and detected by lasers, the contact with the transducers is fully avoided. It is more than natural that the mode identification can be easily implemented in this scheme. The fully non-contact arrangement seems also to be optimal for the use at elevated or low temperatures.

   There is, however, another way how to perform RUS experiments in a fully non-contact regime. It is the *electromagnetic acoustic resonance method (EMAR)* [37], where a solenoidal coil and a static magnetic field are used to induce a induce Lorentz forces on specimen surfaces without using any mechanical contacts. More over, this is method sometimes called mode-selective, which means that particular sets of vibration modes can be selectively excited and detected by changing the direction of the applied magnetic field. The EMAR measurements are, on the other hand, fully dependent on the conductivity and other electromagnetic properties of the tested specimen, and cannot be counted among general experimental techniques of RUS.

All the experimental data presented in this chapter were obtained by the fully non-contact RUS technique. The elastic vibrations in the specimen were excited by sequences of pulses of an unfocused infrared laser beam (Nd:YAG, General Photonics Corporation TWO-45Q, nominal wavelength 1064nm) from the side of the specimen. The vibrations were recorded by scanning red-light laser vibrometer (Polytec OFV-2570 equipped by a scanning unit consisting of two dielectric mirrors on motorized positional stages) on the upper surface of the specimen.

2.2. The Forward Problem

The term *forward problem of RUS* is, in general, used for the evaluation of eigenfrequencies and eigenmodes of free vibrations of an elastic specimen of given geometry and known elastic coefficients. To solve the forward problem, the natural starting point is to formulate the energetic quantities of a dynamically deformed elastic specimen. For given displacement field $\mathbf{u}(\mathbf{x}, t)$ and its time derivative $\dot{\mathbf{u}}(\mathbf{x}, t)$, and for given density ρ and the elastic coefficients C_{ijkl}, these quantities are: the kinetic energy

$$K(\dot{\mathbf{u}}(\mathbf{x}, t), t) = \frac{1}{2}\int_V \rho \dot{u}_i \dot{u}_i \mathrm{d}V, \tag{2}$$

the potential (stored, elastic) energy

$$P(\mathbf{u}(\mathbf{x}, t), t) = \frac{1}{2}\int_V C_{ijkl}\frac{\partial u_i}{\partial x_j}\frac{\partial u_k}{\partial x_l}\mathrm{d}V, \tag{3}$$

and the Lagrangian energy

$$L(\mathbf{u}(\mathbf{x},t),\dot{\mathbf{u}}(\mathbf{x},t),t) = K(\dot{\mathbf{u}}(\mathbf{x},t),t) - P(\mathbf{u}(\mathbf{x},t),t), \tag{4}$$

which is their difference. All the volumetric integrations are meant over the whole volume of the specimen V.

As we are searching for harmonic solutions (eigenvibrations), an assumption about the form of the displacement field can be done by stating

$$\mathbf{u}(\mathbf{x},t) = \mathbf{u}(\mathbf{x})\cos(\omega t) \qquad \text{and} \qquad \dot{\mathbf{u}}(\mathbf{x},t) = -\mathbf{u}(\mathbf{x})\sin(\omega t). \tag{5}$$

This simplifies the Lagrangian energy into

$$L(\mathbf{u}(\mathbf{x}),t) = \frac{1}{2}\int_V \left[\omega^2 \rho u_i u_i \sin^2(\omega t) - C_{ijkl}\frac{\partial u_i}{\partial x_j}\frac{\partial u_k}{\partial x_l}\cos^2(\omega t)\right] \mathrm{d}V. \tag{6}$$

One of the basic properties of the Lagrangian energy is that it follows stationary paths between each two given time points[1]. In our case of $L = L(\mathbf{u}(\mathbf{x}),t)$ it means that

$$\delta\int_{t_1}^{t_2} L(\mathbf{u}(\mathbf{x}),t)\mathrm{d}t = \int_{t_1}^{t_2} L(\mathbf{u}(\mathbf{x})+\delta\mathbf{u}(\mathbf{x}),t)\mathrm{d}t - \int_{t_1}^{t_2} L(\mathbf{u}(\mathbf{x}),t)\mathrm{d}t = 0, \tag{7}$$

for arbitrary but given t_1 and t_2, where $t_1 < t_2$. By choosing these time points such that $t_2 = t_1 + 2\pi/\omega$, and taking the equality

$$\int_0^{2\pi/\omega} \cos^2(\omega t) = \int_0^{2\pi/\omega} \sin^2(\omega t) \tag{8}$$

into account, the time coordinate can be simply eliminated, and we arrive another variational condition:

$$0 = \delta\frac{1}{2}\int_V \underbrace{\left[\omega^2 \rho u_i u_i - C_{ijkl}\frac{\partial u_i}{\partial x_j}\frac{\partial u_k}{\partial x_l}\right]}_{\stackrel{\text{def.}}{=}\lambda(\mathbf{u}(\mathbf{x}))} \mathrm{d}V \stackrel{\text{def.}}{=} \delta\Lambda(\mathbf{u}(\mathbf{x})), \tag{9}$$

where we have defined two new quantities: the *time-averaged Lagrangian energy* $\Lambda(\mathbf{u}(\mathbf{x}))$ and its density $\lambda(\mathbf{u}(\mathbf{x}))$. Following the basic theorems of the calculus of variation, we can express the variation of $\Lambda(\mathbf{u}(\mathbf{x}))$ explicitly as

$$\delta\Lambda(\mathbf{u}(\mathbf{x})) = \frac{1}{2}\int_V \left[\frac{\partial\lambda}{\partial u_i} - \frac{\mathrm{d}}{\mathrm{d}x_j}\left(\frac{\partial\lambda}{\partial(\frac{\partial u_i}{\partial x_j})}\right)\right]\delta u_i \mathrm{d}V + \int_S n_j \frac{\partial\lambda}{\partial(\frac{\partial u_i}{\partial x_j})}\delta u_i \mathrm{d}S, \tag{10}$$

where S is the surface of the specimen and $\mathbf{n}$ its outer normal. As the variations δu_i are arbitrary, we can require that

$$\frac{\partial\lambda}{\partial u_i} - \frac{\mathrm{d}}{\mathrm{d}x_j}\left(\frac{\partial\lambda}{\partial(\frac{\partial u_i}{\partial x_j})}\right) = \rho\omega^2 u_i + C_{ijkl}\frac{\partial^2 u_k}{\partial x_j \partial x_l} = 0 \quad \text{almost everywhere in } V \tag{11}$$

[1] So-called *principle of stationary (or minimal) Lagrangian action* or *Hamiltonian principle*, e.g. [31].

$$\text{and} \qquad n_j \frac{\partial \lambda}{\partial(\frac{\partial u_i}{\partial x_j})} = n_j C_{ijkl} \frac{\partial u_k}{\partial x_l} = 0 \qquad \text{almost everywhere on } S. \tag{12}$$

Obviously, (11) are the equations of steady waves in the considered continuum and (12) are the conditions of a free surface. Thus, by $\delta\Lambda = 0$, we obtain the solutions of the elastodynamic equation (11) for boundary conditions (12), which is exactly what we are searching for – resonant vibrations of an unconstrained specimen.

Let us now try to construct a displacement field $\mathbf{u}(\mathbf{x})$ such that it minimizes $\Lambda(\mathbf{u}(\mathbf{x}))$. If the specimen is, for example, a rectangular parallelepiped of dimensions $d_1 \times d_2 \times d_3$, the Lagrangian is

$$\Lambda = \frac{1}{2} \int_{-d_1/2}^{d_1/2} \int_{-d_2/2}^{d_2/2} \int_{-d_3/2}^{d_3/2} \left[\rho\omega^2 u_i^2(\mathbf{x}) - C_{ijkl} \frac{\partial u_i}{\partial x_j}(\mathbf{x}) \frac{\partial u_k}{\partial x_l}(\mathbf{x}) \right] \mathrm{d}x_1 \mathrm{d}x_2 \mathrm{d}x_3, \tag{13}$$

where the coordinate system $\mathbf{x}$ is chosen such that it has its origin in the center of the specimen and that the edges of lengths d_1, d_2 and d_3 are oriented along the axes x_1, x_2 and x_3. As utilized by the original RPR algorithm [13] and adopted by the pioneering works of RUS [1, 2], the variation of this Lagrangian with respect to $\mathbf{u}(\mathbf{x})$ can be approximated by derivatives of it with respect to the coefficients of polynomial expansions of $\mathbf{u}(\mathbf{x})$ (so-called variational Ritz method). In the other words, for the solution expected in an approximative form

$$u_i(\mathbf{x}) = \sum_{K=1}^{N} \alpha_{[K,i]} \Psi_K(\mathbf{x}), \tag{14}$$

where $\Psi_K(\mathbf{x})$ is some properly chosen functional (e.g. polynomial) basis, the condition $\delta\Lambda(\mathbf{u}(\mathbf{x})) = 0$ is satisfied whenever

$$\frac{\partial \Lambda(\alpha_{[1,1]}, \ldots, \alpha_{[N,1]}, \alpha_{[1,2]}, \ldots, \alpha_{[N,2]}, \alpha_{[1,3]}, \ldots, \alpha_{[N,3]})}{\partial \alpha_{[K,i]}} = 0 \tag{15}$$

for all $[K, i] \in [\{1, 2, \ldots, N\}, \{1, 2, 3\}]$

Although *any* polynomial basis (e.g. $\Psi_{abc} = x_1^a x_2^b x_3^c$ used in [2]) could be suitable for such approximation, it is advantageous here, for the case of the rectangular parallelepiped with mutually comparable dimensions, to take [17]

$$\Psi_{abc} = P_a(\frac{2x_1}{d_1}) P_b(\frac{2x_2}{d_2}) P_c(\frac{2x_3}{d_3}) \tag{16}$$

for

$$a,\ b,\ c = 0, 1, 2, 3 \ldots \qquad a + b + c \leq N, \tag{17}$$

where $P_n(x)$ is the normalized Legendre polynomial of degree n defined as

$$P_n(x) = \frac{\sqrt{(2n+1)/2}}{2^n n!} \left[\frac{\mathrm{d}^n}{\mathrm{d}x^n} (x^2 - 1)^n \right]. \tag{18}$$

The corresponding stationary condition for Lagrangian Λ leads to a symmetric eigenvalue problem

$$\left(\omega^2 \mathbf{E}_{[abc,i][def,j]} - \Gamma_{[abc,i][def,j]} \right) \alpha_{[abc,i]} = 0, \tag{19}$$

with

$$\mathbf{E}_{[abc,i][def,j]} = \delta_{ij}\delta_{ad}\delta_{be}\delta_{cf}, \tag{20}$$

and

$$\Gamma_{[abc,i][def,j]} = \frac{1}{\rho}\frac{8}{d_1 d_2 d_3} C_{ijkl} \int_{-d_1/2}^{d_1/2}\int_{-d_2/2}^{d_2/2}\int_{-d_3/2}^{d_3/2} \frac{\partial \Psi_{abc}}{\partial x_k}\frac{\partial \Psi_{def}}{\partial x_l} \mathrm{d}x_1 \mathrm{d}x_2 \mathrm{d}x_3. \tag{21}$$

Let it be pointed out that for a general polynomial basis, the matrix $\mathbf{E}$ is not unitary, which significantly complicates the solution of the eigenvalue problem.

The integrations in matrix Γ can be done analytically (using a symbolic software) and the eigenvalue problem for this matrix is usually solved by an appropriate numerical algorithm (e.g. the Cholesky method [32]).

Three remarks are to be done here before proceeding to the description of the inverse procedure:

1. For reasonably precise polynomial approximation (14), the matrices Γ are huge and their construction by (21) consumes unacceptable portions of the computation time if it must be done again and again during the optimizing process. For this reason, it is beneficial to notice that this matrix is linearly dependent on C_{ijkl}, regardless of how nontrivial this dependence is.

   Consequently, the derivatives $\partial\Gamma_{mn}/\partial C_{ijkl}$ are independent on C_{ijkl} and can be computed *a priori*. In each run of the optimizing process, the matrix Γ can be, thus, quickly constructed as

$$\Gamma_{mn} = \sum_{ijkl} \frac{\partial \Gamma_{mn}}{\partial C_{ijkl}} C_{ijkl}. \tag{22}$$

   The summation here is carried only over the independent elastic coefficients C_{ijkl}, which means that e.g. for the cubic symmetry the matrices Γ are always obtained as linear combinations of three matrices: $\partial\Gamma/\partial c_{11}$, $\partial\Gamma/\partial c_{12}$ and $\partial\Gamma/\partial c_{44}$.

2. The above described algorithm for solution of the forward problem can be easily modified for a nonrectangular parallelepiped by transforming the Lagrangian (13) into an oblique coordinate system chosen paraxially with the edges of the specimen. For such oblique system $\mathbf{y}$ related to the natural system of the anisotropy $\mathbf{x}$ by a linear relation

$$\mathbf{y} = \mathbf{B}\mathbf{x}, \tag{23}$$

   the Lagrangian transforms into [49]

$$\Lambda = \frac{1}{2}\int_{-d_1/2}^{d_1/2}\int_{-d_2/2}^{d_2/2}\int_{-d_3/2}^{d_3/2} \left[\frac{\rho}{\det \mathbf{B}}\omega^2 u_i^2(\mathbf{y}) - T_{ijkl}\frac{\partial u_i}{\partial y_j}(\mathbf{y})\frac{\partial u_k}{\partial y_l}(\mathbf{y})\right] \mathrm{d}y_1 \mathrm{d}y_2 \mathrm{d}y_3, \tag{24}$$

   where

$$T_{ijkl} = \frac{1}{\det \mathbf{B}} C_{ipko} B_{jp} B_{lo}. \tag{25}$$

   As expressions (13) and (24) are formally identical, the eigenfrequencies of a general parallelepiped can be then evaluated using exactly the same variational procedure as for the rectangular one.

3. The direct problem can be significantly simplified by the symmetry. If the specimen is, for example, a rectangular parallelepiped cut exactly along the principal axes of an orthorhombic (or higher) symmetry, the variational problem (9) has an orthorhombic symmetry, which is the highest common symmetry of the specimen and the material. Consequently, the solutions (eigenmodes) can be expected to inherit this class of symmetry, as they fulfill the new balance laws following from it (according to the Noether's principle). In such case, it is clear that the functions $u_i(\mathbf{x})$ must be either even or odd with respect to all cartesian coordinates x_j with (in contrary) either odd or even partial derivatives $\partial u_i/\partial x_j$. The overall vibrations of such specimen can be, thus, fully characterized by displacement fields in the octant $[0, d_1/2] \times [0, d_2/2] \times [0, d_3/2]$. The way how can this symmetry be utilized for in the numeric calculation of the eigenfrequencies and eigenmodes is in detail discussed in [2], where the matrix Γ is shown to split into eight independent matrices, each one corresponding to one possible symmetry of the resultant vibrational modes.

   There is, moreover, another consequence of such symmetry of the specimen, which is perhaps even more important that the simplification of the forward problem. Consider, for example, an even mode of vibrations, where (for all i)

$$
\begin{aligned}
u_i(x_1 = \frac{d_1}{2}, x_2, x_3) &= u_i(x_1 = -\frac{d_1}{2}, x_2, x_3), \\
u_i(x_1, x_2 = \frac{d_2}{2}, x_3) &= u_i(x_1, x_2 = -\frac{d_2}{2}, x_3), \\
u_i(x_1, x_2, x_3 = \frac{d_3}{2}) &= u_i(x_1, x_2, x_3 = -\frac{d_3}{2}),
\end{aligned} \tag{26}
$$

   and let this mode correspond to the eigenfrequency ω. Then, another mode can be easily constructed by duplication, i.e. by taking

$$
\mathbf{u}^D(\frac{1}{2}\mathbf{x}) = \mathbf{u}\left(\mathbf{x} - [\frac{d_1}{2}, \frac{d_2}{2}, \frac{d_3}{2}]\right), \tag{27}
$$

   in the first octant (i.e. in $[0, d_1/2] \times [0, d_2/2] \times [0, d_3/2]$), and by periodic repetition of this displacement field in the remaining seven eights of the specimen. The duplicated mode obviously fulfills the variational condition (11) for the eigenfrequency $\omega^D = 2\omega$. Similarly, the input even mode (26) can be triplicated with frequency 3ω, quadruplicated with 4ω, or generally n-multiplied with frequency $n\omega$. Such multiplication procedure can be done also for odd modes, where

$$
\begin{aligned}
u_i(x_1 = \frac{d_1}{2}, x_2, x_3) &= -u_i(x_1 = -\frac{d_1}{2}, x_2, x_3), \\
u_i(x_1, x_2 = \frac{d_2}{2}, x_3) &= -u_i(x_1, x_2 = -\frac{d_2}{2}, x_3), \\
u_i(x_1, x_2, x_3 = \frac{d_3}{2}) &= -u_i(x_1, x_2, x_3 = -\frac{d_3}{2}),
\end{aligned} \tag{28}
$$

   or for any mode being alternatively even or odd in individual cartesian coordinates x_i, as it follows from the symmetry of the system with respect to mirror reflections

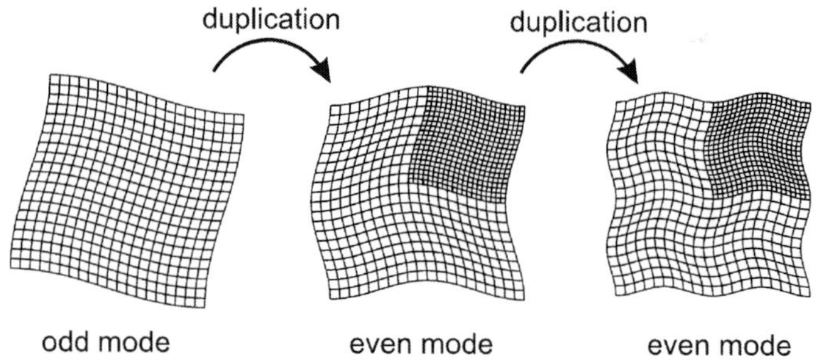

Figure 2. Construction of higher modes of free vibrations by duplication (a 2D model). The initial odd mode is duplicated into an even one by choosing proper phases of the mode in particular quadrants. Further duplication results in the even modes, which are obtained without any phase changes.

$x_i \rightarrow -x_i$ valid in the considered orthorhombic specimen. The difference between the duplication for the even and the odd modes is shown in Fig.2. To summarize, we can say that with each mode of frequency ω, the spectrum of an rectangular specimen aligned with the orthorhombic symmetry of the material contains also all multiplications of this frequency. In the case of generally oriented or non-rectangular parallelepipeds, no such general conclusion can be done. However, whenever the specimen has a shape suitable for spatial repetition, there can be some modes able to be multiplied within the spectrum.

2.3. Inverse Determination of Elastic Coefficients

As it was already mentioned, the inverse procedure is the key point of the all RUS techniques. Within this procedure, the aim is to determine the unknown elastic coefficients, and what we have on disposal are the experimentally obtained resonant spectra and a numerical algorithm for solution of the forward problem.

The obvious way how to obtain the optimal elastic coefficient is to minimize the difference (1) or any nondecreasing function of it. Migliori et al. [2] used

$$\Delta(c_{ij}) = \sum_{p=1}^{n} \frac{\left(f_p - f_p^{\text{calc.}}(c_{ij})\right)^2}{f_p^2}, \tag{29}$$

which reflects the fact that the higher frequencies are due to the approximation (16) less accurately determined and are, thus, involved in (29) with lower weights.

Let us see now how easily can this natural weighting turn into a drawback. Let us consider that our specimen is a thin cylindrical rod made of oak wood (cut along the grain), and let us admit such rod to vibrate both in the longitudinal and the torsional regime (but not in the flexural modes). For the longitudinal vibrations, the steady wave equation (11) can be written as

$$\rho\omega^2 u + E\frac{\mathrm{d}^2 u}{\mathrm{d}x^2} = 0, \tag{30}$$

where u is the axial displacement field and E is the Young modulus in the axial direction of the rod, whereas for the torsional vibrations as

$$\rho\omega^2\theta + G_\perp \frac{\mathrm{d}^2\theta}{\mathrm{d}x^2} = 0, \tag{31}$$

where θ is the torsion angle and $G_\perp$ is the shear modulus for shears in the planes normal to the grain. As the oak wood has approximately $E/G_\perp = 30$, we can estimate that the resonant frequency of the first longitudinal mode is more than five times higher than the resonant frequency of the first torsional mode, and is taken into (29) with a thirty times lower weight. In the other words, the error function (29) is similarly sensitive to an 1% experimental error in the first torsional mode and to a 30% error in the first longitudinal mode. By a particular choice of material (large difference between E and $G_\perp$) and of geometry (1D rod with allowed axially symmetric modes of vibrations) we have disabled a reliable determination of E by any inverse procedure based on minimization of the error function (29).

For generally anisotropic, three-dimensional specimens, the way how the information on the elastic coefficients is distributed in the resonant spectrum is extremely complicated. Similarly to the above discussed example, the lowest resonant frequencies of such specimen correspond to the *soft modes* and contain the information only about the smallest (which means mostly shear) coefficients or their combinations. By moving in the spectrum upwards, some modes corresponding to the other (i.e. *harder*) coefficients may appear, but the major part could be still related to the soft ones (e.g. the spectrum may contain 2ω, 3ω, etc. frequencies of multiple modes discussed for highly symmetric specimens in the third conclusive remark in the previous section.)

The question is how to determine which coefficients can be accurately determined from such resonant spectrum and which can not. On this purpose, let $\alpha_{1,2,\dots,n}$ be the eigenvectors corresponding to measured eigenfrequencies $f^{\mathrm{exp}}_{1,2\dots,n} = \omega^{\mathrm{exp}}_{1,2\dots,n}/2\pi$, and let $C_{k=1,\dots,m}$ be the set of independent elastic coefficients. The derivative of the frequency f^{exp}_p with respect to the constant C_k can be expressed as (using a formula from the perturbation theory, for further details, check [32])

$$\frac{\partial f^{\mathrm{exp}}_p}{\partial C_k} = \frac{\alpha^{\mathrm{T}}_p \frac{\partial \Gamma}{\partial C_k} \alpha_p}{8\pi^2 f^{\mathrm{exp}}_p}, \tag{32}$$

where no summation over p on the right-hand-side is applied. Consequently, the rate of sensitivity of measured spectrum to the k-th elastic coefficient C_k can be taken as a sum of squares of such derivatives over the whole spectrum, i.e.

$$S^2_k = \sum_{p=1,\dots,n} \left(\frac{\partial f^{\mathrm{exp}}_p}{\partial C_k}\right)^2 = \sum_{p=1,\dots,n} \left(\frac{\alpha^{\mathrm{T}}_p \frac{\partial \Gamma}{\partial C_k} \alpha_p}{8\pi^2 f^{\mathrm{exp}}_p}\right)^2. \tag{33}$$

Our aim is to evaluate the elastic constants (or their combinations) which are most accurately determined by the inverse procedure. Let us, for simplicity, consider that what we are trying to find are the linear combinations C^*_k, related to the original set of elastic coefficients C_k by linear equations

$$C_k = \Phi_{kl} C^*_l. \tag{34}$$

To evaluate the sensitivity to such combinations, it is necessary to transform (33) into

$$S_l^{*2} = \sum_{p=1,\dots,n} \left(\frac{\partial f_p^{\text{exp}}}{\partial C_l^*}\right)^2 = \sum_{p=1,\dots,n} \left(\Phi_{kl}\frac{\alpha_p^{\text{T}} \frac{\partial \Gamma}{\partial C_k} \alpha_p}{8\pi^2 f_p^{\text{exp}}}\right)^2 = \Phi_{\cdot l}\mathbf{G}^{\text{T}}\mathbf{G}\Phi_{\cdot l}^{\text{T}}. \qquad (35)$$

where

$$\Phi_{\cdot j} = (\Phi_{1l}, \Phi_{2l}, \dots \Phi_{nl}) \qquad \text{and} \qquad \mathbf{G} = \begin{pmatrix} \frac{\partial f_1^{\text{exp}}}{\partial C_1} & \cdots & \frac{\partial f_1^{\text{exp}}}{\partial C_m} \\ \vdots & \ddots & \vdots \\ \frac{\partial f_n^{\text{exp}}}{\partial C_1} & \cdots & \frac{\partial f_n^{\text{exp}}}{\partial C_m} \end{pmatrix}. \qquad (36)$$

The matrix $\mathbf{G}^{\text{T}}\mathbf{G}$ is symmetric and positive definite, and its eigenvectors can be, thus, chosen to form an orthogonal normalized system. By sorting this eigenvalues in a decreasing order and choosing $\Phi_{\cdot l}$ to be the l-th eigenvector, we obtain linear combinations C_l^* sorted by sensitivity.

Consider now that we are able to construct an inverse procedure reflecting somehow the way how the information on the unknown elastic coefficients is distributed in the spectrum, i.e. a procedure based on a minimization of

$$\Delta(c_{ij}) = \sum_{p=1}^{n} w_p \left(f_p - f_p^{\text{calc.}}(c_{ij})\right)^2, \qquad (37)$$

where w_p are weights tuned such that they do not bias the inversion procedure in the way demonstrated by the above discussed example of the 1D oak wood rod. Another problem can than arise in the optimization itself, especially when only rough initial guesses of the elastic coefficients are on disposal. In the first applications of RUS [2], the calculated and experimentally obtained resonances were arranged into sums like (37) by simple ordering, i.e. the first calculated frequency was subtracted from the first experimentally obtained, the second from the second, etc. The effect which such association of frequencies can have on the minimized function can be clearly seen on the following example.

Consider again a cylindrical rod with axially symmetric modes of vibrations allowed. For this time, let the rod be made of polycrystaline copper, with $E = 110$GPa and $G = 63$GPa, which is a quite small difference ($E/G = 1.75$). As the resonances containing the information on the value of E and on the value of G are close to each other, and as we can evaluate these resonance explicitly (higher frequencies are not disturbed by higher numerical errors), we can safely choose $w_p = 1$ for all p. On the left-hand-side of Fig.3, a contour plot of sum (1) in dependence on E and G is shown, evaluated by simply comparing the first fourteen evaluated resonant frequencies with first fourteen obtainable experimentally (but evaluated here for the correct values of G and E). The sum is obviously unsuitable for minimization. Not only that there multiple minima appearing on $\Delta(E, G)$ (without distinguishing between the longitudinal and torsional modes, the values of E and G are fully interchangeable), but the function $\Delta(E, G)$ is far from smooth, which precludes a meaningful use of any gradient search method.

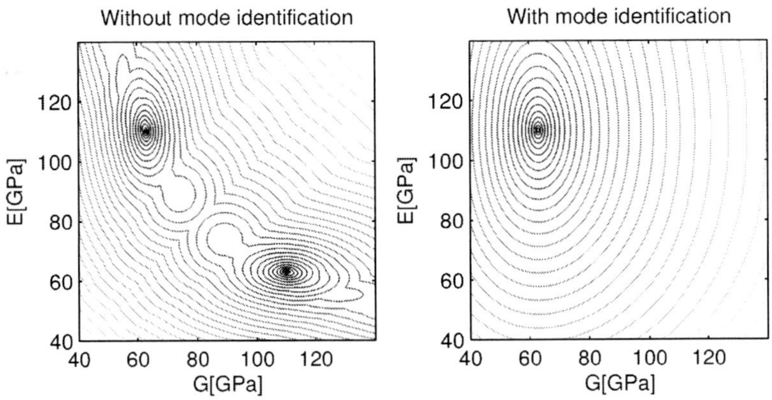

Figure 3. Comparison of the error function $\Delta(E, G)$ without (on the left) and with (on the right) the mode identification.

In the right-hand-side of Fig.3, the same (not weighted) sum is plotted with the *mode identification*, i.e. with the knowledge of which resonances are torsional and which are longitudinal involved. After such improvement, the function $\Delta(E, G)$ becomes fully smooth, with one well defined minimum corresponding to correct values of E and G. For the error function (29) weighted by $1/f_p^2$, the contour plots look nearly the same.

The problem of mode identification was successfully solved by Ogi et al. [17], who scanned one of the surfaces of the vibrating specimen by a laser interferometer. From the projection of displacement patterns into that surface, Ogi et al. were able to identify all the observed modes and arrange the resonances in sum (1) in a correct way. Similar approach was later adopted by Landa et. al [12] and extended by automatic identification of modes, where the measured displacements are fitted with the same order of Legendre polynomials, which enables a reliable comparison to the calculated eigemodes.

However, for a generally oriented, three-dimensional anisotropic specimen, the mode association is possible only for relatively accurate initial guesses of the sought elastic coefficients and for known class and orientation of the anisotropy. That is another factor which complicates a general use of RUS and makes it disadvantageous in comparison with pulse-echo measurements: The overall character of the anisotropy can be directly seen neither from the resonant spectra nor from the shapes of the eigenmodes (which mix the symmetry of the material with the symmetry of the specimen), but, without the knowledge of this character, the mode association, and, consequently, the reliable determination of the elastic coefficients can be, in some cases, close to impossible.

Providing that we have tuned the weights such that the sum (37) does not *a priori* suppress the information on any of the sought coefficients and that we have sufficiently accurate initial estimates of the coefficients to identify all the modes involved in the inversion, the inverse procedure could be expected to converge to correct results. The accuracy of these results is given by the experimental error in the input resonant frequencies, by the accuracy of determination of the geometry and orientation of the specimen, by how accurately the frequencies are evaluated within the forward problem (combining the accuracy of the Ritz method with the accuracy of the numerical algorithm used for the solution of the eigenvalue problem (19)), and by the accuracy of the chosen search algorithm (i.e. how accurately the minimum is localized). Such mixing of experimental and numeric errors in the resultant accuracy of the outputs of RUS nearly precludes any direct determination of the accuracy of the method itself. It can be, however, guessed e.g. from Monte-Carlo simulations the input data disturbed randomly within experimentally reasonable ranges, but such approach always enables it to be guessed only for one specific material and one specific geometry of the specimen.

Among the general ways how to increase the accuracy of the RUS measurements, at least the following three points are worth mentioning here:

1. As far as the preparation and choice of the specimens is concerned, it is always better to use the **general bars with mutually undividable dimensions** rather than the exact cubes or tetragons, **general parallelepipeds** rather than rectangular ones, and **specimens with general orientation** rather than those cut along the principal axes of the material. The reason is clear from the discussion of the resonant spectra of highly symmetric specimens (see the previous section). The lower the symmetry of the specimen is (meaning both the shape itself and the specimen's orientation to the symmetry of the material), the more general vibrational modes can be expected to appear. Let it be pointed out here that the modes constructed by multiplication contain exactly the same information on the elastic coefficients as the basic modes they were constructed from. Moreover, for the highly symmetric specimens, some of the modes can be degenerated (i.e. two or more modes can have the same resonant frequency), which significantly complicates the mode identification.

2. The **increase of the degree of the polynomial approximation** (14) can, naturally, increase the accuracy of the obtained results. However, as the dimension of matrix Γ (without any simplification of Γ by possible symmetries considered) is related to the degree of the polynomial approximation by

$$\mathrm{rank}(\Gamma) = \frac{(N+1)(N+2)(N+3)}{2}, \tag{38}$$

each increase of N is penalized by a dramatic increase of computation time. Fig.4 shows comparison of eigenfrequencies computed for three different degrees N: 12, 14 and 16. The dimensions of the matrix Γ were 1365, 2040 and 2907 respectively. The solution taken as referential here (i.e. the solution which the resonances calculated for different N are compared to) was obtained by finite element method with very large number of degrees of freedom (10^5). This figure also illustrates another important feature: Since the plotted quantity here is

$$\Delta f_p = \frac{f_p^{\text{calc.}}(N) - f_p}{f_p}, \tag{39}$$

it can be easily seen that the frequencies for low N are generally higher than both those for higher N and those calculated by finite elements. In other words, the less flexible the approximation is, the more significant shift of the calculated spectrum upwards can be expected.

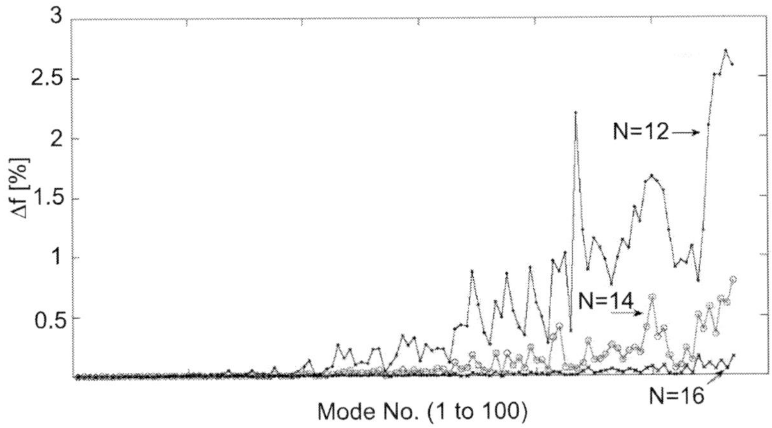

Figure 4. The effect of the degree N of the polynomial approximation on the accuracy of the evaluated resonant frequencies.

The choice of the polynomial approximation should also reflect the shape of the specimen, i.e. if the specimen is a thin plate or shell, another (lower) degree of the polynomials should be used to approximate the displacement field distribution along the normals to midplane of the specimen than along the curves inside the midplane.

3. For the search for the optimal coefficients, it is usually beneficial to **decompose the inversion algorithm into particular iterative steps**. The following architecture of inversion procedure was proposed in [32]:

   (a) We consider a parallelepiped sample with given material symmetry, density ρ and shape. Utilizing the 1st-order homogeneity of Γ with respect to the all independent elastic coefficients C_k, we compute the matrices $\partial\Gamma/\partial C_k$ via (21) by setting $C_k = 1$ and $C_j = 0$ for all $j \neq k$.

   (b) We take an initial guess of the constants $C_k^{(0)}$ (e.g. elastic constants of similar materials found in literature), complete the matrix Γ using equations (22), and calculate its eigenvalues and eigenvectors.

   (c) We compute the surface distributions of the displacement field in the surface of the specimen which is scanned by the laser vibrometer. By comparison of computed and experimentally measured distributions, we try to associate measured and computed spectra. If our constants are far from the correct ones (as usually happens for the initial guesses), only the first few modes are associated.

   (d) Using the equations (35) and (36), we compute the matrix $\mathbf{G}^{\mathrm{T}}\mathbf{G}$ with frequencies f_j^{exp} and the eigenvectors α_j^{as} associated to them and determine the linear combinations C_k^* sorted by sensitivity. (The eigenvectors α_j^{as} are not exactly equal to the experimental eigenvectors and thus, the derivatives $\partial f_j^{\mathrm{exp}}/\partial C_k$ used in (35) are approximate only.)

(e) We minimize the error function

$$\Delta(C_k^*) = \sum_{p_{\text{assoc.}}} \left(f_p^{cal}(C_k^*) - f_p^{exp} \right)^2, \tag{40}$$

where the summation over $p_{\text{assoc.}}$ means that only the associated modes are taken into account. In [32], (as well as for the results presented in this chapter,) the minimization of (40) is done by a gradient (Levenberg-Marquardt) method, which provides a fast and straightforward convergence. The efficiency of the inverse determination of the elastic coefficients is improved by deriving the analytical expression of the gradient and the Hessian of the error function using formulae from perturbation theory

$$\frac{\partial \omega_j^2}{\partial C_k^*} = \alpha_j^{\mathrm{T}} \frac{\partial \Gamma}{\partial C_k^*} \alpha_j, \tag{41}$$

$$\frac{\partial^2 \omega_j^2}{\partial C_k^* \partial C_l^*} = \alpha_j^{\mathrm{T}} \frac{\partial \Gamma}{\partial C_k^*} \frac{\partial \alpha_j}{\partial C_l^*} + \left(\frac{\partial \alpha_j}{\partial C_l^*} \right)^{\mathrm{T}} \frac{\partial \Gamma}{\partial C_k^*} \alpha_j, \tag{42}$$

where

$$\frac{\partial \alpha_j}{\partial C_l^*} = \sum_{\substack{i \\ \omega_i \neq \omega_j}} \frac{\alpha_i^{\mathrm{T}} \frac{\partial \Gamma}{\partial C_l^*} \alpha_j}{\omega_j^2 - \omega_i^2} \alpha_i. \tag{43}$$

(f) We estimate the accuracy of evaluated constants C_k^* from the expression

$$\kappa_k = \sqrt{\frac{\sum_{p_{\text{assoc.}}} \left(f_p^{cal} - f_p^{exp} \right)^2}{\sum_{p_{\text{assoc.}}} \left(\frac{\partial f_p^{cal}}{\partial C_k^*} \right)^2}} \approx \frac{\sqrt{\Delta(C_k^*)}}{S_k^*}, \tag{44}$$

where the last approximative equality is exactly satisfied only if $f_p^{\text{cal}} = f_p^{\text{exp}}$ for all associated modes.

The combinations C_k^* with a low value of κ_k (under some chosen threshold) are accepted, the rest is replaced by proper linear combinations of the initial guesses. From such set of C_k^* the original independent coefficients $C_k^{(1)}$ are computed by equations (33). Then, we return to the (b) step of the procedure with new constants $C_k^{(1)}$ and repeat this process until we match all the measured resonance modes and fit their frequencies.

This algorithm using the parameters C_k^* sorted by accuracy κ_k as sought variables (rather than the original constants C_k) enables us to perform the optimization gradually, i.e. to evaluate the most accurately determinable combination C_1^* as the first one (with the rest being fixed), then to evaluate C_1^* and C_2^*, etc., until a chosen sensitivity level is reached. Such inverse procedure is more robust than the classical method optimizing all original constants C_k together [32].

There are, of course, more ways in which the RUS technique can be improved. The accuracy of the input spectra can be improved by taking the attenuation in the material into account, i.e. by considering the effect which the attenuation has on the position of the maxima of the resonance peaks. The experiments can be done in an evacuated chamber to minimize the damping of the specimen by air. However, such particular improvements can slightly correct the obtained results, but undoubtedly cannot help the method to overcome its applicability limits given by material, and exceed, therefor, out of the frame of this chapter.

3. RUS Measurements Close to Applicability Limits

Let us now proceed to particular applications of the RUS method for determination of elastic coefficients of various anisotropic solids. The aim here will not be to illustrate the reliability and certainty of RUS when applied to materials for which this method was already verified and provides accurate and easily interpretable results. Quite on the contrary, the issues described in the following text are chosen such that the suitability of the RUS method for their solution is, on the first look, questionable.

In this section, the focus will be laid on the cases where the applicability of RUS is embarrassed due to the properties of the examined material itself. Namely, the particular issues solved within this section will be:

1. the case of **extremely strong anisotropy**, for which the specimen tends to vibrate in the softest modes related to the softest elastic coefficients, and the spectrum, consequently, does not contain sufficient information on the harder ones

2. the case of **weak anisotropy of unknown class of symmetry and unknown orientation**, where the inverse procedure leads to 21-dimensional optimization, as the material must be considered as fully triclinic

3. the case of **temperature dependent attenuation** where the required information (thermal dependencies of elastic coefficients) are shaded by simultaneous changes in the quality of the measured spectrum.

As it was mentioned in the introduction, the modifications of the RUS method described within this chapter cannot be understood as any kind of universal recipes valid for wider classes of similar problems. They are suitable for the particular materials, particular shapes of the specimens and particular experimental arrangements only; they illustrate, according to the main message this chapter aims to bring, the diversity and variability of the world of RUS.

3.1. Extremely strong anisotropy: Single crystals of Cu-Al-Ni

The first example of application of RUS discussed here will be the determination of elastic coefficients of a single crystal of the Cu-Al-Ni shape memory alloy. In the high-temperature *austenitic phase*, the crystal of this alloy has a cubic symmetry. Upon cooling and after applying proper mechanical loading (for details, see [38]), the material can be transformed into the low-temperature *martensitic phase*, which is orthorhombic. The elasticity of the cubic austenite can be fully described by three independent elastic coefficients c_{11}, c_{12} and c_{44}, which are, for the axes set parallel to the principal direction of the material, arranged in the matrix of elastic coefficients in the following way:

$$c_{ij} = \begin{pmatrix} c_{11} & c_{12} & c_{12} & 0 & 0 & 0 \\ c_{12} & c_{11} & c_{12} & 0 & 0 & 0 \\ c_{12} & c_{12} & c_{11} & 0 & 0 & 0 \\ 0 & 0 & 0 & c_{44} & 0 & 0 \\ 0 & 0 & 0 & 0 & c_{44} & 0 \\ 0 & 0 & 0 & 0 & 0 & c_{44} \end{pmatrix}. \tag{45}$$

After being transformed into martensite, the symmetry of the material decreases into the orthorhombic symmetry. Elasticity of such material can be, then, fully described by nine independent elastic coefficients, which, in natural coordinates of this system, form matrix

$$c_{ij} = \begin{pmatrix} c_{11} & c_{12} & c_{13} & 0 & 0 & 0 \\ c_{12} & c_{22} & c_{23} & 0 & 0 & 0 \\ c_{13} & c_{23} & c_{33} & 0 & 0 & 0 \\ 0 & 0 & 0 & c_{44} & 0 & 0 \\ 0 & 0 & 0 & 0 & c_{55} & 0 \\ 0 & 0 & 0 & 0 & 0 & c_{66} \end{pmatrix}. \tag{46}$$

. The coordinate systems in which the elasticity matrices of austenite and martensite adopt forms (45) and (46) are related by the coordination relations following from the nature of the martensitic phase transition between these two phases (e.g. [39]).

Our aim will be to determine all the independent elastic coefficients of both austenite and martensite from RUS measurements on a single specimen (i.e. the same specimen being first in austenite and then in martensite). The solution of this problem is significantly complicated by the strong anisotropy of the material. In the cubic phase, the strength of the anisotropy can be characterized by the *anisotropy factor*

$$A = \frac{2c_{44}}{c_{11} - c_{12}}. \tag{47}$$

This factor is equal to one for an isotropic material, for common single crystals of metals (aluminium, nickel, copper, gold, silicon) this factor ranges between 0.5 and 5. For the austenitic phase of Cu-Al-Ni, this factor is approximately equal to 12.

What does such high value of A mean physically? To understand it, we have to introduce few fundamental terms of the elastodynamics of anisotropic materials, particularly of the elastic wave propagation. The starting point for us will be the *Christoffel equation* (for derivation, see [40] or any similar textbook)

$$(n_j C_{ijkl} n_l - \rho v_\varphi^2 \delta_{ik}) U_k = 0, \tag{48}$$

which relates the phase velocity of the elastic wave v_φ with the direction of propagation $\mathbf{n}$, the elastic coefficients of the material C_{ijkl}, the density ρ and the polarization vector of the wave $\mathbf{U}$; δ_{ik} is the Kronecker's symbol. For given direction $\mathbf{n}$, this equation becomes an eigenvalue problem for the so-called *Christoffel matrix*

$$\Gamma_{ik} = n_j C_{ijkl} n_l, \tag{49}$$

which can be solved by finding the roots of secular equation

$$\det(\Gamma_{ik} - \rho v_\varphi^2 \delta_{ik}) = 0. \tag{50}$$

As it follows from definition (49), the Christoffel matrix Γ_{ik} is symmetric and positively definite, therefore its eigenvalues ρv_φ^2 are always real positive and its eigenvectors $\mathbf{U}$ create an orthogonal triplet.

In an isotropic case, the largest eigenvalue $\rho(v_\varphi^{\mathrm{L}})^2$ correspond to an eigenvector $\mathbf{U}^{\mathrm{L}}$ parallel to the direction $\mathbf{n}$, where the superscript L denotes the *longitudintal mode*. The remaining two solutions of (48) coincide in one wave of a *transverse mode* elliptically polarized in plane normal to $\mathbf{n}$, which travels along the direction of $\mathbf{n}$ at phase velocity v_φ^{T}.

In the anisotropic case, the solutions of the Christoffel equation are much more general. The secular equation (50) has, in general, no degenerated roots and the corresponding polarization directions are neither parallel nor normal to the direction $\mathbf{n}$. It is said that the planar waves in each direction can propagate at three different wave modes; one quasi-longitudinal (qL) mode and two quasi-transverse (qT^1, qT^2) modes, withal the qL mode is the one with polarization vector $\mathbf{U}^{\mathrm{qL}}$ closest to the direction of propagation $\mathbf{n}$.

In principal directions of a cubic material, the phase velocities can be expressed analytically. In particular, in the [110] direction of a cubic crystal, the phase velocities are

$$v_\varphi^{\mathrm{qL}} = \sqrt{\frac{c_{11} + c_{12} + 2c_{44}}{2\rho}} \stackrel{\mathrm{def}}{=} \sqrt{\frac{c_L}{\rho}}, \qquad v_\varphi^{\mathrm{qT}^1} = \sqrt{\frac{c_{44}}{\rho}}$$

$$\text{and} \qquad v_\varphi^{\mathrm{qT}^2} = \sqrt{\frac{c_{11} - c_{12}}{2\rho}} \stackrel{\mathrm{def}}{=} \sqrt{\frac{c'}{\rho}}. \tag{51}$$

The anisotropy factor A relates, thus, the phase velocities of qT^1 and qT^2 modes in this direction by relation

$$v_\varphi^{\mathrm{qT}^1} = \sqrt{A} v_\varphi^{\mathrm{qT}^2}. \tag{52}$$

So, for A being approximately equal to twelve, we get

$$v_\varphi^{\mathrm{qT}^1} = 2\sqrt{3} v_\varphi^{\mathrm{qT}^2}. \tag{53}$$

Moreover, as the cubic materials typically have[2] $v_\varphi^{\mathrm{qL}} \approx 2 v_\varphi^{\mathrm{qT}^1}$, we can write

$$v_\varphi^{\mathrm{qL}} = 2 v_\varphi^{\mathrm{qT}^1} = 4\sqrt{3} v_\varphi^{\mathrm{qT}^2}. \tag{54}$$

In other words, if the specimen of such material was a 1D rod (similar to the oak wood rod considered in the introduction) cut along the [110] direction, the first resonant frequency

[2] for the [110] direction in the examined single crystal of Cu-Al-Ni, this ratio is even higher, approximately equal to 2.2

corresponding to c' would be approximately two times lower than the first one corresponding to c_{44} and nearly seven times lower than the first one corresponding to c_L. With the multiplication of the modes taken into account, we can approximately say that 70% of the spectrum of such rod would contain the information on c' whereas only 10% on c_L. With all the resonant frequencies determined with the same accuracy, we can, thus, expect the coefficient c_L to be obtainable with seven times higher experimental error than c'.

In a real 3D specimen, the information on particular elastic coefficients is in the spectrum distributed even less uniformly, as it is illustrated by the following experimental example. The examined specimen was a parallelepiped of the austenite with dimensions 5.11 x 6.16 x 5.22mm and orientations [0.53; -0.80; -0.26], [0.81; 0.58; -0.07], [0.21; -0.15; 0.96], i.e. with two directions close to [1-10] and [110] principal axes and the third one close to [001]. The RUS measurement was performed in the frequency range 0.1-0.6MHz which resulted in 65 identified modes, which seems to be an inappropriately large number for identification of three independent coefficients. The inverse procedure was run in the multi-stage modification (see the third item of the concluding discussion in subsection 1.3) to obtain an orthogonal set of combinations of elastic coefficients sorted by accuracy. The resulting combinations $C^*_{k=1,2,3}$ are listed in Tab.1. Unsurprisingly, the most accurately determined combination is nearly exactly equivalent to c' and the second one to c_{44}. As the combinations C^*_k are orthogonal by definition, the last combination cannot be equivalent to c_L, but it (similarly to c_L) contains the information on $c_{11} + c_{12}$.

It is obvious that the inaccuracy in C^*_3 is unacceptably high. The resonant spectrum simply does not contain sufficient information on the c_L coefficient, which is related to longitudinal modes of propagation. The applicability of the RUS method is here restricted to accurate determination od coefficients c' and c_{44}. The full anisotropy cannot be captured, unless even more higher modes are taken into account.

Let us, on the other hand, make a short comparison of the RUS method with the pulse-echo measurements, where the elastic coefficients are obtained from values of phase velocities by inverting relations similar to (51). Let us consider a specimen being cut exactly along the [110], [1-10] and [001] directions. In the the [110] and [1-10] directions, the velocities are given by (51), in the [001] directions, the relations are

$$v_\varphi^{\mathrm{qL}} = \sqrt{\frac{c_{11}}{\rho}}, \qquad v_\varphi^{\mathrm{qT}^1} = v_\varphi^{\mathrm{qT}^2} = \sqrt{\frac{c_{44}}{\rho}}. \tag{55}$$

Obviously, the coefficients c_{11} and c_{44} can be conveniently determined from measurements of longitudinal and shear wave phase velocities in direction [001]. Then, it is sufficient to

Table 1. Combinations of elastic coefficients for the austenitic phase of SMA and the accuracies which are the combinations determined with.

k	Combination	C^*_k [GPa]	κ_k [GPa]
1	$0.71c_{11} - 0.70c_{12} + 0.07c_{44}$	18.11	0.14
2	$0.01c_{11} + 0.1c_{12} + 0.99c_{44}$	108.38	2.14
3	$0.7c_{11} + 0.71c_{12} - 0.08c_{44}$	169.46	34.94

measure the qL velocity in direction [110] to evaluate c_{12} as $2\rho\left(v^{\mathrm{qL}}_{\varphi[110]}\right)^2 - 2c_{44} - c_{11}$. But how does it look like with the accuracy?

Consider that we obtain the phase velocities in forms $v^{\mathrm{qL}}_{\varphi[001]} \pm \delta v^{\mathrm{qL}}_{\varphi[001]}$, $v^{\mathrm{qT}}_{\varphi[001]} \pm \delta v^{\mathrm{qT}}_{\varphi[001]}$ and $v^{\mathrm{qL}}_{\varphi[110]} \pm \delta v^{\mathrm{qL}}_{\varphi[110]}$. Using the known rules for recalculation of experimental errors between quantities, we can write that

$$c_{11} = \rho\left(v^{\mathrm{qL}}_{\varphi[001]}\right)^2 \pm 2\rho v^{\mathrm{qL}}_{\varphi[001]}\delta v^{\mathrm{qL}}_{\varphi[001]}, \tag{56}$$

$$c_{44} = \rho\left(v^{\mathrm{qT}}_{\varphi[001]}\right)^2 \pm 2\rho v^{\mathrm{qT}}_{\varphi[001]}\delta v^{\mathrm{qT}}_{\varphi[001]}, \tag{57}$$

and finally

$$\begin{aligned} c_{12} &= \rho\left(v^{\mathrm{qL}}_{\varphi[110]}\right)^2 - 2\rho\left(v^{\mathrm{qT}}_{\varphi[001]}\right)^2 - \rho\left(v^{\mathrm{qL}}_{\varphi[001]}\right)^2 \\ &\pm 2\rho\left(v^{\mathrm{qL}}_{\varphi[110]}\delta v^{\mathrm{qL}}_{\varphi[110]} + 2v^{\mathrm{qT}}_{\varphi[001]}\delta v^{\mathrm{qT}}_{\varphi[001]} + v^{\mathrm{qL}}_{\varphi[001]}\delta v^{\mathrm{qL}}_{\varphi[001]}\right). \end{aligned} \tag{58}$$

For c' evaluated now as $c_{11} - c_{12}$, the experimental accuracy is

$$\delta c' = \rho\left(v^{\mathrm{qL}}_{\varphi[110]}\delta v^{\mathrm{qL}}_{\varphi[110]} + 2v^{\mathrm{qT}}_{\varphi[001]}\delta v^{\mathrm{qT}}_{\varphi[001]} + 2v^{\mathrm{qL}}_{\varphi[001]}\delta v^{\mathrm{qL}}_{\varphi[001]}\right). \tag{59}$$

If we take into account that

$$v^{\mathrm{qL}}_{\varphi[110]} \approx \sqrt{2}v^{\mathrm{qL}}_{\varphi[001]} \approx 2v^{\mathrm{qT}}_{\varphi[001]} \tag{60}$$

and consider that all the phase velocities were determined with the same accuracy, i.e. that

$$\delta v^{\mathrm{qL}}_{\varphi[110]} \approx \delta v^{\mathrm{qL}}_{\varphi[001]} \approx \delta v^{\mathrm{qT}}_{\varphi[001]}, \tag{61}$$

we can conclude that c' is from the pulse-echo measurements determined with $(4 + 2\sqrt{2})$ times lower accuracy than c_{44}. As c' is approximately five times smaller than c_{44}, the pulse echo measurements at the level of 1% experimental error in c_{44} results in nearly 35% error in determination of c'.

If the coefficient c' is not determined from the above procedure but from direct measurements of the $v^{\mathrm{qT}^2}_{\varphi[110]}$ velocity, the accuracy can be incomparably higher. However, the experimental determination of this velocity is extremely complicated. In paper [38], the authors examined variously oriented single crystals of Cu-Al-Ni by pulse echo methods, but were not able to detect reliably this wave in any direction. Similarly, Stipcich et al. [41] were not able to detect any reliable echo for this velocity in Ni-Mn-Ga although their specimen was cut exactly along the (110) planes. On the other hand, this wave was repeatedly measured in materials with lower anisotropy factors, such as Co-Al-Ni ([42], A =4) or Cu-Al-Mn ([43], A =5.4). The reasons are numerous (e.g. strong magneto-elastic attenuation as discussed in subsection 3.3), but the most important of them probably is that for such strong anisotropy, the qT^2 mode is extremely strongly affected by the *directional dispersion*, i.e. by the fact that the anisotropy focuses the energy carried by qT^2-waves to few preferred directions whereas the others (e.g. the [110] direction) become energetically suppressed (more details on this effect can be found, again, in [40] or any similar textbook).

Table 2. Combinations (dominating terms only) of elastic coefficients of martensite of Cu-Al-Ni sorted by accuracy they can be determined with.

k	Dominating terms	C_k^* [GPa]	κ_k [GPa]
1	$c_{11} + c_{22} + 2c_{55} - 2c_{12}$	51.93	0.37
2	$-c_{11} - c_{22} + 2c_{55} + c_{12}$	18.34	0.90
3	$2c_{44} + 2c_{66}$	115.92	2.77
4	$2c_{44} - 2c_{66} - c_{23}$	10.98	3.56
5	$2c_{33} - c_{23} - 2c_{13}$	8.02	5.75
6	$c_{11} + c_{44} + 2c_{23} - c_{13} - c_{12}$	25.63	5.87
7	$2c_{11} - 2c_{22} + c_{33} - c_{23} + c_{13}$	111.60	17.08
8	$2c_{11} + c_{22} - 1c_{33} + 2c_{12}$	182.55	53.35
9	$c_{22} + 2c_{33} + c_{23} + c_{13}$	302.09	73.54

For c' determined as the difference between experimentally obtained c_{11} and c_{12}, the experimental error of this coefficient (59) is similarly unacceptable as the error in C_3^* obtained from RUS measurements. This lead us to the idea, that the RUS and pulse-echo measurements can be, in some sense, complementary to each other, and that their combination could be a good way how to determine the elastic coefficients of strongly anisotropic materials. This idea will become essential in the second part of this subsection, where we will try to determine the elastic properties of the same specimen but after being transformed into a single variant of martensite.

In the monovariant of martensite, the shape of the specimen was a non-rectangular parallelepiped (5.14mm×5.92mm×5.39mm) with face normals [-0.71; 0.58; -0.37], [0.32; 0.84; 0.42], [0.55; 0.23; -0.80] in the natural coordinate system of the orthorhombic crystal lattice of martensite. RUS measurement was performed in the frequency range 0.1-0.8MHz, 70 resonances were involved in the inversion procedure, which was, again, run in the multi-stage modification with the result obtained in a form of orthogonal combinations of elastic coefficients sorted by accuracy. These results are outlined in Tab.2, where the combinations are reduced to few dominating terms to highlight the particular character of each of them. Obviously, the combinations Nos.7$\div$9 are extremely inaccurately determined. The question is, whether especially these three combinations can be determined by pulse-echo measurements, i.e. whether the RUS and pulse-echo measurements are really complementary to each other. As the specimen in martensite has a quite general crystallographic orientation and the class of symmetry is quite low (orthorhombic), no analytic formulae can be derived to relate the phase velocities in the direction normal to the parallelepiped's faces with the elastic coefficients. These relations must me sought numerically, namely by analyzing which of the combinations from Tab.2 depend sensitively on values of which phase velocities. On this purpose, the partial derivatives of particular phase velocities with respect to the combinations $\partial v_\varphi / \partial C_k^*$ were approximated by finite differences

$$\frac{\partial v_\varphi}{\partial C_k^*} \approx \frac{v_\varphi(C_1^*, \ldots, C_k^* + \delta C_k^*, \ldots, C_9^*) - v_\varphi(C_1^*, \ldots, C_k^*, \ldots, C_9^*)}{\delta C_k^*}, \tag{62}$$

and these differences were evaluated for all modes of propagation (qL, qT^1, qT^2) and three possible directions (normals to the parallelepiped's faces). The result is graphically embodied in Fig.5, from which one can conclude that the relation between the determined combinations and the phase velocities measurable by pulse-echo methods is rather nontrivial. What is undisputable is the fact that the first (the most accurately determined) combination

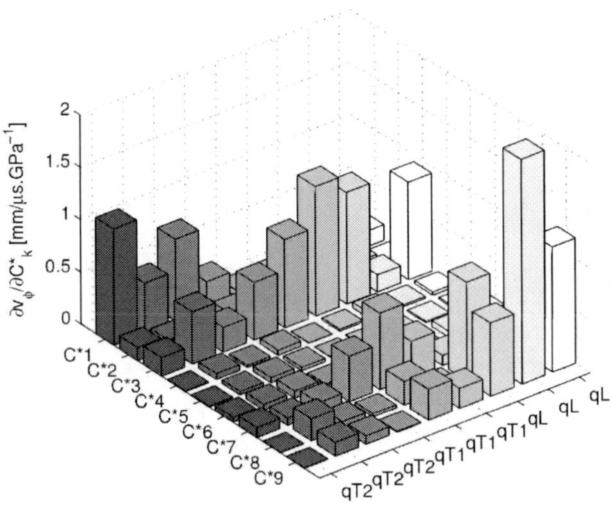

Figure 5. Sensitivity of values of the phase velocities to the combinations of elastic coefficients C_k^*. Labels qL, qT1 and qT2 denote the modes of propagation. For each mode, three velocities are considered, corresponding to three directions given by the normals to the specimen's faces.

C_1^* is somehow related to the qT^2 mode of propagation whereas the last (the least accurately determined) combination C_9^* to the qL mode. This is an evident similarity with the previous case of the austenitic specimen. On the other hand, the combinations cannot be strictly divided by their correspondence to particular modes of propagation, as the combination No.3 is sensitively dependent to all modes of propagation. On the other hand, the combinations Nos.4÷6 do not significantly correspond to any of the modes of propagation. The question is how the above findings can improve the inverse evaluation of the elastic coefficients. The quasi-longitudinal velocities v_φ can be easily measured using the pulse-echo technique, so this additional information may be used for more accurate determination of combinations Nos.7÷9. The most natural approach seems to be the direct involvement of the phase velocities into the optimizing process, i.e. adding a term

$$\Delta_\varphi = \sum_{n=1}^{3} (v_\varphi^{\mathrm{cal}}(C_{ijkl}) - v_\varphi^{\mathrm{exp}})^2 \tag{63}$$

to the error function (1). However, such extension of the error function disables the construction of linear combinations, and the whole optimization process becomes more complicated. Another possibility is to take the combinations No.1÷6 as precisely determined from the RUS measurements, and Nos. 7÷9. as independent. After the combinations

Table 3. Elastic coefficients of the martensitic phase of Cu-Al-Ni. The results obtained by RUS and by combination of RUS and pulse-echo measurements of qL velocities is compared to the results of [38] obtained by pulse-echo measurements on various specimens.

Source	c_{11} [GPa]	c_{22} [GPa]	c_{33} [GPa]	c_{44} [GPa]	c_{55} [GPa]	c_{66} [GPa]	c_{23} [GPa]	c_{13} [GPa]	c_{12} [GPa]
RUS	187.17 $\pm$16.47	149.18 $\pm$15.00	236.38 $\pm$27.10	69.40 $\pm$2.96	22.62 $\pm$0.82	63.01 $\pm$3.85	93.07 $\pm$16.43	65.03 $\pm$17.77	140.18 $\pm$14.77
RUS & qL	185.38 $\pm$1.83	147.55 $\pm$1.87	229.89 $\pm$0.84	71.02 $\pm$1.85	22.93 $\pm$0.53	63.74 $\pm$1.69	97.35 $\pm$2.14	74.11 $\pm$1.51	138.96 $\pm$1.01
[38]	184.46 $\pm$1.12	151.45 $\pm$0.75	238.58 $\pm$1.87	66.39 $\pm$0.21	22.85 $\pm$0.18	60.55 $\pm$0.40	86.83 $\pm$1.05	70.09 $\pm$1.07	140.41 $\pm$0.77

Nos.1$\div$6 are determined from the frequency spectrum, the remaining combinations can be easily fitted to the longitudinal phase velocities. This approach enables us to reach the average difference between measured and calculated qL phase velocities to be smaller than 10^{-6}mm/μs without disturbing the optimality of the fit between the measured and calculated spectra. Tab.3 compares the accuracies of individual independent coefficients evaluated without and with involving the qL phase velocities in the algorithm. In the first row, the estimates of the experimental errors were obtained from κ_k by the linear relation

$$\delta C_k = |\Phi_{kl}| \frac{\kappa_l}{3}, \tag{64}$$

where Φ_{kl} are the coefficients of the combinations (34) and the factor $\frac{1}{3}$ reflects the fact that the $[C_k^* - \kappa_k; C_k^* + \kappa_k]$ is assumed as a 3σ-interval. In the second row, a similar procedure was applied, but with the 3σ intervals of combinations Nos.7$\div$9 determined by Monte Carlo simulation with the requirement of agreement in qL phase velocities being better than 10^{-6}mm/μs. The impact of involvement of the qL phase velocities into the inverse algorithm is dramatic: The experimental errors in coefficients c_{33}, c_{13}, and c_{12} are reduced more ten times. The smallest improvement is in the coefficients c_{44}, c_{55} and c_{66}, i.e. in the shear coefficients closely related to the first (softest) eigenmodes and already accurately determined by RUS.

In Tab.3 the results from RUS are also compared to these obtained from pulse-echo in fifteen different directions (see [38] for more detail). The accuracies of the coefficients from [38] seem to be slightly higher (except of c_{33}). However, they were obtained by Monte Carlo simulations with chosen input errors in the values of phase velocities and in the specimen orientation, and are, thus, artificial in some sense; the experimental errors of the results of RUS were, on contrary, evaluated directly from the nature of the method by definition (44). Moreover, it must be pointed out that the results from RUS were obtained on one specimen only, whereas in [38], five differently oriented specimens were used.

In general, we can conclude that the RUS method can be applied to materials with extremely strong anisotropy, but the results cannot be expected to have satisfactory accuracy unless some additional information is involved, e.g. the values of qL phase velocities in given directions of the material. For the cubic austenite, the RUS method and the inversion from phase velocities are fully complementary – the first cannot accurately determine the c_L coefficient, the latter the c' coefficient. In the case of the orthorhombic martensitic phase, the situation is more complicated, but the combination of RUS and pulse-echo measurements can result in acceptably accurate determination of all independent elastic coefficients.

3.2. Weak (Averaged) Anisotropy: Elastic Properties of Finely Grained Materials Processed by ECAP

Consider now a quite different problem. Let the specimen be a parallelepiped again, but not a single crystal, where the orientations can be accurately determined from X-ray measurements, but a *textured polycrystal*, where the anisotropy has a statistical, averaged character which can be given by both the preferred crystallographic orientation of the grains and the microstructure (i.e. the pattern of grain boundaries). In such case, the principal axes of the symmetry (if there are any) are completely unknown, as well as the class of the symmetry[3]. To characterize the elasticity of the material, a full triclinic description with 21 independent elastic coefficients must be used in the first step; the character and orientation of the anisotropy can be only estimated *a posteriori* by finding some cartesian system in which the tensor C_{ijkl} has the highest symmetry. The only reasonable assumption we can do about such anisotropy is that to expect it to be weak, i.e. to be describable as a small perturbation of the isotropic elasticity of an untextured polycrystal with isotropic microstructure. Obviously, the RUS technique is unreplaceable here, since the information obtainable from pulse-echo measurements (three triples of phase velocities in directions normal to the faces of the specimen) can never be sufficient for determination of all 21 constants. Let it be pointed out that this problem is completely different from the case discussed by Sarrao et al. [16], who have shown that the RUS method can be suitable for determination of crystallographic orientations of small single crystals. The main difference is that Sarrao et al. knew *a priori* the class of symmetry of the examined material, which enabled them to extend the inverse procedure by involving the orientation of the principal axes as additional sought unknown variables.

As an illustrative example, we will evaluate the elastic properties of a polycrystal of copper processed by *equal channel angular pressing* (ECAP) [44–47]. ECAP technology is a method for manufacturing of fully dense nanoscopically grained materials, based on subjecting the material to repeated plastic deformation by moving a workpiece several times through a die containing two intersecting channels of identical cross-sections. During each processing cycle, the grains become finer. The first pass of the workpiece induces the pattern in grain boundaries, which rotates for a small angle during every pass.

The specimen used in our experiments was a $3\times5\times7$mm rectangular parallelepiped, cut from the workpiece after the first route through the processing die. The one pass only was chosen because the material in this state has the weakest anisotropy – the microstructure of the grain boundaries does not play such important role as after many passes where the grains are much smaller (and the volume fraction of the grain boundaries, thus, much higher). The density of the specimen was expected to be the same as for common polycrystalline copper

[3]Both the class and the orientation of the anisotropy can be sometimes approximately guessed from optical or EBSD microscopy of the grain texture, but as our primary aim is to check the power of the RUS method to solve this problem, we will not take any such additional information into account.

Table 4. Combinations of elastic coefficients (dominating terms only) for a weakly anisotropic, nanograined polycrystal of copper sorted by accuracy.

k	Dominant coefficients	C_k^* [GPa]	κ_k [GPa]
1.	$c_{11}+c_{22}+c_{33}+2c_{44}+2c_{55}+2c_{66}-c_{23}-c_{13}-c_{12}$	119.47	0.18
2.	$c_{33}+c_{44}+2c_{55}-3c_{66}-c_{13}-c_{46}+c_{56}$	23.29	0.32
⋮	⋮	⋮	⋮
16.	$c_{22}+c_{33}-c_{13}-2c_{25}+2c_{35}+c_{24}-c_{26}-c_{34}+c_{36}+c_{45}$	22.06	7.30
17.	$-c_{22}+c_{33}-c_{44}-c_{23}+c_{13}-c_{12}+c_{25}-c_{35}+2c_{14}+2c_{24}-2c_{34}$	45.48	9.64
18.	$c_{11}+2c_{33}+c_{23}+c_{13}-c_{14}-c_{16}-c_{24}-c_{26}-c_{34}-2c_{36}$	1.83	22.15
19.	$c_{22}+c_{12}+c_{14}-c_{16}+2c_{24}-c_{26}+2c_{34}-c_{36}$	183.06	40.20
20.	$c_{11}+c_{22}+c_{33}+c_{13}+c_{12}+c_{15}+c_{25}+c_{35}$ $+c_{46}+c_{14}+c_{16}+c_{24}+c_{26}+c_{34}+c_{36}$	70.10	49.41
21.	$c_{11}+c_{22}+c_{13}+c_{12}-2c_{15}-2c_{25}-2c_{35}+c_{14}+c_{24}+c_{36}$	182.14	58.83

(8.96 g.cm^{-1}). This specimen was investigated by the common RUS procedure, taking first 100 resonances and the shapes of corresponding eigenmodes as an input. Similarly as for the previous case of extremely strong anisotropy, the multi-stage inverse algorithm was applied, and thus, the result was obtained in a form of linear combinations of the sought elastic coefficients sorted by accuracy. As the initial guesses, isotropic elastic coefficients of polycrystalline copper were taken. These guesses enabled reliable identification most of the input modes.

Tab.4 includes the first (i.e. the most accurately determined) two and the last (i.e. the least accurately determined) six combinations. In this table, not the full combinations are listed (each having 21 terms) but only few dominating terms are shown to highlight the overall character of each combination. Obviously, the first of these combinations has a strong physical meaning. As our material is nearly isotropic, we can consider for a while that

$$c_{11} \approx c_{22} \approx c_{33}, \tag{65}$$

$$c_{12} \approx c_{23} \approx c_{13}, \tag{66}$$

$$\text{and } c_{44} \approx c_{55} \approx c_{66} \approx \frac{c_{11}-c_{12}}{2}. \tag{67}$$

Then, the combination No.1 is proportional to c_{44}, which is the the shear modulus of the isotropic continuum.

The relation between the first combination and the shear velocities is clear even when a full anisotropy is considered. Being written in form

$$C_1^* \sim c_{44}+c_{55}+c_{66}+\frac{c_{11}-c_{12}}{2}+\frac{c_{22}-c_{23}}{2}+\frac{c_{33}-c_{13}}{2}, \tag{68}$$

this combination can be approximately understood as an average value of squares of shear velocities in various principal directions. Similar discussion can be done also for the second combination, where, for the isotropic continuum ($c_{46} \approx c_{56} \approx 0$), the dependence remains on c_{44}, c_{55}, c_{66} and $\frac{c_{33}-c_{13}}{2}$ only, which are all equal to the shear modulus again.

The last four combinations (Nos. 18÷21) are extremely inaccurately determined. For this reason we decided to tune these combinations by pulse-echo measurement in a similar way as it was done in the previous subsection for the strong anisotropy. Seven phase velocities (three quasilongitudinal and four quasitransverse) were taken into account. After such correction, the resultant triclinic elasticity with coefficients

$$c_{ij} =$$

$$= \begin{pmatrix} 199.70 \pm 0.79 & 111.50 \pm 0.92 & 102.65 \pm 0.80 & 5.39 \pm 1.54 & 4.64 \pm 1.09 & -0.68 \pm 0.76 \\ & 196.87 \pm 1.10 & 109.08 \pm 0.67 & 6.33 \pm 1.45 & 3.40 \pm 1.54 & -1.85 \pm 0.62 \\ & & 200.83 \pm 1.46 & 5.53 \pm 1.39 & 2.06 \pm 1.43 & -3.90 \pm 1.20 \\ & & & 42.29 \pm 0.42 & 1.69 \pm 0.95 & 0.19 \pm 0.87 \\ & \text{symm.} & & & 42.65 \pm 0.42 & 0.01 \pm 0.55 \\ & & & & & 42.62 \pm 0.42 \end{pmatrix} \text{GPa} \tag{69}$$

approximated all these phase velocities with errors lower than $0.05\text{mm}.\mu\text{s}^{-1}$. Similarly as in the case of the martensite of the Cu-Al-Ni, the estimates of experimental errors were recalculated from the known linear relations between C_k^* and c_{ij}

Let us now try to identify the class and orientation of the anisotropy to lower the number of independent coefficients to the essential minimum. As the material is nearly isotropic, the surfaces of the phase velocity (qL, qT^1, qT^2) are close to spheres. On contrary, the difference between the quasishear velocity surfaces $\text{qT}^1 - \text{qT}^2$ (which is identically equal to zero in the isotropic material) copies sensitively the symmetry of the material, and can be, thus, used for its identification.

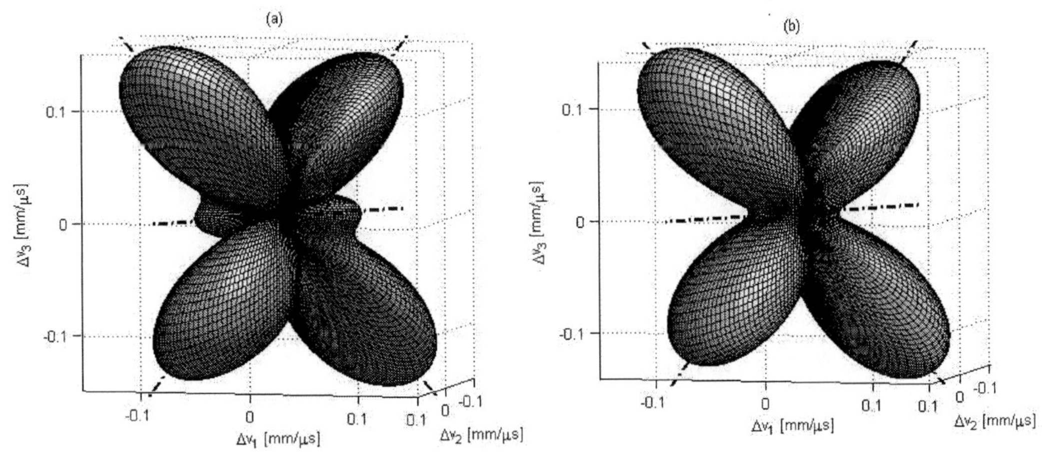

Figure 6. Surfaces of difference between qT velocities: (a) fully triclinic anisotropy determined by RUS method in combination with pulse-echo measurements of qL phase velocities; (b) orthorhombic approximation of the material.

In Fig.6(a), the surface of the difference

$$\Delta v(\mathbf{n}) = v^{\text{qT}^1}(\mathbf{n}) - v^{\text{qT}^2}(\mathbf{n}) \tag{70}$$

is plotted in the axes given by the edges of the specimen. Obviously, this surface has higher class of symmetry than fully triclinic. Three mutually orthogonal axes can be easily identified, having a general orientation to the edges of the specimen. After rotating the matrix (69) onto these orthogonal axes and setting the coefficients

$$c_{14} = c_{15} = c_{16} = c_{24} = c_{25} = c_{26} = c_{34} = c_{35} = c_{36} = c_{45} = c_{46} = c_{56} = 0, \quad (71)$$

we obtain an orthorhombic system with the elastic coefficients (in the rotated axes)

$$c_{ij} = $$

$$= \begin{pmatrix} 203.74 \pm 1.64 & 106.23 \pm 0.67 & 116.79 \pm 1.63 & 0 & 0 & 0 \\ 106.23 & 182.97 \pm 1.35 & 105.64 \pm 0.90 & 0 & 0 & 0 \\ 116.79 & 105.64 & 199.82 \pm 0.29 & 0 & 0 & 0 \\ & & & 41.26 \pm 0.40 & 0 & 0 \\ & \text{symm.} & & & 42.31 \pm 0.26 & 0 \\ & & & & & 49.42 \pm 0.17 \end{pmatrix} \text{GPa}, \quad (72)$$

where the experimental errors were determined by simply rotating the matrix of accuracies from (69) into the coordinates of the orthorhombic symmetry. The $\Delta v(\mathbf{n})$ surface for this system is plotted in Fig.6(b). Fig.7 shows how accurately this orthorhombic system approximates the properties of the original triclinic system. The compared quantities here are the *slowness vector components*, i.e. the reciprocals of the phase velocities. Similarly

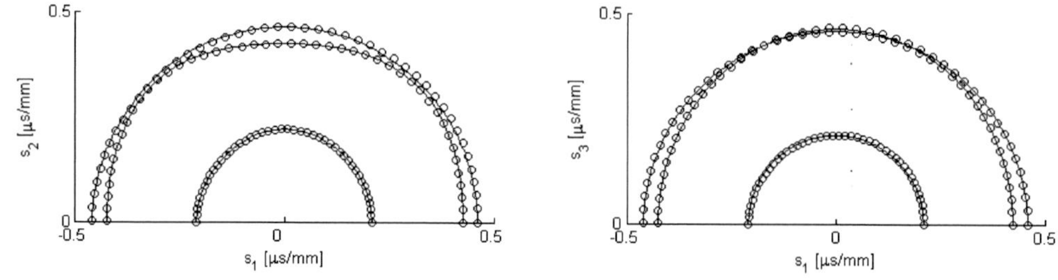

Figure 7. Principal cuts (by x_1x_2 and x_1x_3 planes) of slowness surfaces for the examined material. The circles correspond to fully triclinic anisotropy determined by RUS method in combination with pulse-echo measurements of qL phase velocities, the solid lines to the orthorhombic approximation of the material.

good agreement can be seen between the resonant spectra and the shapes of the corresponding eigenmodes for the originally considered full triclinic anisotropy and the identified orthorhombic system. In Fig.8, such comparison is shown. Only for few modes (e.g. mode No.30 in Fig.8) some difference between the shapes of modes evaluated for the triclinic and the orthorhombic symmetry can be seen. The values of resonances are, however, matched with an excellent agreement.

We can conclude that the RUS method (in combination with pulse-echo measurements) is able to identify the class and orientation of anisotropy in weakly anisotropic materials. However, the above outlined procedure can be applied only for nearly isotropic materials, where the isotropy can be used as the initial guess for the inverse algorithm, which enables the mode association.

Mode information	Experimental pattern	Evaluated for triclinic	Evaluated for orthorhombic
mode No.69 f_{exp} = 671.2kHz f_{tric} = 670.9kHz f_{ort} = 669.0kHz	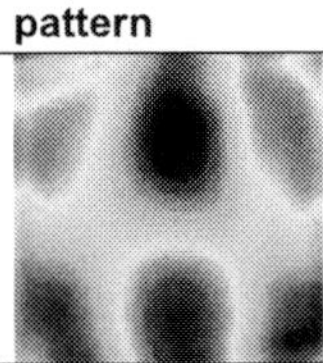	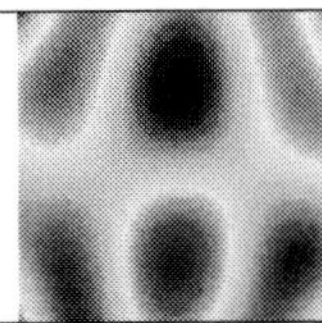	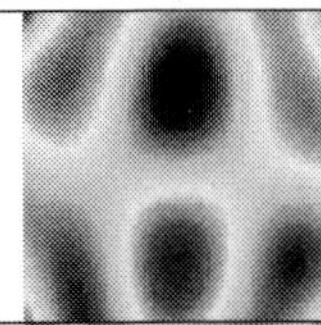
mode No.77 f_{exp} = 708.4kHz f_{tric} = 709.6kHz f_{ort} = 708.6kHz	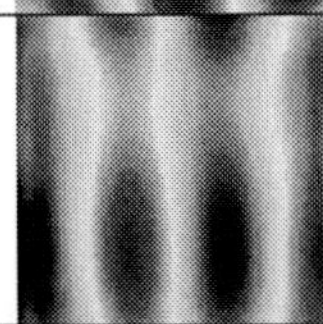	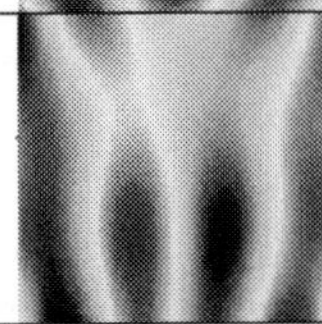	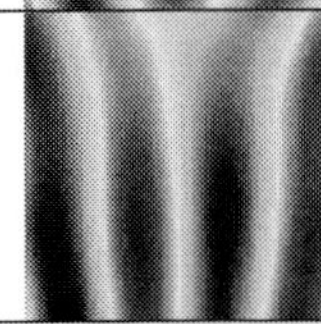
mode No.30 f_{exp} = 495.3kHz f_{tric} = 496.6kHz f_{ort} = 496.7kHz	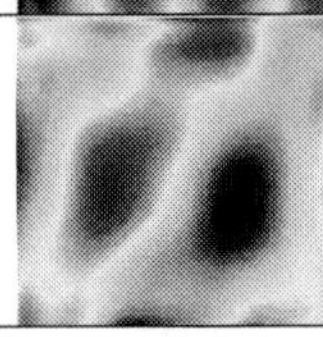	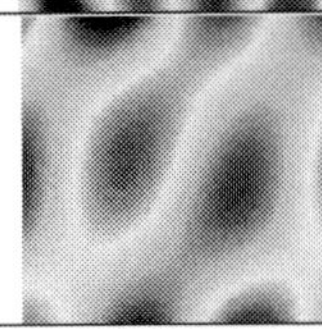	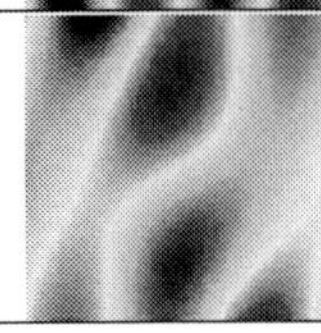

Figure 8. Examples of comparison between measured and evaluated displacement patterns for particular eigenmodes; $f_{\rm exp}$ denotes the experimentally determined eigenfrequency, $f_{\rm tric}$ the corresponding frequency evaluated for the full triclinic matrix (69), $f_{\rm ort}$ for the orthorhombic approximation (72).

There is one more remark to be done here concerning the RUS investigation of weak averaged anisotropies, i.e. anisotropies induced not by the crystal lattice but by an oriented microstructure. In the above investigated case of polycrystalline copper processed by ECAP, the texture is nanoscopic, which means that all its characteristic dimensions can be considered as incomparably smaller than the dimensions of the specimen as well as than the wavelengths of the all modes involved in the inverse procedure. However, it is important to understand what the expression *incomparably smaller* exactly means, i.e. how coarse microstructuring will not limit the applicability of RUS.

For simplicity, let us return to our well-tried example of a 1D specimen. The following numerical example gives an illustration of how the coarsening of the microstructure can significantly affect the measurements: A one-dimensional string of 7.2mm in length was considered, consisting of 48, 24, and 12 elements of unit mass density and alternating bending wave velocities $v_\varphi^A = 1\text{mm}/\mu\text{s}$ and $v_\varphi^B = 3\text{mm}/\mu\text{s}$. The volume fraction of the component B was chosen as $\mu = 2/3$. The eigenfrequencies of bending modes of such spring were determined using the COMSOL Multiphysics$^{\rm TM}$ eigenvalue solver with the spring meshed by 1152 Lagrangian-quadratic finite elements. The stability of the solution was checked by decreasing the FEM mesh density down to 288 elements on the string. The results are shown in Fig.9. For a homogeneous spring, the spectrum should linearly increase, as the equation of the steady waves (11) here has form

$$\rho\omega^2 u + S_{A+B}\frac{\mathrm{d}^2 u}{\mathrm{d}x^2} = 0, \tag{73}$$

where S_{A+B} is a bending stiffness obtained by homogenization of the bending properties of elements of material A and B. However, the results of the numeric simulation show

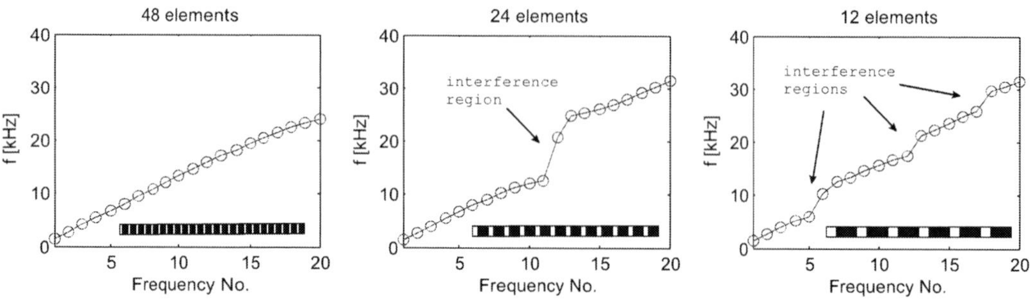

Figure 9. Effect of texture coarsening on eigenfrequencies for a 1D string composed of decreasing number of elements.

something quite different. Even for the spring consisting of 48 elements, the evaluated spectrum deviates from a linear trend (but remains, for the displayed first 20 eigenfrequencies, increasing smoothly). For 24 elements, a significant jump in a spectrum appears at about 15kHz. Further coarsening (to 12 elements) reveals that such jumps form a periodic serration on the increase of the spectrum, appearing in the regions where the eigenmodes and the structure interfere. We can conclude that even for the ratio between the dimension of the specimen and the characteristic length of the microstructure being equal to 50, the application of RUS for the determination of homogenized (averaged) elastic coefficients may be questionable.

In the more general 3D case, where the eigenmodes, even if they have eigenfrequencies close to each other, may geometrically differ in such way that one of them strongly interferes with the microstructure and the second in not influenced at all, the effect of coarsening on the resultant frequency spectrum becomes more complex. In [49], the effect of the interference between the eigenmodes of vibrations and the microstructure is illustrated for martensitic microstructures of Cu-Al-Ni, i.e. for geometrically ordered mixtures of variously oriented single variants of martensite analyzed in the previous section. For evaluation of the homogenized elastic coefficients of such ordered mixtures, an energetic algorithm described in [48] is used. The conclusion in [49] is that as soon as the thicknesses of particular laminas in the microstructure start to be higher than 100μm, the resonant spectra of a parallelepiped having few millimeters in each dimension exhibit similar interference effects as the above discussed 1D spring.

3.3. Thermal Dependencies of Elastic Coefficients in Media with Strong Magneto-Acoustic Attenuation: Single Crystal of Ni-Mn-Ga

Whereas the above two discussed applications of RUS were rather illustrative (chosen to have extremely strong or extremely weak anisotropy), which enabled their findings to be formulated in clear conclusions, the last case described within this section will concern an application of this method to an extremely complex material, coupling more physical phenomena together. The main aim here will be to show how the RUS applicability can be limited by ultrasound attenuation in the material. However, the findings from the first two subsections will be also utilized.

The internal friction in materials and resulting ultrasound attenuation belong among the most natural limiting factors for RUS measurements. Although the theoretical works dedicated to how the effect of attenuation can be avoided ([36] and the list of references therein) or even exploited for determination of damping parameters [26] are numerous, the essential question is always the same: How to identify the individual resonance peaks within the attenuated spectrum?

Consider now a part of a spectrum containing N resonant frequencies. For f_n being the resonant frequency of the n–th mode (n between 1 and N) with amplitude A_n and phase ϕ_n, the amplitude of a spectrum can be approximated by function

$$F(f_n, A_n, \phi_n, \mathrm{FWHM}_n) = \sum_{n=1}^{N} \frac{A_n \frac{\mathrm{FWHM}_n}{2} e^{\mathrm{i}\phi_n}}{\mathrm{i}(f - f_n) + \frac{\mathrm{FWHM}_n}{2}}, \quad (74)$$

where FWHM_n means *full width at half maximum* of the n–th mode, which is one of the possible parameters to measure the attenuation. In Fig.10, the ways how the spectrum can be biased by an increase of FWHM parameters are shown on illustrative synthetic spectra (FWHM here is considered as the same for all plotted peaks). The left–hand–side of the figure shows how a peak of lower amplitude can get completely overlapped by a near peak of higher amplitude. On the right, a junction of two peaks located close to each other is shown, providing that the amplitudes of the peaks are comparable. Obviously, it is nearly impossible to decompose the attenuated spectrum into individual peaks without at least an approximative knowledge about the number and the locations of the resonant frequencies contained in it.

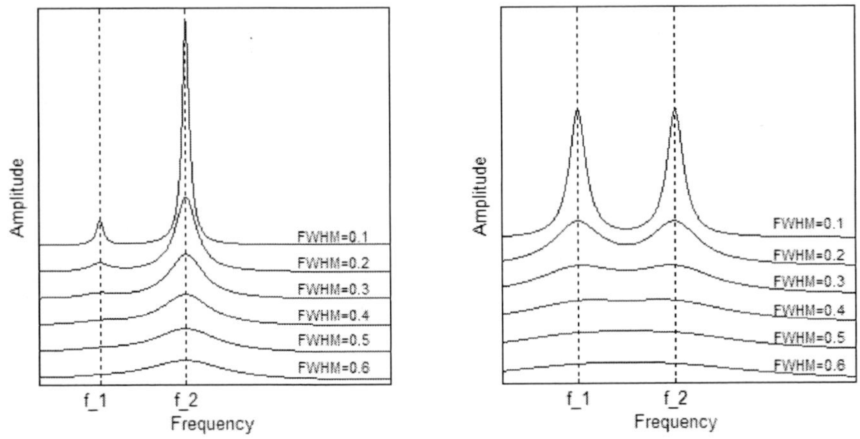

Figure 10. The effect of increasing attenuation on the spectrum: Disappearance of smaller peaks (on the left) and junction of neighboring peaks with comparable amplitudes (on the right).

In the following example, the attenuation combines with an extremely strong cubic anisotropy (comparable to that investigated in section 3.1), which makes the application of RUS even more complicated. On the other hand, the findings from the section 3.1 will be shown as very helpful in this case, and will enable us to determine quite accurately the c' coefficient, although only few peaks from the spectrum will be identified. The examined specimen is a 7.7×5.6×4.4mm rectangular parallelepiped (cut approximately along the principal {100} planes) of a near stoichiometric Ni-Mn-Ga alloy, which exhibits extremely strong magneto-elastic attenuation in the temperature interval between the premartensitic transition temperature ($T_{\text{pM}} \approx 257$K) and the Curie point ($T_{\text{C}} \approx 385$K). In [50], this attenuation has been investigated by combination of ultrasonic methods (pulse–echo measurements of longitudinal waves in [100] and [110] directions and the RUS). The results have shown that this attenuation is strongly anisotropic (see Fig.11 for an outline):

- The attenuation of longitudinal waves in the [100] direction is completely unaffected by the T_{pM} temperature. It increases towards some maximal attenuation, and then it falls down till the Curie point is reached, where it changes significantly its slope.
- The attenuation of longitudinal waves in the [110] direction jumps discontinuously at the T_{pM} temperature and increases steeply to the maximum (which is at slightly higher temperature than for the [100] direction). With further increase of the temperature, the attenuation slowly decreases, but seems to be fully unaffected by the T_{C} temperature.
- The attenuation of the first mode detected within the spectrum obtained by RUS measurements exhibits a significant change of the slope at T_{C} and steeply increases with further decrease of the temperature. However, at about $T = 280$K, further measurements (in the fully non-contact regime) were disabled by water vapor condensation at the faces of the specimen.

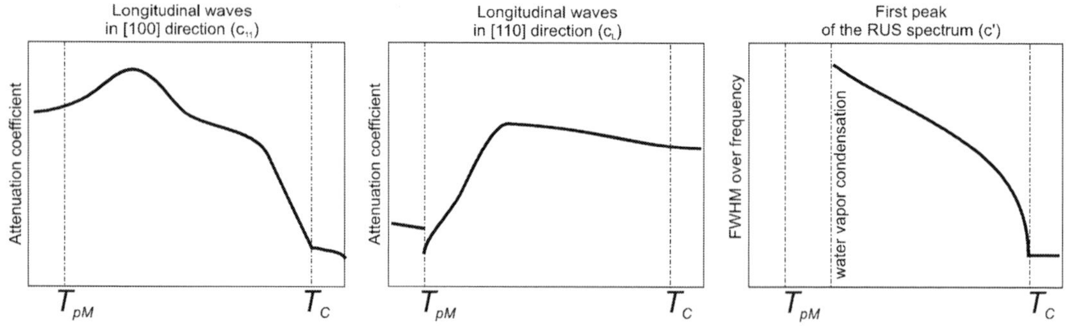

Figure 11. Anisotropic character of the attenuation in the Ni-Mn-Ga single crystal; sketched after the results of [50].

Here, we will focus on the RUS measurements only. The effect which the magneto-elastic attenuation has on the spectrum is illustrated by Fig.12, where the spectra in frequency band 50–170kHz are plotted for three different temperatures: Above the Curie point (i.e. at 389K), the spectrum has a good quality and all the resonant frequencies can be reliably determined. Below the Curie point (353K), the quality of the spectrum significantly decreases. Individual peaks start overlapping and merging. This effect is even more evident at 317K, where the spectrum is such strongly attenuated that the identification of individual resonances with acceptable accuracy is close to impossible. Another effect illustrated by Fig.12 is the drift of the whole spectrum with the temperature. Upon cooling, the first resonant frequency moves from 76.1kHz above the Curie point down to 60.1kHz at 317K. The rest of the spectrum (or at least the part of the spectrum shown in Fig.12) drifts in a similar way. Such dramatic changes of the resonances with the temperature cannot be ascribed to the thermal expansion of the specimen, which is about 18.10^{-6}m.K^{-1} [51] (which means that the shifts in the resonances should be also at about 10^{-5} level). The resonant spectra

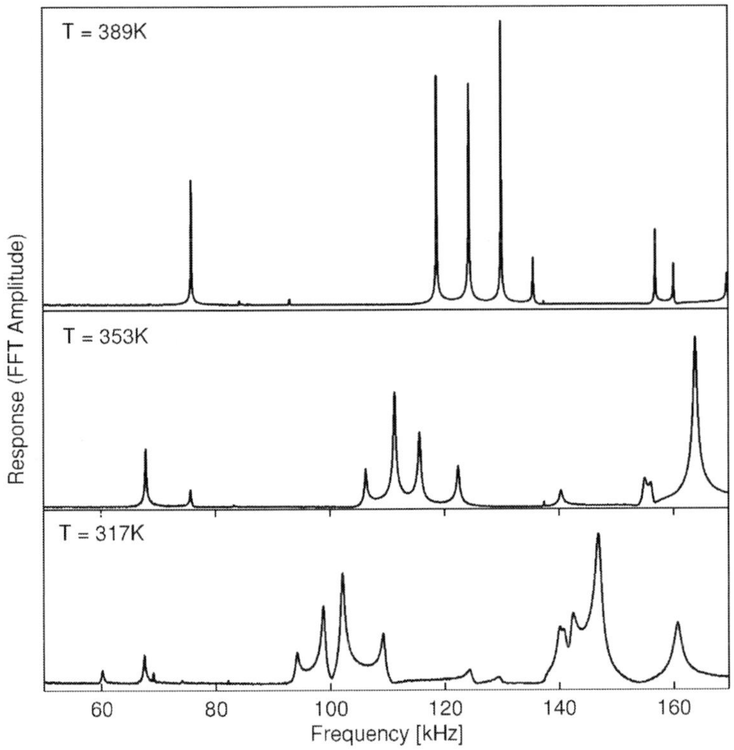

Figure 12. Illustration of the attenuation increase below the Curie point (outputs of the RUS measurements at different temperatures).

of the specimen were recorded during cooling from above the Curie point down to the T_{pM} temperature. The strategy was following:

- At 393K (i.e. safely above the Curie point), the specimen was scanned in the full 20×20 grid to identify accurately the resonant frequencies as well as the shapes of 37 vibrational modes. Such information was sufficient for determination of the c' and c_{44} coefficients.

- Then, the specimen was heated up to 398K, from where the temperature was decreased in successive steps of approximately −5K till a 280K temperature was reached, where further measurements were disabled by water vapor condensation at the faces of the specimen. At each temperature, the surface of the specimen was scanned by a sparse 3×3 grid, which was not sufficient for the identification of the shapes of the vibrational modes, but enabled reliable determination of the spectra.

At 393K, the full scan data were used for determination of the elastic coefficients by the same inverse procedure as applied in section 3.1 to austenite of Cu-Al-Ni, so the results were, again, obtained in a form of linear combinations of elastic coefficients sorted by accuracy. These are listed in Tab.5. Similarly to the case of Cu-Al-Ni (see Tab.1 for comparison), the first combination has nearly exactly the meaning of c', the second of c_{44} and the third of $c_{11} + c_{12}$.

After transforming back from these linear combinations, the reliable results are $c' = (6.9 \pm 0.1)$GPa and $c_{44} = (97.7 \pm 4.7)$GPa, individual values of c_{11}, c_{12} or c_{L} cannot be, naturally, determined. Our aim, however, was the determination of the elastic coefficients

below the Curie point, namely in the vicinity of the premartensitic transitions. As the spectra there were extremely attenuated, only the first few peaks were determinable by fitting the chosen interval of the spectrum by function of form (74), where the parameters A_n, ϕ_n, f_n and FWHM_n were determined by numeric optimization (simplex search method). The analysis was performed in successive steps, starting at the highest temperature (above the Cure point), and then fitting the spectra at lower and lower temperatures. As initial guesses for the search at each temperature, the values obtained in the previous step (i.e. at the previous, higher temperature) were used. This enabled the first two peaks in the spectrum to be accurately traced down through the whole temperature interval (down to $T = 280$K), and four more peaks to be traced down to $T = 353$K (below this temperature, the algorithm was able to localize the peaks and fit the amplitude of the spectrum, but the evaluated phases ϕ_n were not agreeing sufficiently with the experimental results and the FWHMs were not obtainable with sufficient accuracy).

In comparison to the number of resonant frequencies involved in the inverse procedure in the previous cases (up to 100 peaks), such data (two or six peaks) seem to be completely insufficient for the determination of any of the elastic coefficients. However, we can remind us our previous findings about the sensitivity of individual elastic coefficients to individual modes of vibrations for extremely strong cubic anisotropy, according which should the first peaks in the spectrum be nearly explicitly dependent on the c' coefficient only. But how can be such assumption utilized in this case?

If we assume that the shapes of the first few eigenmodes do not change with the temperature ($\alpha_j \neq \alpha_j(T)$), we can express the thermal dependencies of corresponding resonant

Table 5. Combinations of elastic coefficients for the single crystal of austenitic Ni-Mn-Ga above the Curie point.

k	Combination	C_k^* [GPa]	κ_k [GPa]
1	$0.71c_{11} - 0.71c_{12} + 0.02c_{44}$	12.35	0.01
2	$-0.01c_{11} + 0.02c_{12} + 0.99c_{44}$	98.104	4.72
3	$0.71c_{11} + 0.71c_{12}$	232.32	193.32

frequencies as

$$\frac{\partial f_j}{\partial T} = \frac{\alpha_j^{\mathrm{T}} \frac{\partial \Gamma}{\partial T} \alpha_j}{8\pi^2 f_j} = \frac{\alpha_j^{\mathrm{T}} \frac{\partial \Gamma}{\partial C_k} \alpha_j}{8\pi^2 f_j} \frac{\partial C_k}{\partial T}, \tag{75}$$

where the effect of the thermal expansivity is fully neglected. As we have already discussed, the partial derivatives $\partial\Gamma/\partial C_k$ are independent on C_k. For this reason, we can understand (75) as a linear relation between $f_j \frac{\partial f_j}{\partial T}$ (no Einstein's summation law applied) and the thermal derivatives of elastic coefficients. Looking back in Tab.5, we can indubitably assume that the dominant dependence is on c' only[4] and write

$$f_j \frac{\partial f_j}{\partial T} = K_j \frac{\partial c'}{\partial T}, \tag{76}$$

where K_j are constants. Then, with the exact knowledge of c' at some T_0 (the full scan at the temperature above the Curie point) and with thermal dependencies of f_j approximated by differences

$$\left.\frac{\partial f_j}{\partial T}\right|_{T=T_i} \approx \frac{f_j(T_{i+1}) - f_j(T_i)}{T_{i+1} - T_i} \tag{77}$$

we can get an estimate of the change of c' with decreasing temperature

$$\Delta c'(T_k) = \sum_{i=1}^{k} \frac{1}{K_j} f_j(T_i) \frac{f_j(T_{i+1}) - f_j(T_i)}{T_{i+1} - T_i} \tag{78}$$

for every mode (i.e. for every j). In Fig.13, the thermal dependencies of f_j and the corresponding values of $\Delta c'$ are plotted. Near the Curie point, the $\Delta c'$s evaluated for all modes are in an excellent agreement, small divergence appears between $\Delta c'$ determined from mode No.1 and mode No.2 at lower temperatures, which can be ascribed to slight changes of the shapes od the first two modes (i.e. of vectors α_1 and α_2) with the decreasing temperature. However, the difference is still incomparably smaller than the theoretically estimated experimental error of this coefficient (59) for the pulse-echo measurements, so we can suppose that the RUS method here is still more suitable for determination of this coefficient than the pulse-echo methods are.

It is rather complicated to summarize the findings of this subsection into any general conclusion. There is still a lot to be improved within the RUS analysis of Ni-Mn-Ga (or similar) single crystals. Especially the better knowledge of the micromagnetic mechanism of the attenuation is lacking, which might enable us to utilize the RUS results for deeper analysis of this magneto-elastic phenomenon. From the point of view of applicability limits of RUS, we have shown the way how the magneto-elastic attenuation complicates the determination of elastic coefficients by RUS measurements, in the sense that only first few peaks can be identified far below the Curie point, which precludes reliable determination

[4] Indeed, when expressed numerically, the ratios are

$$\alpha_j^{\mathrm{T}} \frac{\partial \Gamma}{\partial C_1^*} \alpha_j \approx 50 \alpha_j^{\mathrm{T}} \frac{\partial \Gamma}{\partial C_2^*} \alpha_j \approx 5.10^3 \alpha_j^{\mathrm{T}} \frac{\partial \Gamma}{\partial C_3^*} \alpha_j .$$

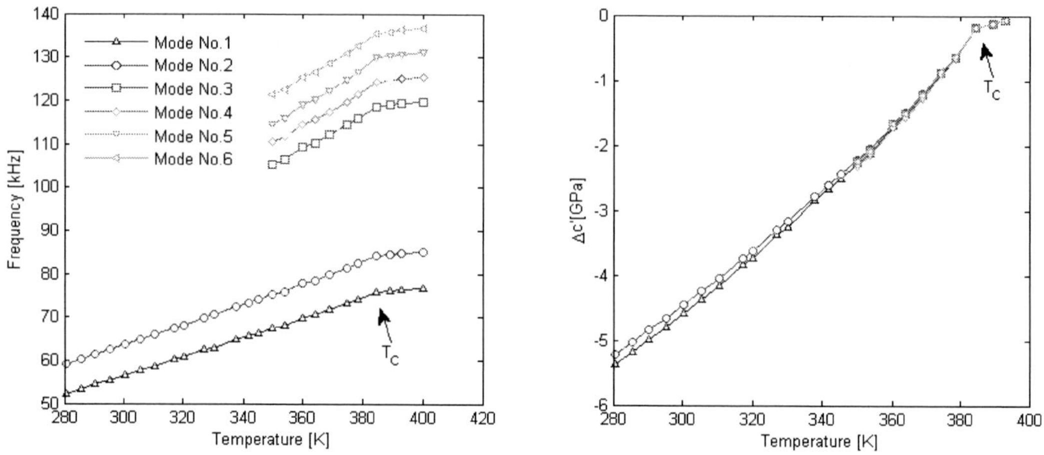

Figure 13. Thermal dependencies of locations of the first six peaks in the spectrum (on the left) and the changes of c' evaluated from them.

of all elastic coefficients, especially if the material is as strongly anisotropic as Ni-Mn-Ga. Hereto, it is necessary to point out that no enormous effort was made to analyze thoroughly the spectra (maybe more peaks and their thermal dependencies could have been identified throughout the spectrum) – the aim was to show that the knowledge of the theoretical background of RUS enables us to obtain at least some information on the elastic coefficients, even though the input information is minimal. We have, thus, shown that the knowledge of thermal dependencies of first few resonant frequencies can be sufficient for determination of the one elastic coefficient which is closely related to the corresponding modes.

The RUS analysis of Ni-Mn-Ga single crystals represents here the case when the the applicability of this method is complicated by combination of many different factors (strong anisotropy, significant thermal dependencies of elastic coefficients, temperature-dependent attenuation). Such case illustrates at the same time the main demerit and the strength of the RUS method: Although the nature of the method itself requires sophisticated postprocessing of the experimental data (which must be, moreover, modified for each particular material, etc.), the sought elastic coefficients can be obtained with high accuracy, providing that the resonant spectra contain sufficient information about them.

4. Conclusion

This chapter brings a survey through main ideas of the RUS method for determination of elastic coefficients of anisotropic solids, with a focus on the limitations of this method given by the properties of the examined materials. Throughout the text, at least five essential questions were open, regarding the applicability a reliability of this method. By formulating these questions explicitly and answering them based on the findings outlined in the chapter, we can summarize the whole content of the text as follows:

Q1. Does the resonant spectrum of free elastic vibrations of a small specimen of known geometry always contain sufficient information on the elasticity of the

material? The answer is, surprisingly, not. Some additional information is always required to associate the values of resonant frequencies to individual vibrational modes (see Fig.3 and the discussion around there). This information may have a form of the shapes of the eigenmodes determined by a scanning laser interferometer, or it may by simply brought by very accurate initial guesses of the elastic coefficients, for which the shapes are very similar to the real ones.

Q2. Is there any well-founded way how to estimate the accuracy of the results of RUS? If yes, can this accuracy be improved by increasing the number of frequencies involved in the inverse procedure? Yes, the accuracy can be estimated by formulas like (44), which are following fully from the nature of the method. On the other hand, the increase of the accuracy by involving more and more frequencies in the inverse procedure is not axiomatic. In the example of the wooden rod (see the beginning of subsection 2.3.), the ratio between number of frequencies in the spectrum corresponding to the longitudinal and to the torsional modes never exceeds the value $\sqrt{G_\perp/E}$ as the all the higher modes are given only by multiplication (see Fig.2 and the discussion above it) of the lower ones.

Q3. Is the applicability of RUS anyhow limited by the strength of the anisotropy of the examined material? Yes, in some sense. As we have seen both for the single crystals of Cu-Al-Ni (subsection 3.1) and for Ni-Mn-Ga (subsection 3.3), the RUS method is not able to determine accurately the coefficients corresponding to the hardest vibrational modes, if the anisotropy is sufficiently strong. However, as it was shown for both the austenitic phase and the martensitic single variant of Cu-Al-Ni, the RUS measurement can be properly complemented with pulse-echo measurements, which are most suitable for determination of the hardest modes related to the quasi-longitudinal velocities. For extremely weak anisotropy (subsection 3.2), no limitations were found, especially after the RUS measurements were, again, complemented by pulse-echo measurements.

Q4. Is the applicability of RUS anyhow limited by the number of sought independent elastic coefficients? Here, the answer is definitely not (providing that reasonable initial guesses are available). The example of the nanocrystalline copper manufactured by the ECAP method (subsection 3.2) showed that the RUS method can be reliably applied for determination of all 21 elastic coefficients, or even for the detection of the symmetry class of the material. The only problems can be encountered when the material is periodically microstructured (Fig.9), where the microstructure can, at certain frequencies, interfere with the vibrational modes.

Q5. Can the RUS method be easily modified for particular materials with special, more complex properties? Although this question is quite general, we can say that the answer might be yes. The adaptability of RUS was illustrated in subsection 3.3 for the magneto-elastically attenuated single crystal of the Ni-Mn-Ga alloy. The positive answer can be also supported by many of the references listed below. On the other hand, there are many challenging issues for the RUS method not solved yet. Continuously graded material, metallic foams, metamaterials or nanoscale objects are few topic of those to which the RUS community turns now and which will check the real adaptability of this method. The limitations which will be found for such highly advanced applications of RUS cannot be foretold yet.

Acknowledgement

Authors wish to express their thanks to V. Novák and J. Kopeček (Institute of Physics ASCR, v.v.i.) for providing specimens from CuAlNi, M. Janeček (Faculty of Mathematics and Physics, Charles University) for Cu specimens prepared by ECAP technique, and I. Aaltio (Laboratory of Materials Science, Helsinki University of Technology, Finland) for NiMnGa specimens.

The work was supported by the project No. 101/06/0768, and post-doc. project No.202/09/P164 of the Czech Science Foundation (CSF), the project A200100627 of the Grant Agency of ASCR, the institutional project of IT ASCR , v.v.i., CEZ:AV0Z20760514, and from the research center 1M06031 of the ministry of education of the Czech Republic.

References

[1] Stekel, A.; Sarrao, J.L.; Bell, T.M.; Lei, M.; Leisure, R.G.; Visscher, W.M.; Migliori, A. *J. Acoust. Soc. Am.* 1992, Vol.2I, 663–668.

[2] Migliori, A.; Sarrao, J.L.; Visscher, W. M.; Bell, T.M.; Lei, M.; Fisk, Z.; Leisure, R.G. *Physica B* 1993, Vol.183 (1-2), 1–24.

[3] Willis, F.; Leisure, R.G.; Kanashiro, T. *Phys. Rev. B* 1996, Vol.54, 9077–9085.

[4] Sarrao, J.L.; Mandrus, D.; Migliori, A.; Fisk, Z.; Bucher, E. *Physica B* 1994, Vol.199-200, 478–479.

[5] Ulrich, T.J.; Darling, T.W. *Geophys. Res. Lett.*, 2001, Vol.28, 2293–2296.

[6] Kim, Y.H.; Song, S.-J.; Kwon, S.-D.; Cheong, Y.-M.; Jung, H.-K. *Ultrasonics* 2004, Vol.42, 551–555.

[7] Guo, H.; Lal, A. *Proceedings of the IEEE Ultrasonics Symposium* 2001, 863–866.

[8] Araki, W.; Kamikozawa, T.; Adachi, T. *NDT&E Int.* 2008, Vol.41, 82–87.

[9] Li, G.; Lamberton, G.; Gladden, J.R. submitted to *Phys. Rev. B*.

[10] Fan, Y.; Tysoe, B.; Sim, J.; Mirkhani, K.; Sinclair, A.N.; Honarvar, F.; Sildva, H.;Szecket, A,; Hardwick, R. *Ultrasonics* 2003, Vol.41, 369–375.

[11] M. K. Fig, MSc Thesis, Montana State University 2005.

[12] Landa, M.; Sedlák, P.; Šittner, P.; Seiner, H.; Novák, V.: *Mater. Sci. Eng. A* 2007, vol.462, 320–324.

[13] Demarest, H. H. Jr. *J. Acoust. Soc. Am.* 1971, Vol.49, 768–775.

[14] Ohno I. *J. Phys. Earth* 1976, Vol.24, 355–379.

[15] Maynard, *J. Phys. Today* 1996, Vol.49, 26–31.

[16] Sarrao, J.L.; Chen, S.R.; Visscher, W.M.; Lei, M.; Kocks, U.F.; Migliori, A. *Rev. Sci. Instrum.* 1994, Vol.65, 2139–2140.

[17] Ogi H.; Sato K.; Asada T.; Hirao M. J. Acoust. Soc. Am. 2002, Vol.112, 2553–2557.

[18] Lei, M.; Sarrao, J.L.; Visscher, W.M.; Bell, T.M.; Thompson, J.D.; Migliori, A.; Welp, U.W.; Veal, B.W. *Phys. Rev. B* 1993, Vol.47, pp. 6154–6156.

[19] Sarrao, J.L.; Mandrus, D.; Migliori, A.; Fisk, Z.; Tanaka, I.; Kojima, H.; Canfield, P.C.; Kodali, P.D. *Phys. Rev. B* 1994, Vol.50, 13125–13131.

[20] Spoor, P.S.; Maynard, J.D.; Kortan, A.R. *Phys. Rev. Lett.* 1995, Vol.75, 3462–3465.

[21] Foster, K.; Leisure, R.G.; Shaklee, J.B.; Kim, J.Y.; Kelton, K.F. *Phys. Rev. Lett.* 1999, Vol. 59, 11132–11135.

[22] Nakamura, N.; Ogi, H.; Hirao, M. *Acta Mat.* 2004, Vol.52, 765–771.

[23] Ogi, H.; Nakamura, N.; Hirao, M. *Fatigue Fract. Eng. M.* 2005, Vol.28, 657–663.

[24] Ikeda, R.; Tanei, H.; Nakamura, N.; Ogi, H.; Hirao, M.; Sawabe, A.; Takemoto, M. *Diam. Relat. Mater.* 2006, Vol.15, 729–734.

[25] Ledbetter, H.; Leisure, R.G.; Migliori, A.; Betts, J.; Ogi, H. J. *Appl. Phys.* 2004, Vol.96, 6201–6206.

[26] Leisure, R.G.; Foster, K.; Hightower, J.E.; Agosta, D.S. *Mater. Sic. Eng. A.* 2004, Vol.370, 34–40

[27] Ogi, H.; Fukunaga, M.; Hirao, M., Ledbetter, H. *Phys. Rev B* 2004, Vol.69, 241041–241048.

[28] Muller, M.; Sutin, A.; Guyer, R.; Talmant, M.; Laugier, P.; Johnson, P.A. 2005 *J. Acous. Soc. Am.* 2005, Vol. 118, 3946–3952.

[29] Payan, C.; Garnier, V.; Moysan, J.; Johnson, P.A. *J. Acous. Soc. Am.* 2007, Vol.121, EL125–EL130.

[30] Muller, M.; Tencate, J.A.; Darling, T.W.; Sutin, A.; Guyer, R.A.; Talmant, M.; Laugier, P.; Johnson, P.A. *Ultrasonics* 2006, Vol.44(SUPPL.), e245–e249

[31] Fung, Y.C.; Tong, P.: *Classical and computational solid mechanics.* Singapore: World Scientific Publishing, 2001.

[32] Sedlák, P., PhD Thesis, CTU in Prague, 2008.

[33] Yoneda, A. *Earth Planets Space* 2002, Vol.54, 763–770.

[34] Tian, J.; Ogi, H.; Hirao M. *Appl. Phys. Lett.* 2005, Vol.87, 204107-1–204107-3.

[35] Radovic, M.; Barsoum, M.W.; Ganguly, A.; Zhen, T.; Finkel, P.; Kalidindi, S.R.; Lara-Curzio, E. *Acta Mater.* 2006, Vol.54, 2757–2767.

[36] Zadler, B.J., PhD Thesis, Colorado School of Mines, 2005.

[37] Ogi, H.; Ledbetter, H.; Kim, S.; Hirao, M. *J. Acoust. Soc. Am.* 1992, Vol.106, 660–665.

[38] Sedlák, P.; Seiner, H.; Landa, M.; Novák, V.; Šittner, P.; Manosa, L. *Acta Mater.*, 2005, Vol.53, 3643–3661.

[39] Bhattacharya, K. *Microstructure of Martensite*, Oxford: Oxford University Press, 2003.

[40] Auld, B.A. *Acoustic fields and waves solids*, Vol.1, New York: John Wiley and Sons, 1973.

[41] Stipcich, M.; Manosa, L.; Planes A. *Phys. Rev. B* 2004, Vol.70 054115-1–054115-5.

[42] Bennett, B.W.; Shannette G.W. *Acoust. Lett.* 1980, Vol.4, 99–104.

[43] Saunders, G.A.; Liu, H.J.; Bach, H. *J. Mat. Sci.* 1998, Vol.33, 4589–4594.

[44] Kim,H.S.; Ryu W.S.; Janecek, M.; Baik, S. Ch.; Estrin, Y. *Adv. Eng Mater.* 2005, Vol.7, 43–46.

[45] Kim, H. S. *Mater. Sci. and Eng. A* 2006, Vol.430, 346–349.

[46] Yoon, S.Ch.; Seo M.H.; Kim, H.S. *Scripta Mater.* 2006, Vol.55, 159–162.

[47] Kim, H.S.; Estrin, Y. *Mater. Sci. Eng. A* 2005, Vol.410–411, 285–289.

[48] Landa, M.; Sedlák, P.; Seiner, H.; Heller, L.; Bicanová, L.; Šittner, P.' Novák, V. Accepted to *Applied Physics A* (2008).

[49] Seiner, H. PhD Thesis, CTU in Prague, 2008.

[50] Seiner, H. Bicanová, L.; Sedlák, P.; Landa, M.; Heller, L.; Hannula, S.-P. submitted to *Mater. Sci. Eng. A.* (special issue 'Proceedings of 15th ICIFMS 2008').

[51] Rudajevova, A. *Kovove Mater.* 2008, Vol.46, 71–76.

In: Horizons in World Physics, Volume 268
Editors: M. Everett and L. Pedroza, pp. 137-180
ISBN: 978-1-60692-861-5

Chapter 4

HYBRID INTEGRATED EXTERNAL CAVITY LASERS BASED ON SILICA PLANAR WAVEGUIDE GRATING

Kyung Shik Lee[1] and Jeong Hwan Song[2]
[1]School of Information and Communication Engineering,
Sungkyunkwan University, Suwon, South Korea
[2]Integrated Photonics Group, Tyndall National Institute, Cork, Ireland

Abstract

Grating technology has been well matured, and therefore successfully utilized to select the oscillation wavelength and polarization properties of conventional laser diodes. This is because the lasing characteristics of the laser can be easily controlled by simply writing a specific grating directly in a waveguide cavity. Grating and hybrid integration techniques are presented. Several types of hybrid integrated external cavity lasers (ECLs) demonstrated with a variety of silica waveguide gratings on a Si-substrate are discussed. They include highly polarized ECLs, a dual-wavelength ECL and ECLs with high sidemode suppression ratio.

1. Introduction

It has been well known that a variety of gratings can be easily written by UV beam in photosensitive Ge-doped silica waveguides. By applying the grating technology to silica planar lightwave circuit (PLC), researchers have demonstrated a number of grating-based silica waveguide devices. The grating-based PLC devices include add-drop filters, external cavity lasers (ECLs), optical sensors and arrayed waveguide grating (AWG) devices. One of the important PLC devices is the external cavity laser, which is hybrid integrated on a silica PLC platform with a UV-written Bragg grating. The hybrid integrated ECL is a promising key component for the future wavelength division multiplexing (WDM) networks and optical sensing systems. This is because, compared with other types of ECLs, this PLC-type ECL has some advantages. The lasing wavelength of the ECL, which is determined by the Bragg wavelength of the silica wavelength grating, is very stable. It has also advantage of the capacity for high-density integration. In addition, the polarization and spectral properties of the laser beam can be easily chosen by simply selecting a proper type of grating to write, because the grating is polarization- and wavelength-sensitive.

In this chapter, we first discuss how photosensitivity in Ge-doped silica waveguides is improved and gratings are formed in the silica waveguides. Next, we consider the hybrid integration technology, by which one can fabricate the hybrid integrated ECLs. Also, we review the recent results on the grating-based hybrid integrated ECLs developed by employing a variety of the silica waveguide gratings. In the discussion, some techniques on how to control the high degree of sidemode suppression ratio, the state of the polarization and the lasing modes are included.

2. Silica Waveguide Bragg Grating Technology

Since the first observation of the refractive index change in Ge-doped silica fibers in 1978 [1], there has been a great interest in the photosensitivity in optical waveguides and the fabrication of Bragg gratings within the core of the optical waveguides such as optical fiber and silica planar waveguide. The capability of fabricating Bragg gratings in these photosensitive waveguides has dramatically changed the field of fiber optic telecommunications. Over the last 15 years, a number of researchers have investigated not only the fundamentals but also the applications of the gratings in the areas of telecommunication and sensor systems [2-6]. And now the Bragg grating technologies are well matured. This section overviews the writing techniques of the Bragg gratings and the Hydrogenation of increasing photosensitivity as well as the growth characteristics of the Bragg gratings in the optical waveguides.

2.1. Bragg Grating Formations

Bragg gratings in optical waveguides are one of the core technologies and have emerged since fiber optic telecommunications expanded rapidly. The initial discovery was showed that an intense periodic pattern of ultra violet could create a refractive index pattern. Further works by Meltz *et al.* [7] developed a side writing technique so that a pattern of any period could be written in the optical waveguides and the Bragg grating could be used as a filter at the telecommunication band of wavelengths. Bragg grating filters have the potential to be extremely low cost and high quality. The devices can perform primary functions such as reflecting and filtering with high efficiency. UV written Bragg gratings are relatively easy to fabricate and are becoming increasingly inexpensive to manufacture.

Bragg gratings were firstly fabricated using internal inscribing. Nowadays, two main laser-writing techniques are in use for the fabrication of Bragg gratings: interferometric and phase mask methods [8, 9].

The interferometric fabrication technique utilized an interferometer that split the incoming UV light into two beams and then recombined two splitting beams to form an interference pattern. This pattern was inducing a refractive index modulation in the optical waveguide core. Splitting can be accomplished using either amplitude or wave-front interferometry. Bragg gratings in optical waveguide can be fabricated by both amplitude-splitting and wave-front-splitting interferometers. An amplitude splitting interferometer is that the power of UV source is divided into two distinct beams by a beamsplitter and two separated beams reflecting via mirrors form interference-periodic patterns on the optical waveguide core as shown in Fig. 1(a)-(c). It requires high coherent UV source for forming periodic patterns. This technique was used for the first side-written gratings by Meltz *et al.* in 1989. The main advantage of an amplitude interferometer is that the grating period can be controlled by changing the angle of the mirror for adjusting the angle between two interfering beams. This control is especially important for the fabrication of WDM (wavelength division multiplexing) devices. Wave-front-splitting is typically achieved by a geometrical dividing of a UV beam using a prism as shown in Fig. (d)-(e). The grating length is limited to mostly half of the UV beam width and the wavelength tunability of the grating is difficult. These restrictions are quite disadvantageous. In contrast to the amplitude splitting technique, the wave-front splitting interferometers can reduce the sensitivity of mechanical and atmospheric vibrations during the grating growth process.

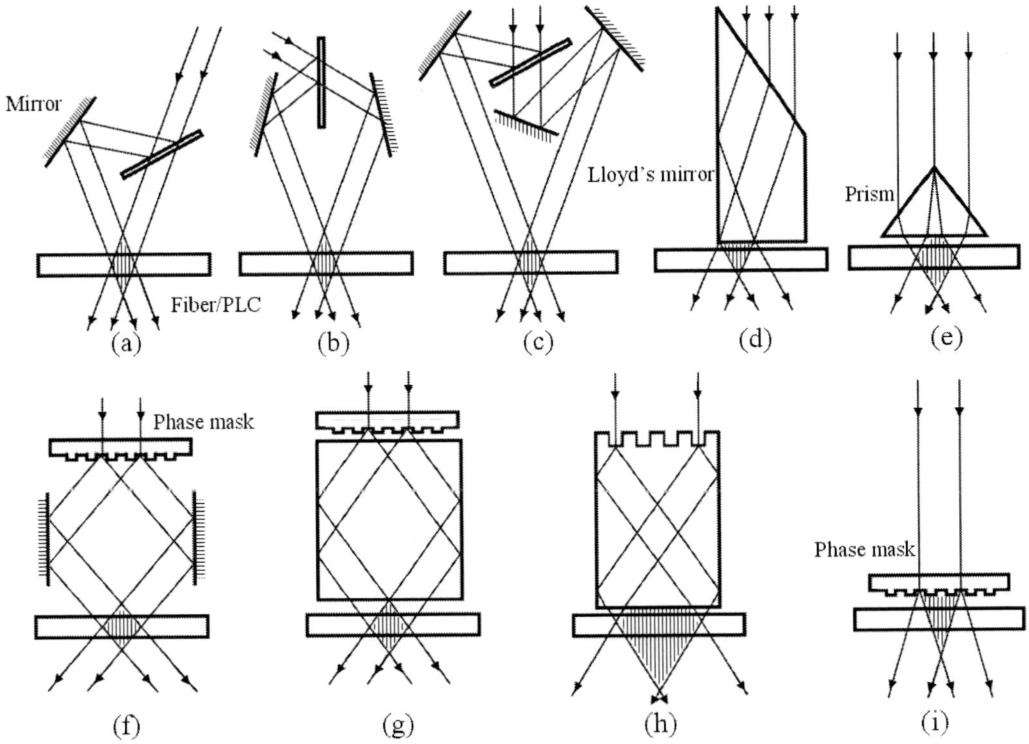

Figure 1. Various methods for fabrications of Fiber/PLC Bragg gratings. (a)Mach-Zehnder interferometer (amplitude-splitting with a mirror), (b)amplitude-splitting interferometer with two mirrors, (c)amplitude-splitting interferometer with three mirrors, (d)Lloyd's mirror interferometer, (e)symmetric prism interferometer, (f)noncontact, interferometric phase mask technique with two mirrors, (g)interferometric phase mask technique with a prism, (h)phase mask combined with a prism block and (i) direct writing technique using phase mask [9].

The Fig. 1(f)-(i) shows various fabrication methods using a phase-mask for Bragg grating in optical waveguides [9]. The phase-mask technique [see Fig. 1(i)] was proposed by Hill *et al.* in 1993. Essentially, the phase-mask is the transmission diffraction grating that recreates a periodic modulation onto the optical waveguide via illumination of UV source. The phase-masks can be made by holographic or lithographic methods. The phase-mask diffracts a few percent of light in the zero-order and the most of light (about 70%) in the +1 and –1 orders. When the phase mask is situated directly on a fiber or planar waveguide, these two diffracted order beams (+1 and –1 orders) interfere to produce a periodic pattern that inscribes a corresponding pattern in the optical waveguide's core material. It is important to mention that the period of the written grating is half of the period of the phase mask ($\Lambda=\Lambda_{mask}/2$) and is independent of the wavelength of UV source used for illumination. Temporal coherence variations due to mechanical and other vibrations do not play as great a role in the phase-mask technique as they do in the interferometric technique and the manufacturing process can be greatly simplified. KrF excimer laser sources are the most common UV sources for fabricating Bragg gratings using a phase mask. The phase mask is placed in close nearness or in contact with an optical waveguide because the UV laser source has quite a short spatial coherence. The separation between fiber or planar waveguide and the phase mask is a critical parameter in writing quality of Bragg gratings. The good spatial coherence is related with a high contrast fringe pattern. When the distance between an optical waveguide and the phase mask is increased the separation between the two interfering beams emerging from the mask is increased. As the distance between the two interfering beams extends the interference fringe contrast will decay. It means that a good quality Bragg grating requires short distance between an optical waveguide and the phase mask. A disadvantage of this technique is that a separate phase mask is required for each different Bragg wavelength since it is not possible to tune by changing the angle between interfering beams (the angle between diffraction orders is constant) or writing wavelength.

2.2. Photosensitivity and Hydrogenation

Since the discovery of photosensitivity and the development of optical waveguide gratings, there have been considerable efforts in understanding the photosensitivity in optical waveguides. In order to understand the photosensitivity in Ge-doped silica waveguides, the mechanisms of photo-transformation from defect precursors to defects by UV laser have been suggested [1-5]. It is widely accepted that the germanium oxygen deficient centers (GODCs), which show 240nm absorption band, are responsible for the photosensitivity in Ge-doped silica optical waveguides.

The core region of an optical waveguide is doped with germanium in order to raise the refractive index above that of pure silica in the cladding region. This provides the optical guiding properties for the optical waveguides. The germanium also creates a quantity of GeO (germanium monoxide) which creates oxygen deficient centers when exposed to UV radiation. These defects cause a rise in the refractive index. In general, this process is slow and is dependent on the intensity of the UV source. To enhance photosensitivity, there are various methods such as high germanium concentration doping and co-doping with both germanium and various co-dopants, which are boron, tin, europium, cerium and so forth, as well as hydrogenation (hydrogen loading)[3]. The index change of high germanium doped fibers is the order of 2.5×10^{-4} by UV irradiation. The boron co-dopanted fiber has photo-induced index change of 7×10^{-4}. By hydrogen loading [10], the process of creating defects can be speeded up by inducing a second photosensitive effect and hydrogen into the core allows the UV radiation to create Si-OH which increases the refractive index of the core in optical waveguides. These two effects combine to create the periodic variation of the refractive index in the core region. This hydrogen loading technique is a simple and effective approach for achieving high photosensitivity in Ge-doped silica optical waveguides. The following equations [11] show the H_2 diffusion equation for the optical fiber core.

$$\frac{C(r,t)}{C_0} = 1 - 2\sum_{n=1}^{\infty} \frac{\exp(-D\alpha_n^2 t) J_0(r\alpha_n)}{a\alpha_n J_1(a\alpha_n)} \tag{1}$$

$$D = 2.83 \times 10^{-4} \exp\left(\frac{-40.19kJ/mole}{RT} \right) \quad [cm^2/\sec] \tag{2}$$

Here, a is the radius of an optical fiber, C_0 is the initial H2 concentration as a function of pressure and temperature. D is the diffusion coefficient of H2. J_0 and J_1 are the Bessel functions and αn is the solution of $J0(a\alpha n)=0$ as well as r is a radial position in the cylindrical coordinates. The R=8.311J/(K-mol) and T is the temperature in degrees Kelvin. The relationship between the concentration of H2 and index change of fiber Bragg gratings has been reported that the index change of ~2×10-5 per the H2 concentration of 100ppm was increased linearly [12, 13]. Another consideration of calculating the H2 diffusion concentration is the case of planar waveguides. The diffusion equation for a planar waveguide can be expressed by [11],

$$\frac{C(l,t)}{C_0} = 1 - \frac{4}{\pi}\sum_{n=0}^{\infty} \frac{(-1)^n}{2n+1} \exp\{-D(2n+1)^2 \pi^2 t / 4l^2\} \cos\frac{(2n+1)\pi x}{2l} \tag{3}$$

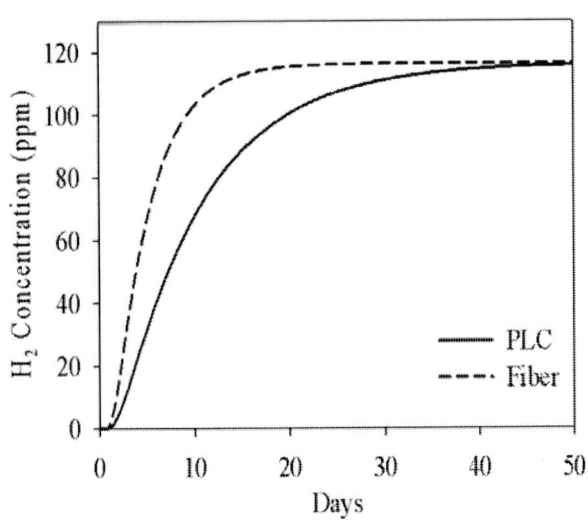

(a)

Figure 2. Continued on next page.

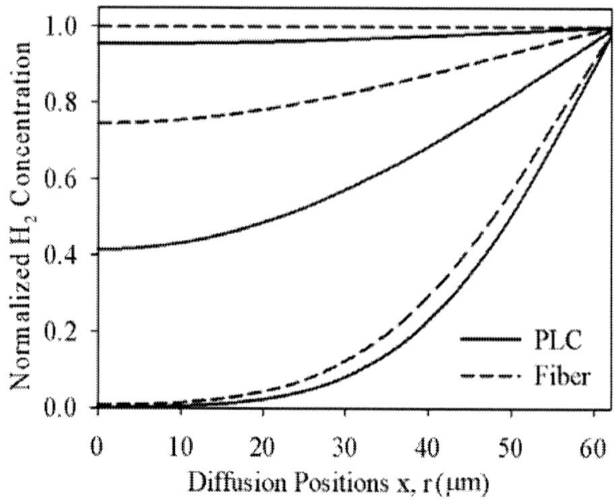

(b)

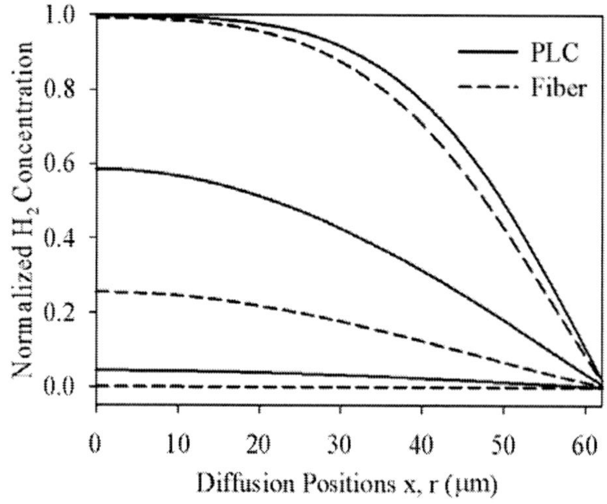

(c)

Figure 2. Diffusive curves of H_2 concentration with different times in PLC and fiber (a): [21°C (room temp.), a=l=62.5μm, 1atm], normalized H_2 concentration of diffusion in (b): [Pairs (solid and dashed) of top, middle and bottom curves are for 30, 7, and 1days, respectively.] and diffusion out (c): [Pairs (solid and dashed) of top, middle and bottom curves are for 1, 7, and 30days, respectively.].

Here, the l is the half-width of the planar waveguide and the x is a position in the rectangular coordinates. For better understanding, assuming that $a=l=62.5\mu m$ at room temperature (21°C) with 1atm of pressure we compared with H_2 concentrations diffused in both optical fiber and planar waveguide, the saturation time of H_2 concentrations diffused in an optical fiber is over 21 days and that of a planar waveguide is over 52 days as shown in Fig. 2(a). Fig. 2(b) and (c) shows the position of in-diffusion and out-diffusion as a function of normalized H_2 concentrations with different times, respectively. Table 1 also shows the summary of simulation parameters and results. Assuming that the values of both a fiber radius and planar waveguide thickness are same, the saturation time of H_2 concentration diffused in the planar waveguide is much longer than that of the optical fiber. However, the thicknesses of the core and cladding in conventional planar waveguides are less than 10μm and 20μm, respectively. In addition, high temperature can accelerate to the diffusive speed of H_2 and high pressure can contribute to the increment of the concentration of H_2 to both the optical fiber and planar waveguide. The H_2 loading time is less than 1 week in the conventional use. According to the diffusion equation for a planar waveguide, 7 days is adequate for H_2 molecules to be saturated in the planar waveguide core, whose thickness is 25μm, at the room temperature as shown in Fig. 3.

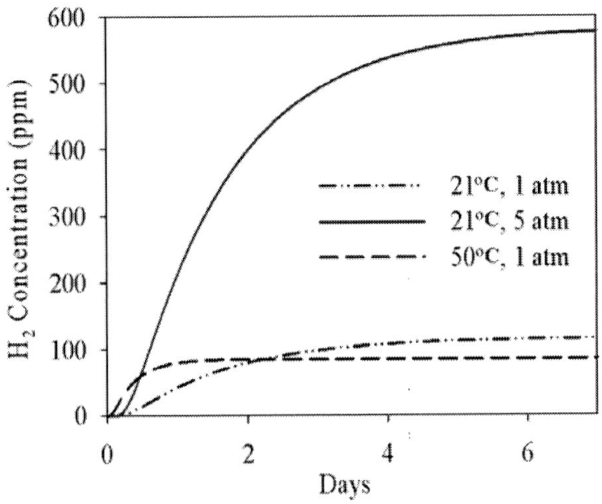

Figure 3. Hydrogenation simulations of the practical PLCs whose thickness of waveguide region (core and clad) is 25μm.

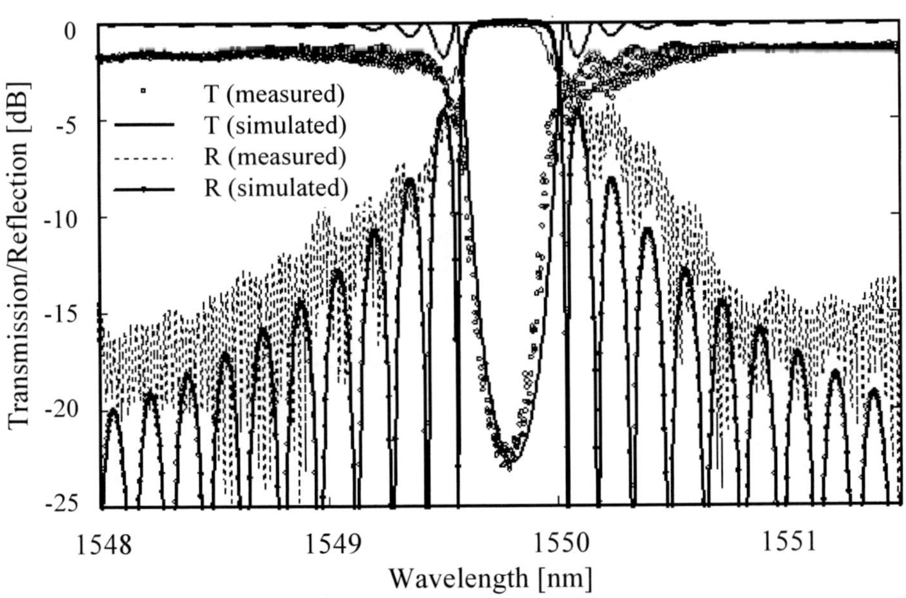

Figure 4. PLC Bragg grating fabricated by phase mask method.

Table 1. Simulation parameters for Hydrogen diffused in and out of a PLC and fiber.

PLC	Optical fiber
$C_0 = P \times 3.37 \exp(1.0416 \times 10^3 / T)$ [ppm]	
$D = 2.83 \times 10^{-4} \exp(-40.19[kJ/mol]/RT)$ [cm^2/s]	
$R = 8.311$ [J/K·mol]	
l=62.5 μm	a=62.5 μm
P =1atm	P =1atm
T=21°C	T=21°C
x=0	r=0
Saturation time = 52 days	Saturation time = 21 days

The example of the PLC Bragg grating fabricated by a phase mask method is shown in Fig. 4. The PLC was soaked in H_2-circumstance over 7 days at room temperature under high pressure (approximately 50atm). The fabrication conditions are the UV power of 135mJ/cm^2, the repetition rate of 5Hz, and the exposure time of 5 minutes. The fabrication conditions and results are summarized in Table 2.

Table 2. Summary of simulation parameters and experimental results of fabricated PLC grating with fabrication conditions.

Parameters	Simulations	Results	Fabrication conditions	
Grating length (L_g)	5mm	5mm	Phase mask length (L_{mask})	5mm
Transmittance (T)	-22.6dB	-21dB	UV power	135mJ/cm^2
Reflectivity (R)	99%	99%	UV exposure time (t_{uv})	5 minutes
Band width (BW)	0.42nm	0.35nm	Repetition rate	5Hz
Bragg wavelength (λ_B)	1549.8nm	1549.78nm		

2.3. Growth Characteristics of PLC Gratings

Bragg grating written in planar lightwave circuits (PLCs) have attracted interest because they are suitable for compaction, cost reduction, and mass production, and offer long-term stability. In this section, we will review the growth characteristics of the PLC gratings, which have recently been reported [14] below.

Gratings can be formed in PLCs using the phase mask technique described in section 2.1. The PLCs are Ge-doped silica planar waveguides, which have intrinsic photosensitivity, and are typically hydrogenated at room temperature and 100atm pressure for more than 4 days to increase the photosensitivity. To write a grating in the PLC waveguide, a UV light beam from a KrF excimer laser (λ=248 nm, 5Hz) was focused by a cylindrical lens to the PLC waveguide through a phase mask. To vary the grating growth parameters, the laser average fluence per pulse Fp should be controlled, and the grating length L be adjusted using a beam shutter. Fig. 5 shows the transmission spectra of the gratings formed in the PLCs at different exposure times. The 5-mm-long Bragg gratings with a pitch Λ of 533.65nm were written by the laser beam of fluence 485mJ/cm^2. As the exposure time t increases, the strength of the Bragg grating became stronger, the bandwidth between the first zeroes of gratings (FWFZ) $2\Delta\lambda$ wider, and the peak wavelength λ_B shifted to longer wavelength. According to the coupled-mode theory [3-5], the peak reflectance R and the bandwidth are determined by

$$R = \tanh^2\left[(\eta\pi L\Delta n)/\lambda\right] \tag{4}$$

and

$$2\Delta\lambda = (\lambda^2/\pi n_{eff}L)\sqrt{(\eta\pi L\Delta n/\lambda)^2+\pi^2}\,, \tag{5}$$

respectively. Here, η is the fraction of the single-mode intensity confined in core, which is estimated to be 0.85 for the gratings, n_{eff} is the effective refractive index of the core and Δn is the refractive index modulation. The index change Δn during the grating growth can be obtained from equation (4) for weak gratings and computed from equation (5) for strong gratings whose reflectivity approaches 100%. Typically, the dependence of Δn on the exposure time t is nonlinear. According to the one-photon absorption model [15] and the power law, Δn as a function of t is expressed by

$$\Delta n = \Delta n_{\max}[\exp(-AIt)] \tag{6}$$

and

$$\Delta n = Ct^b\,, \tag{7}$$

respectively [14], where $\Delta n_{\max}$ is the maximum refractive index modulation and I is the intensity of the UV light. The peak reflectance of the PLC gratings is plotted versus exposure time in Fig. 6. The dashed curve is the fit predicted by the one-photon absorption model with $\Delta n_{\max}=8.6\times10^{-4}$ [14]. The solid curve is obtained by fitting Δn to equation (7) with $C=2.3\times10^{-7}$ and $b=1.53$. Similar to the case of fiber gratings [15], the fit given by equation (6) is poor, but the power-law fit [equation (7)] for the PLC gratings is good. In addition to the 5-mm-long Bragg grating, a 3-mm-long Bragg grating was formed with the same fluence of 485mJ/cm^2. The 1-mm-, 3-mm-, and 5-mm-long gratings were also formed with the laser fluence 294 mJ/cm^2 to fully understand the grating growth characteristics under the different growing conditions listed in Tables 3 and 4. In every case, the growth of the PLC gratings is well described by the power law [14]. The parameters best fitted to equation (7) for five different writing conditions are listed in Table 3. Note that C increases with the laser fluence per pulse F_p. This implies that the laser pulse intensity should be raised to reduce the writing time t of the PLC gratings for mass production. In the grating writing process, the total laser fluence F is also a critical factor, which is defined as the total number of laser pulses times F_p. To see if Δn has a power law dependence on the total fluence F, the calculated Δn values can be fitted to the power law [16] of the form,

$$\Delta n = AF^B\,. \tag{8}$$

The solid curve in Fig. 6 is the fit with $A=8.8\times10-8$ and $B=1.47$ for the grating pitch of 533.65nm. The error bars in Fig. 7 are attributed to the fluctuation of the laser fluence per pulse [14]. The parameters obtained for the five different writing conditions are also listed in Table 3.

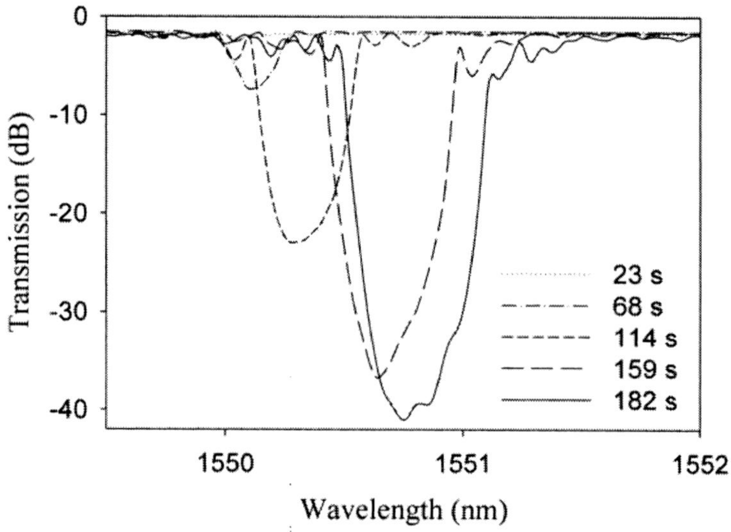

Figure 5. Transmission spectra of PLC gratings at different exposure times [14].

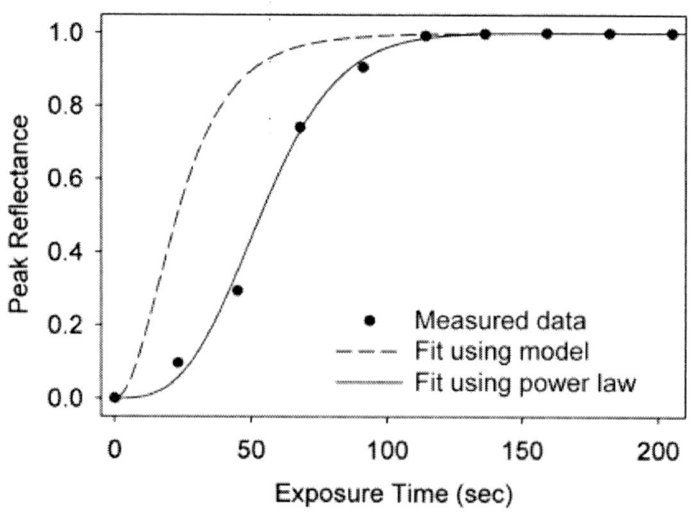

Figure 6. Peak reflectance of a PLC grating as a function of exposure time. Closed circles represent the measured data, and the dashed and solid curves are the fits predicted by the absorption model and the power law, respectively [14].

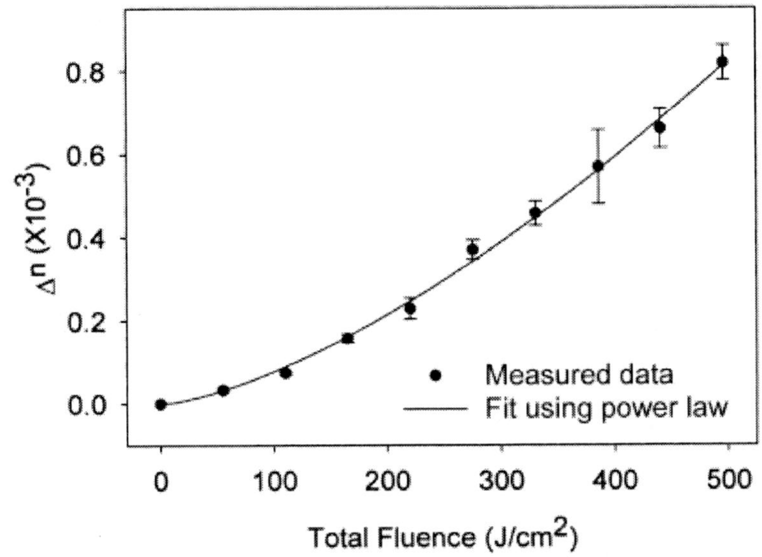

Figure 7. Refractive index change versus the total laser fluence. Closed circles represent the measured data, and the solid curve is the power-law fit with A $=8.8\times10^{-8}$ and B=1.47 [14].

Table 3. Parameters of the power laws for different writing conditions [14]. The average values, A=8.6±0.9×10−8 and B =1.47±0.02.

			Equation (4)		Equation (5)	
Λ (nm)	L (mm)	F_p (mJ/cm^2)	C ($\times10^{-7}$)	b	A ($\times10^{-8}$)	B
	3	294	2.7	1.37	9.5	1.46
533.65		485	3.7	1.44	8.9	1.48
	5	294	1.2	1.53	8.9	1.46
		485	2.3	1.53	8.1	1.48
450.91	1	292	3.0	1.32	7.8	1.45

Table 4. The properties and the writing conditions of the two PLC gratings (L=1 mm) [14]

		First grating	Second grating
Designed grating parameters	R	0.55	0.7
	$2\Delta\lambda$ (nm)	1.8	1.8
	λ_B (nm)	1550.5	1309.5
Measured grating properties	R	0.55	0.73
	$2\Delta\lambda$ (nm)	1.9	1.9
	λ_B (nm)	1550.5	1309.5
Writing conditions	Λ (nm)	533.65	450.91
	F (J/cm^2)	382	473
	F_p (mJ/cm^2)	529	292
	t	144	324

3. Hybrid Integration Technology

The silica-based planar lightwave circuit (PLC) is an optical integrated circuit with the same structure as optical fiber. PLCs are highly stable and reliable because they have no moving parts, and the waveguide material, silica glass, is physically and chemically stable. Moreover, PLCs are suitable for large-scale integration and mass production. PLCs offer easy fiber attachment and a low propagation loss of less than 0.1 dB/cm at an 1.55-μm wavelength [17-20].

Many types of PLC devices, including arrayed waveguide gratings (AWGs), multi/demultiplexers, PLC switches, interleave filters and chromatic dispersion compensators have been developed through sophisticated design and fabrication techniques. The enlargement of wavelength division multiplexing (WDM) technology needs the mass production of complicated pattern-waveguides. The WDM technology allows the fact that optical fibers can carry many wavelengths of light simultaneously without interaction among each wavelength. Several wavelength-selective key components are required to make WDM possible [21]. Wavelength multiplexers allow the insertion of the entire separate wavelength laser signals to be combined into single fiber. In recent years there has been a new interest in broadband access methods based on the optical fiber. The Internet has become popular and the number of people has been requiring broadband access. Moreover, new services, for example high definition television (HDTV), IP (Internet protocol) telephony, video on demand (VOD), interactive gaming or two-way video conferencing, have been developed and they require broader bandwidth. A further growth in the number of people requiring broadband access is forecasted. Ethernet-passive optical network (E-PON) and WDM-PON, which are well known as fiber to the home (FTTH) networks, appear to be the best candidates for the next-generation broadband access networks [22].

For the large-scale mass production of the wavelength multiplexers to be satisfied for the needs of the both network systems, a PLC technology is the most promising candidate for low-cost solutions. The cost reduction of optical components is becoming a key issue and the PLCs on a silicon platform will play a key role for the low cost solution of multifunctional components for both WDM and FTTH network systems [23-26]. The PLCs and integrated optical structures offer the possibility to miniaturize bulk optical functions and they provide completely novel functions.

In this section, we briefly review a PLC technology including fabrication methods. Furthermore, the key technologies for hybrid integration are also introduced such as PLC platform and flip chip bonding.

3.1. Fabrication of Planar Lightwave Circuits

Planar lightwave circuits (PLCs) provide key devices to optical telecommunication systems. Various and complicated optical waveguide devices can be fabricated by this technology and can be mass productive. The processing of PLC-fabrication is involved in forming optical waveguides on planar substrates. All techniques require photolithography, since the waveguide widths are usually on the order of a few microns. There are two representative approaches to planar waveguide formation: flame hydrolysis deposition (FHD) and plasma enhanced chemical vapor deposition (PECVD).

Flame hydrolysis deposition (FHD) [18-20] is one of the main technologies for the fabrication of PLC devices such as directional couplers, splitters, Mach-Zehnder interferometers and arrayed waveguide gratings as well as PLC based-sub-system modules for WDM and FTTH network systems. The first step of the process is to deposit two successive glass particle layers as the undercladding, called also buffer, and core on a silicon substrate or a thermally oxidized (TO) silicon wafer by the flame hydrolysis of $SiCl_4$. This process leads to the deposition of a soot of silica (SiO_2) that for the core layer can be doped by the inclusion of GeO_2, P_2O_5, and B_2O_3, all from their respective halide gases. After sintering of the soot at ~1300°C, glasses of different refractive indices can be obtained. The desired core ridges for the channel waveguides are then defined by photolithography followed by reactive ion etching. The core pattern is finally covered with an overcladding layer in another FHD process. Deposition of sequential layers of silica with different refractive indices leads to the micro-fabrication of buried optical waveguide structures. The deposition is relatively fast (several mm/min), glass composition can be widely changed by co-deposition of different precursors, although this composition can be changed both physically and chemically during the consolidation process. The FHD process is described in Fig. 8. The optical quality of the fabricated waveguides is excellent and the propagation losses are very low. A propagation loss of 0.1dB/cm was obtained in a 2-m-long waveguide with a 2%Δ (Δ=index difference between core and cladding) and a loss of 0.035dB/cm was obtained in a 1.6-m waveguide with a 0.75%Δ. A further loss reduction down to 0.017dB/cm has been achieved in a 10-m-long waveguide with a 0.45%Δ [19].

Channel waveguide devices [24-26] for the sub-system module of FTTH systems are fabricated on silicon substrates by using a combination of flame hydrolysis deposition (FHD) and reactive ion etching (RIE). Thermally grown oxides of 15-μm thickness are used as undercladding layers on which the core layers were deposited by the FHD process. The channel waveguides are formed by the photolithography technique and then covered by uppercladding layers of 25μm thickness. The core is $6.5\times6.5\mu m^2$ in size, the relative index difference between the core and cladding layers is 0.75% and the NA (numerical aperture) is estimated to be 0.178. Atomic emission spectroscopy is used for the measurements of Ge contents of the waveguide and the measured values were about 13.5wt% and 2.1wt% for core and uppercladding, respectively.

Cho *et al.* proposed a precursor-premixed type torch for the FHD process to improve the crosstalk property of an AWG [27]. The torch was composed of several concentric tubes having small inlets for gas supplies. In the conventional type torch (Type A) as shown in Fig. 9, the carrier gas (Ar) containing precursor materials such as $SiCl_4$, $GeCl_4$, $POCl_3$ and BCl_3 is injected from the center tube of the torch, and the fuel gas (H_2) diluted with Argon and the oxidizer gas (O_2) are injected from the outer tubes in that order. The outermost tube acts as a shield that increases flame stability by reducing the entrainment of the ambient air. Since the fuel and oxidizer are kept separated until the exit of the torch, the combustion that results from the diffusion of H_2 and O_2 start to take place in the boundary region between the gas injection tubes for H_2 and O_2. This means that the concentration of H_2O or OH needed for the hydrolysis reaction is highest in this ring-shaped region while the concentrations of precursors are highest in the center region of the torch. Therefore, the generation of fine particles begins in another ring-shaped region farther from the torch exit. The mixing ratios of SiO_2 particles to dopant particles such as GeO_2, P_2O_5 and B_2O_3 are varied within a flame in the radial direction as well as in the streamwise direction, which causes the refractive index variation of the deposited layer, accordingly. Although the overall uniformity of the deposited layer within a substrate can be controlled with the adjustments of the speed profiles of torch traversing and rotation of the turntable supporting substrates, the local non-uniformities caused by non-uniform concentrations within a flame cannot be fully eliminated even at the final stage (consolidated transparent film) in spite of the dopant diffusion in the consolidation process that may reduce the index non-uniformity in the film. To reduce the index fluctuations, a new configuration into the torch of same dimensions was proposed, which is depicted as Type B in Fig. 9. The fuel H_2, dilution gas Ar and precursors are injected together from the center nozzle (I) and the oxidizer O_2 is injected from the outermost nozzle (III). Inert gas Ar is introduced between H_2 and O_2 to prevent undesirable deposition of glass particles on the end face of the torch. Because the fuel and the precursors are injected in premixed state, both the combustion and the hydrolysis reactions are expected to take place nearly at the same region within the flame in the precursor-premixed type torch. Therefore, the spatial concentration distributions of the dopants in the flame that determine the index uniformity of the deposited film can have more uniform profiles than those in a flame of conventional type torch (Type A in Fig. 9). As a result, the AWG crosstalk level was enhanced of 5~8dB by the precursor-premixed type torch.

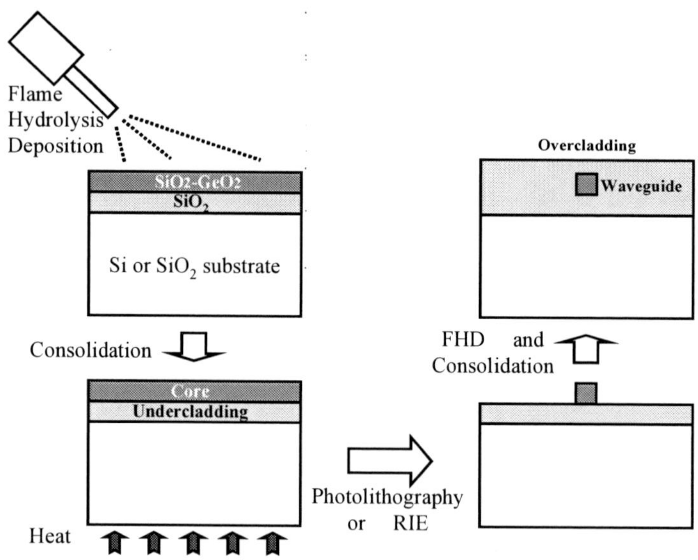

Figure 8. PLC fabrication process using flame hydrolysis deposition (FHD) and photolithography (or RIE).

A plasma enhanced chemical vapor deposition (PECVD) [28] has been primarily used for diffusion masks and passivation. More common applications of this process for microelectronic fabrication started with the introduction of commercial processing equipment. In recent years, new material demands and lower-processing-temperature requirements in large-scale integration circuits, solar energy cells, flat-panel displays, and optoelectronic integrated systems have made plasma-enhanced deposition processes increasingly important. One of the major advantages of the plasma deposition processing, besides its high deposition rate, is the flexibility for depositing films with desirable properties. Films with unique composition and given physical and chemical properties can be obtained by adjusting deposition parameters such as temperature, RF power, pressure, precursor's gas mixture and their ratios. Although the deposition is at relatively low temperatures, to achieve low loss material high temperature annealing is commonly applied. Mechanically and chemically stable thick films can be deposited and easily formed into channel waveguides and waveguide components by plasma etching. The deposition rates can be in the range of 0.15~0.3mm/min and the optical characteristics of the obtained films are qualified. The propagation loss of the waveguide from the PECVD process is less than 0.1dB/cm.

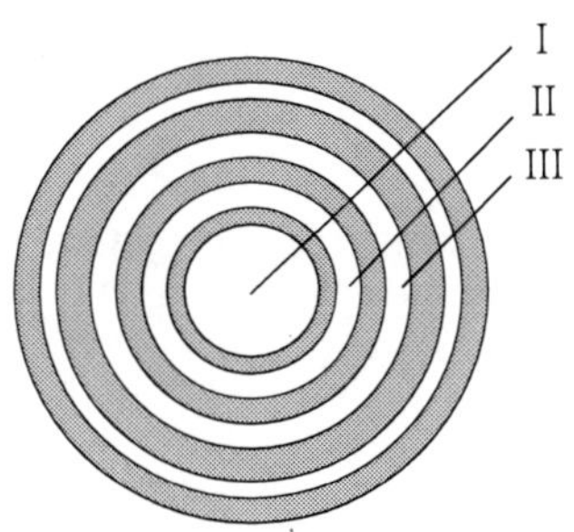

Nozzle	Type A	Type B
I (8mm)	precursor, carrier gas Ar	precursor, carrier gas Ar, H_2 (1.2), Ar (2.0)
II (10mm × 13mm)	H_2 (1.6), Ar (3.2)	Ar (1.0)
III (16mm × 19mm)	O_2 (6.0)	O_2 (4.0)

Figure 9. FHD torch configurations (Type A: conventional, Type B: precursor-premixed, nozzle size: inner diameter × outer diameter, unit of gas flow rate: l/min) [23].

3.2. Hybrid Integration

Hybrid integration and optoelectronic packaging can provide highly functional photonic components and can interconnect real telecommunication systems. These are valuable and practical technologies in the area of the photonics science. Recently the increment of needs for WDM and FTTH components accelerate to develop cost-effective and multifunctional solutions using the hybrid integration. We briefly introduce the key technologies of the hybrid integration in the area of optoelectronic packaging in the following small sections.

3.2.1. PLC Platform

Hybrid integrated components use a PLC platform to connect active devices to passive waveguides. Typical PLC platforms have used the silica-on-terraced-silicon structure [29] as an optical bench and a heat sink of the PLC platform, which incorporates both PLC and electrical wire on a silica layer. The PLC and the OE devices can be aligned by passively placing the devices onto the silicon terrace. The silicon terrace acts as an alignment plane for OE-devices, and the embedded-type silica-waveguides have excellent optical properties suitable for high-performance planar lightwave circuits. The PLC platform consists of a PLC region, an assembly region and a wiring region. In the PLC region, silica waveguide patterns are formed on the ground plane of the terraced silicon substrate. The silica waveguides have various patterns such as AWG, directional coupler, Y-branch and so on for the optical guiding purpose. The silicon terrace acts as both an optical bench for alignment and a heat sink for OE devices in the assembly region. An Au electrode and an AuSn solder pad are formed on a thin passivation layer deposited on silicon terrace. The height from the solder surface to the waveguide core center is designed to be same as the height of the active layer of the OE devices. Au alignment markers are formed on the terrace and they can guide right position of the OE devices by a passive alignment. Therefore, the OE devices can be flip-chip bonded simply by using the alignment markers of the silicon terrace and the OE-devices. In the wiring region, the power lines and signal lines are formed to interconnect the OE devices and the module package. Electrical integrated circuits can be flip-chip bonded in the region as well. The PLC platform using the silicon terrace fabrication process is shown in Fig. 10 [19]. First, an undercladding layer is deposited on a chemically etched Si substrate. Then, the surface is flattened by a mechanical polishing. Second, a thin layer for height adjustment is deposited on the flattened surface by polishing process. Third, the core and overcladding layers are formed. Fourth, a Si terrace is formed by RIE and then a thin SiO passivation layer is formed on the Si terrace. After that, electrodes and alignment marks are simultaneously formed by lifting off the evaporated Au layer. Finally, a thin AuSn solder film (about 2–3μm thick) is evaporated and patterned using the liftoff method. The AuSn film consists of three layers of Au–Sn–Au with a composition of 80 wt.% Au and 20 wt.% Sn for the flip chip bonding of the OE devices.

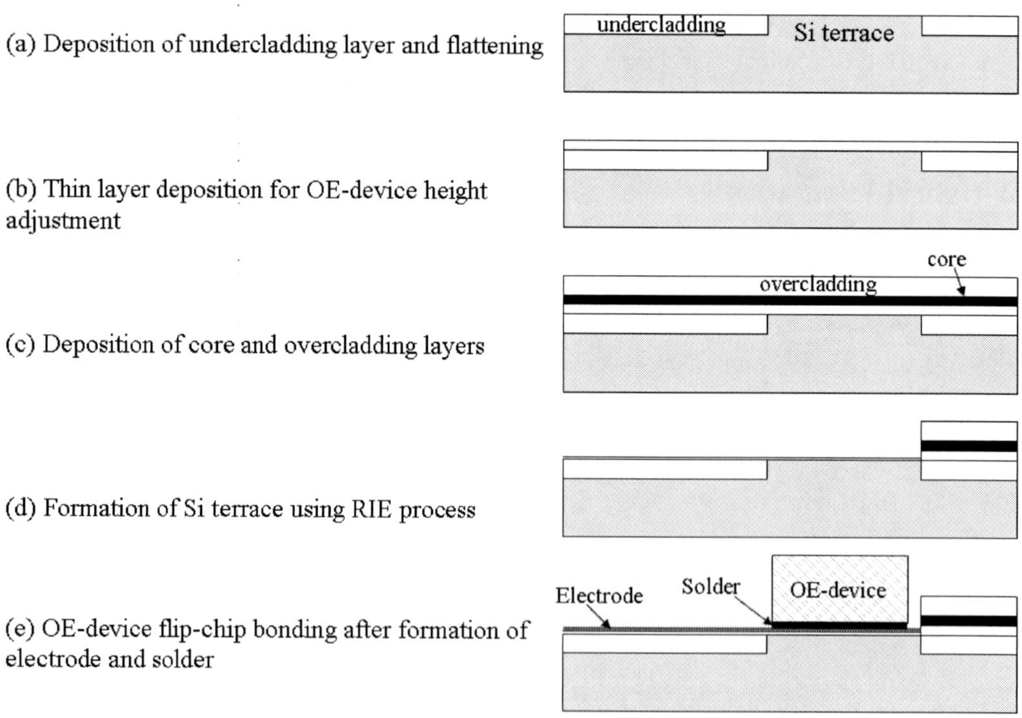

Figure 10. PLC platform fabrication process [19].

There is another approach to fabricate PLC platforms using a terraced silica for hybrid integration [30]. This method is simple with no performance degradation and does not need additional equipments except metallization (evaporating, plating) ones. In addition, there is no polishing process needed unlike the silicon terraced platform. Fig. 11 shows the fabrication sequence of the terraced-silica platform for hybrid integration. Underclad and core silica layers are deposited by the flame hydrolysis deposition (FHD) method on a silicon wafer [29]. Waveguides are patterned by the dry etching process. Alumina layer is deposited as an etch stopper by an E-beam evaporator. Alumina has good selectivity against silica during the deep dry etching process and maintains its integrity after the overcladding process. The overclad silica layer of about 25μm is deposited and sintered at above 1000°C as shown in Fig. 11 (a)-(d). The terraced-silica structure is made by the deep (~30μm) dry etching process. The alumina etch-stop layer process and over etching process are continued to make terraced-silica structures. Over etching fixes the height of terraced-silica while does not change the reference surface. Electrode patterns are fabricated by the evaporation and lift-off method. Solder layer is fabricated by electroplating. Solder height is controlled to be 0.5~1μm higher than the terraced-silica surface. Optoelectronic devices such as LD and PD are bonded using the flip chip bonding (FCB) technique as shown in Fig. 11 (e)-(h).

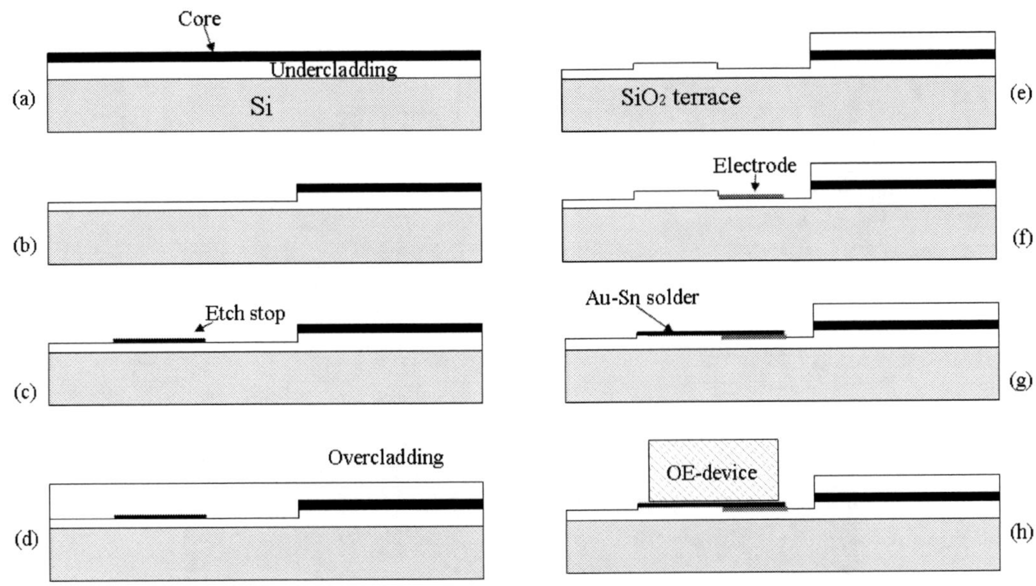

Figure 11. The fabrication sequences of the terraced-silica platform. (a) undercladding and core deposition, (b) waveguide patterning (c) etch-stop layer deposition, (d) overcladding deposition, (e) deep etch, (f) electrode patterning, (g) solder patterning, (h) flip chip bonding of OE-devices [30].

3.2.2. Flip Chip Bonding

After the fabrication of the PLC platform, laser diode (LD) and other device can be bonded on it with high accuracy. For this bonding process, flip chip bonding (FCB) technique and accurate flip chip bonder are needed [30, 31]. One of the main issues for the platform is the lateral and vertical position errors from the misalignment between active devices and waveguides. Lateral position errors result from the lithography and the FCB process. In the case of the terraced silica structure, a waveguide alignment mark which is made simultaneously when making the waveguide pattern is used and the top surface of the silicon or silica terraced structure are brought into a contact with the bottom surface of optoelectronic devices. Therefore, the lateral position error due to the lithography error can be reduced significantly and hence the lateral position error becomes dominated by the FCB error. The bonding conditions and sensing errors of the FCB equipment can affect this position error. For the accurate vertical alignment, the position of a terrace surface relative to the waveguide center must be controlled precisely. In the silica terraced structure, deep etching process and the solder height do not cause this positional error because the etch-stop layer prevents the etching of terraced region and the fabricated terraced-silica also acts as a mechanical stop. Therefore, the vertical position could be controlled within 0.5μm.

High coupling efficiency can be achieved with the help of accurate alignment marks in the waveguide and in an LD. Fig. 12(a) show the align marks for the PLC platform. It is composed of a cross hair and four small squares. The position offset by the FCB can be measured with an infrared microscope. As Fig. 12(a) indicates, the align marker is misaligned longitudinally. The FC bonder should be calibrated for the offset value (here, 8μm).

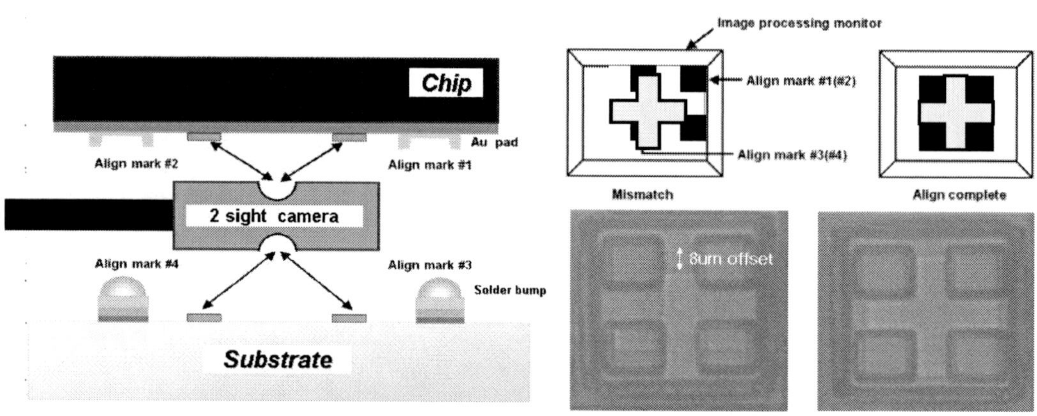

(a)

(b)

Figure 12. (a) Index matching method for flip chip bonding and IR (infra-red) inspection of a LD on the PLC platform after flip chip bonding (before and after calibration), (b) the SEM photo of LD bonded on the platform [30].

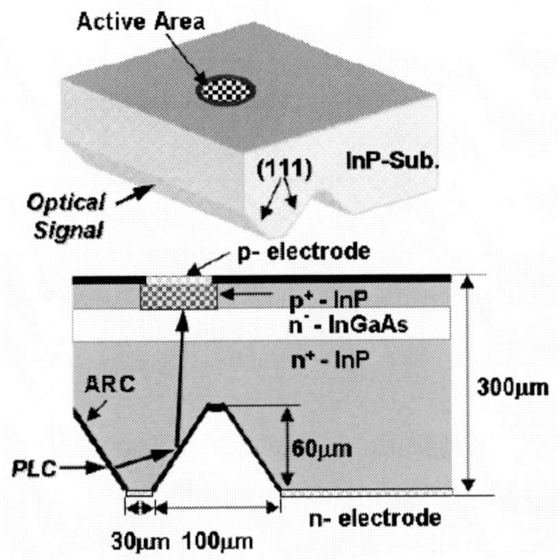

Figure 13. The structure of the RMF-PD [32].

To minimize the mode-size mismatch between the LD and the waveguide, the far field angle (FFA) of the LD should be reduced. A spot-size converter or mode adapter integrated LD (SSC-LD) with a tapering ridge waveguide [18] has often been used. The typical FFA of SSC LD is less than 13×15°. The bonded SSC-LD on the platform is shown in Fig. 12(b). To improve the coupling efficiency between the waveguide and a photodiode (PD), the reflecting mirror facet (RMF) PD can be used [32]. It is composed of two v-grooves and a conventional vertical PD as shown in Fig. 13. The two adjacent v-grooves convert a horizontal beam to nearly a vertical beam. After the FCB, the LD showed good reliability and high coupling efficiency [30]. These hybrid integration techniques will be used to fabricate the hybrid integrated ECLs, which will be discussed in the next sections 4 and 5.

4. Hybrid Integrated External Cavity Lasers Based on Wavelength Selective Grating

The external cavity semiconductor laser (ECL) is a key device for the wavelength division multiplexing (WDM) systems and access area networks. One of the simple but effective methods for controlling the spectral and modal properties of the ECL is the hybrid integration technique combined with the well-developed grating technology. In this hybrid integration for constructing an ECL, typically, a Bragg grating is first integrated in a PLC waveguide and then an FP-LD is aligned with the PLC waveguide as shown in Fig 14. Because the PLC grating written in the silica PLC waveguide acts as a mirror in this hybrid integrated ECL, the oscillation wavelength and the number of lasing mode depend on the reflection spectra of the PLC waveguide grating.

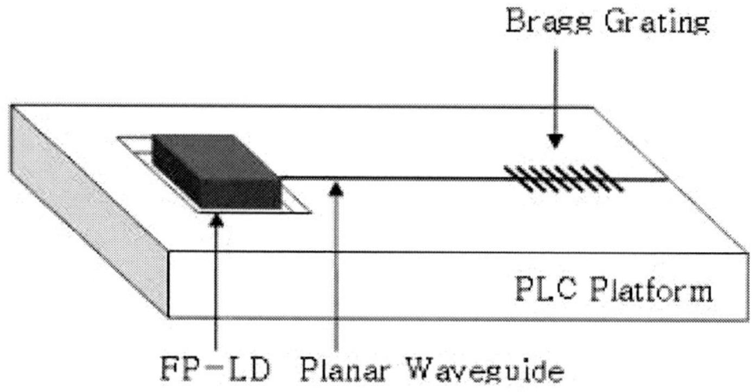

Figure 14. Schematic diagram of hybrid integrated ECL.

4.1. Hybrid Integrated Single Wavelength External Cavity Laser

Single wavelength, dual wavelength or multiwavelength ECLs are very important in WDM systems and sensing systems. They can be cost-effectively constructed by using the hybrid integration technique (see Fig 14). The hybrid integrated ECL exhibits single wavelength or multiwavelength oscillation depending on the types of the Bragg grating integrated in the PLC. If a Bragg grating with one specific Bragg wavelength is selected as the external cavity mirror, the ECL will oscillate at the single wavelength corresponding to the Bragg wavelength. A SSC-LD is often used for the FP-LD in this hybrid integrated single wavelength ECL to achieve low coupling loss [14, 18]. The rear facet of the FP-LD is coated with 80% of reflection and the front facet is antireflection (AR) coated with 1% of reflection for reducing the parasitic reflections [14].

Ge-doped silica waveguide is used for the PLC waveguide and hydrogenated at room temperature with 100atm pressure to enhance photosensitivity. The PLC Bragg gratings can be well written by the phase mask technique described in section 2.1. The transmission spectra of the PLC gratings [14] grown by the method are plotted (solid curves) in Fig. 15(a) and Fig. 16(a). The first PLC grating was written with a phase mask whose pitch is 533.65nm at a repetition rate of 5Hz and at the UV fluence of 529mJ/cm^2 for t=144s [14]. Meanwhile, the second PLC grating was written with a phase mask of 450.91nm pitch at the laser fluence of 292mJ/cm^2 for t=324s at 5Hz. The length of the grating was set to 1mm. The Bandwidth and the reflectance of the first PLC grating, whose Bragg wavelength is ~1550nm, were 1.8nm and 0.55, respectively. The second PLC grating was designed at ~1309nm [14]. In order that the bandwidth and reflectance of the 1mm-long PLC grating were 1.8nm and 0.7, respectively, the modulation index required for the grating was 5.9×10^{-4}. Then from the power law equation (8) with $A=7.8\times10^{-8}$ and B=1.45, the total fluence of 473J/cm^2 is given. Thus, the second grating was formed at the UV fluence of 529mJ/cm^2 for t=144s. Note that the measured bandwidths and reflectance of the two gratings are closely matched to the prediction (dotted line).

The oscillation spectrum of the hybrid integrated ECL with the first PLC grating is shown in Fig. 15(b). The FP-LD exhibiting multimode oscillation, shown in the inset in Fig. 15(b), was stabilized at single longitudinal mode with side mode suppression ratio (SMSR) of ~40dB after using the PLC grating [14]. Similarly, Fig. 16(b) shows the oscillation spectrum of the hybrid integrated single longitudinal mode ECL with the second PLC grating, while that of the FP-LD without the grating is displayed in the inset. It indicates that the FP-LD exhibiting multimode oscillation was stabilized at single longitudinal mode by employing the PLC Bragg grating as an external cavity mirror. The lower SMSR value, compared with the case in Fig. 16(b), was attributed to the low coupling ratio between the FP-LD and the PLC waveguide [14].

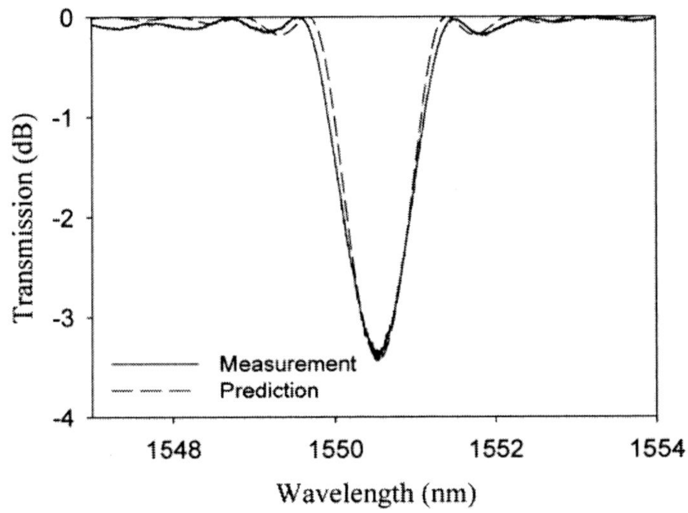

(a)

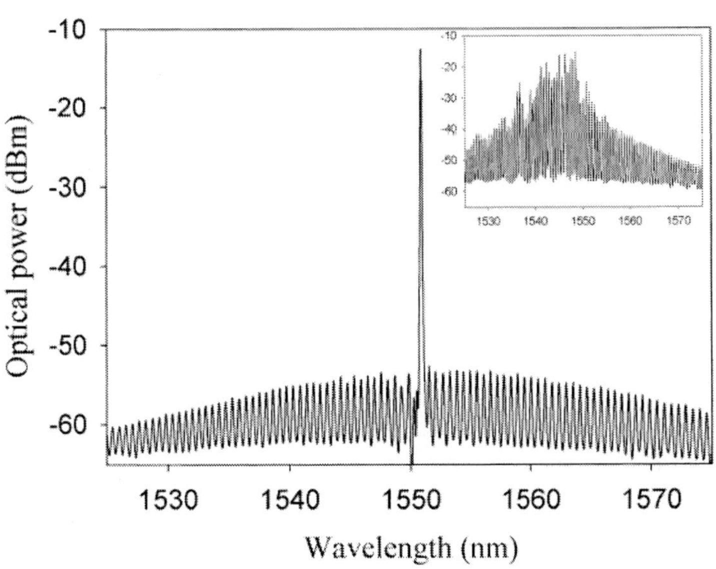

(b)

Figure 15. (a) Transmission spectra of the first PLC grating (λB=1550.5 nm), where solid line is the measured curve, which is well matched to the dashed curve predicted by the coupled-mode theory, and (b) oscillation spectra of an ECL with the first PLC grating. The inset shows multimode oscillation of the FP-LD without the PLC grating [14].

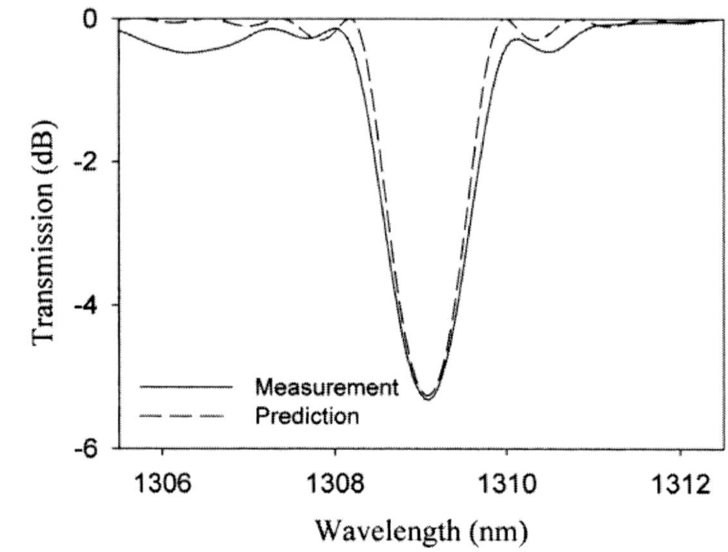

(a)

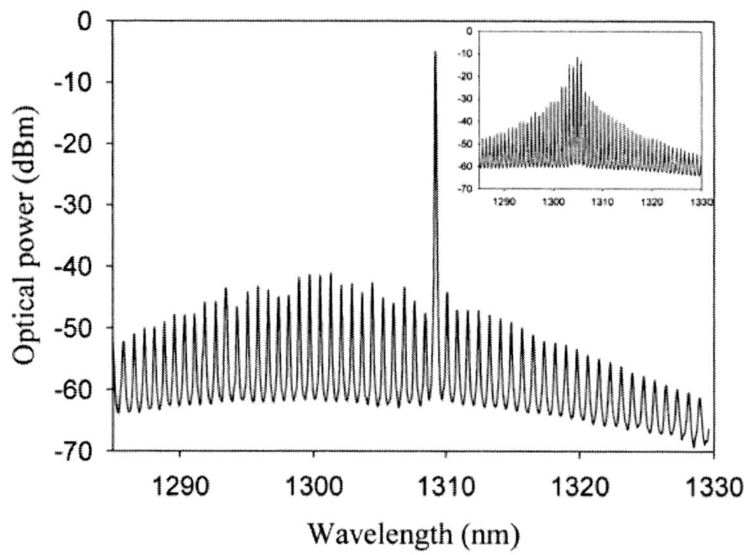

(b)

Figure 16. (a) Transmission spectra of the second PLC grating (λ_B=1309.5nm), and (b) oscillation spectra of an ECL with the second PLC grating [14].

Also, to fabricate the hybrid integrated ECL, one can bond the FP-LD to one PLC waveguide using the flip chip bonding technique and couple it to a printed circuit board (PCB) for easy connection with the LD driver [33]. And a Bragg grating can be integrated in another PLC waveguide. Then, the two PLC waveguides can be aligned using an aligner and bonded together using an index matching oil. Loss measured between the two PLC waveguides is about 0.5-1dB [33].

4.2. External Cavity Lasers with High-Sidemode Suppression Ratio Using Grating-Assisted Directional Coupler

In WDM systems, external cavity lasers (ECLs) with Bragg gratings as external reflectors are attractive candidates as a light source because of their high wavelength stability. In addition to the wavelength stability, the sidemode suppression ratio (SMSR) is one of the most important parameters of this type of light sources. Low SMSR values may lead to intolerable intensity noise through the mode partition noise or large levels of crosstalk in the WDM systems [34, 35]. The SMSR values of the conventional distributed feedback laser (DFB) and monolithic semiconductor laser were usually around 40dB or less [36-39]. The SMSR can be improved by simply adding an additional optical filter to filter out the sidemode spontaneous emission of the laser. But in this method, additional cavity loss and cost result in a problem.

In order to improve the SMSR of ECL, an ECL with a grating-assisted directional coupler has recently been reported [40]. We will discuss the principle and the experimental results of the ECL with a grating-assisted directional coupler below.

4.2.1. Principle of Operation

In a conventional ECL structure, the laser output is emitted passing through the Bragg grating with a narrow reflection bandwidth. If the light reflected from the Bragg grating is taken as the laser output, the SMSR of the ECL can be improved because only the main laser cavity mode within the reflection band of the grating is reflected from the grating meanwhile the laser sidemodes are transmitted through it. Thus, the scheme shown in Fig. 17(a) was applied [41] to achieve high SMSR. However, this setup is disadvantageous because of the large external cavity loss due to the loss of the laser output through the port 2.

Fig. 17(b) shows a schematic diagram of the method to improve the SMSR of the ECL by using a grating-assisted directional coupler as an external mirror. The grating-assisted directional coupler is the coupler device with a Bragg grating written in the part of its coupling region, and is often used as an add-drop filter for wavelength division multiplexing (WDM) applications [42, 43]. By using the grating-assisted directional coupler instead of a 3dB coupler used in Fig. 17(a), the SMSR of the ECL can be improved without any additional devices such as an optical circulator without suffering from additional cavity loss.

The coupling region of the grating-assisted directional coupler can be divided into three regions as shown in Fig. 17(b) [42]. Without the Bragg grating written in region II, the regions I, II and III act as the coupling regions of the conventional directional couplers. However, with the Bragg grating in region II, the light in the reflection band selected by the Bragg grating is reflected in region II. Then, a part of the reflected light is dropped to the drop port and the remaining part of the light is returned back to the input port through the directional coupling mechanism provided by regions II and I. In the scheme [40], the light returned to the input port acts as the feedback light of the conventional ECL, while the light dropped to the drop port acts as the output light. Thus, if the reflectivity of the Bragg grating is high, most of the light within the reflection band will not pass through the grating-assisted coupler. As a result, the scheme [40] suffers from little external cavity loss, unlike the scheme with the 3dB coupler in Fig. 17(a).

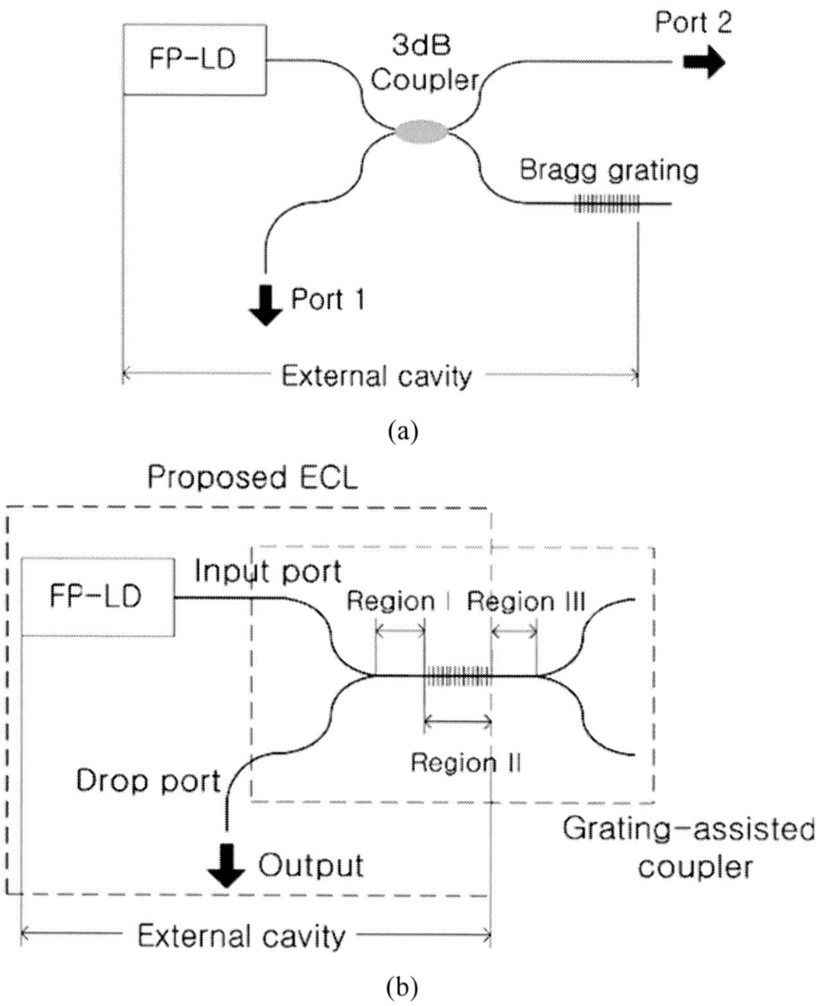

Figure 17. Schematic diagrams of the ECL with high-SMSR composed with (a) a 3dB directional coupler and a Bragg grating, and (b) a grating-assisted coupler [40].

In the conventional ECL structure, the feedback light power is determined by the reflectivity of the external mirror. Therefore, the power ratio R of the light returned to the input port to the total reflected light in the proposed scheme can be treated as the reflectivity of the external mirror of the conventional ECL [40]. The reflectivity of the external mirror affects the resonance in the external cavity and the output power of the ECL. For example, low reflectivity of the external mirror may cause no resonance in the external cavity, whereas high reflectivity may cause low output power. So, the optimization of the reflectivity of the external mirror is very important for the realization of the ECL of high performance [40]. In general, the Bragg gratings with the reflectivity in the range of 50~95% are usually used in order for stable feedback action and good output characteristics [14, 36-38]. The power ratio R of the grating-assisted coupler is determined by the parameters of the Bragg grating and the coupler. Especially, when the strength of the Bragg grating is strong, the power ratio R depends mainly on the length of the region I [42]. Therefore, the power ratio R should be adjusted properly by controlling the length of the region I, in order for the grating-assisted coupler to act as the external mirror, whose reflectivity is appropriate for the stable feedback action [40].

4.2.2. Experimental Results

The reflectivity of the Bragg grating for the grating-assisted coupler was designed to be 99% to minimize the cavity loss of the ECL [40]. The lengths of the Bragg grating and the region I were set to 1.5mm and 2.4mm, respectively. With these parameters, the power ratio R results in the reflectivity of 55%. The grating-assisted directional coupler was formed by writing a Bragg grating in the coupling region of a PLC directional coupler using the phase mask technique [14]. An optical circulator with insertion loss of 0.5dB was also used to measure the reflection spectrum of the light returned to the input port. The reflectivity and the bandwidth of the grating were about 99% and 0.7nm, respectively. The reflection spectra of the lights returned to the input and dropped to the drop port are shown in Fig. 18. The center wavelength of the reflection spectra was about 1549nm. At the center wavelength, the power returned to the input port was higher than the power dropped to the drop port by 0.6dB, indicating that the grating-assisted coupler acts as an external mirror with the reflectivity of 53.4%. To couple the light emitted from the pigtailed fiber of the FP-LD into the input port of the PLC-type grating assisted coupler, a fiber block was used with a coupling loss of about 0.5dB [40]. The cavity mode spacing of the ECL was about 0.01nm.

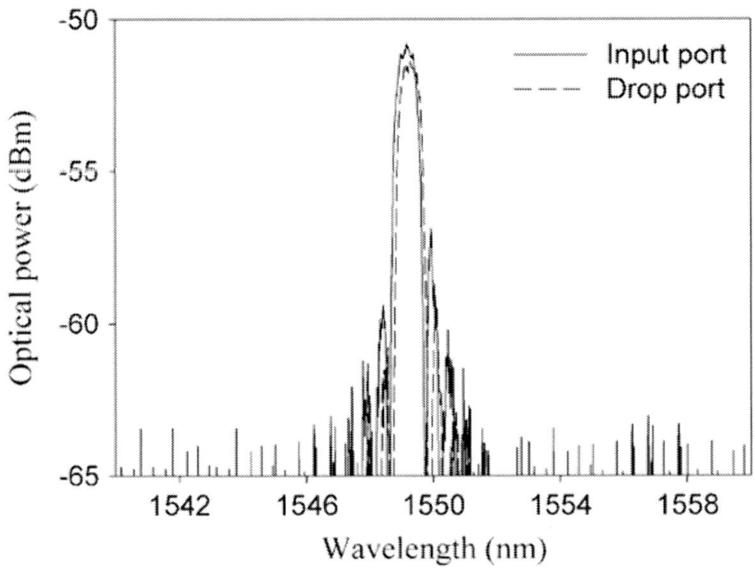

Figure 18. Measured reflection spectra of the lights emitted from the input port (solid curve) and the drop port (dashed curve) of a grating-assisted coupler [40].

Fig. 19 shows the oscillation spectrum of the ECL measured at 25°C and 25mA from the drop port of the grating-assisted coupler. Note that the FP-LD oscillating at multi longitudinal modes (as in the inset of Fig. 19) becomes stabilized at a single-longitudinal mode with high SMSR by the grating-assisted directional coupler. The oscillation wavelength was consistent with the Bragg wavelength of the PLC grating formed in the grating-assisted coupler, indicating that the grating-assisted coupler acted as the external mirror and the reflectivity is appropriate for the stable feedback action. Note that the SMSR of the ECL with a grating-assisted directional coupler is as high as 60dB. This SMSR value is 10 to 20dB higher than those of the conventional ECLs.

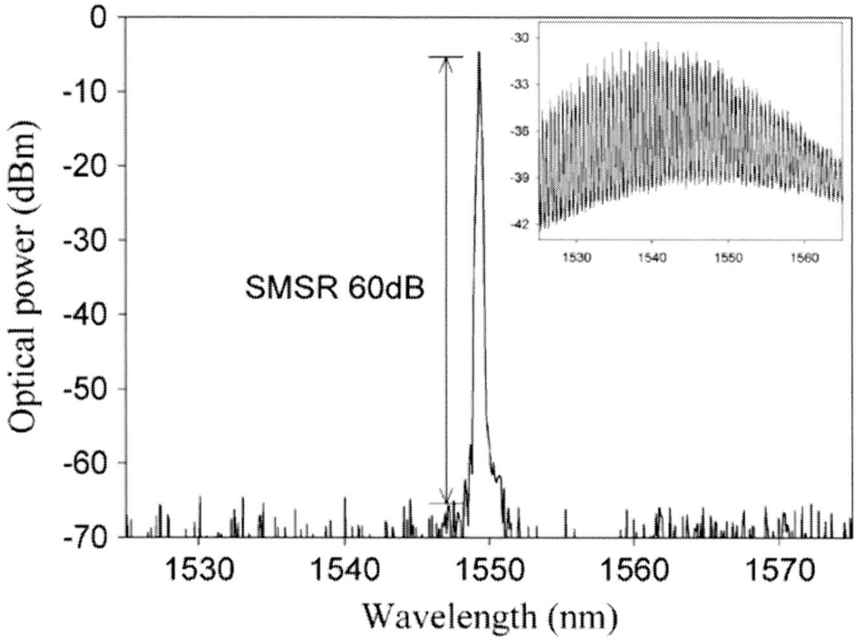

Figure 19. The oscillation spectra of the ECL composed of a grating-assisted coupler. The inset shows multimode-oscillation of the FP-LD used to compose the ECL [40].

4.3. Dual Wavelength Hybrid Integrated External Cavity Laser

Many researchers [44-49] have investigated dual-wavelength lasers (or two-color lasers) for various applications such as Terahertz frequency generation [44-46], dual wavelength interferometry [50], THz-imaging [51] and wavelength switching. For the dual-wavelength laser operation, one has to combine a single laser with one or two external cavities. In these dual-wavelength lasers, one of the key devices is the Bragg grating which acts as an external cavity mirror. The gratings include bulky gratings [44-46], distributed Bragg gratings [47, 48] and fiber Bragg gratings [49]. The distributed Bragg gratings [47, 48] were grown directly in the vertical cavity surface emitting lasers (VCSELs), therefore, the dual-wavelength lasers based on VCSELs with two monolithic gratings are more stable and compact than the dual-wavelength lasers with the other types of gratings. However, this kind of VCSEL-based dual-wavelength laser is expensive to fabricate because of its complex structure. A hybrid-integrated dual-wavelength laser has recently been demonstrated [52] using a sampled Bragg grating formed in a planar lightwave circuit. This kind of hybrid-integrated dual-wavelength laser is a good candidate for the dual-wavelength source because it is cost-effective and has an external cavity, which is short and compact enough for the two-color coherent lasing [44].

In this section, we will only focus on the dual-wavelength source for terahertz beat signal generation. Since a good THz beat signal is necessary starting point for THz radiation generation, it is extremely important to demonstrate a stable hybrid-integrated dual-wavelength laser [52]. This hybrid-integrated ECL exhibiting strong dual-wavelength laser oscillation [52] with Terahertz beat frequency will be discussed in detail below.

4.3.1. Device Fabrication

Fig. 20 shows the schematic of the hybrid integrated dual-wavelength ECL in which an FP-LD and a sampled Bragg grating are integrated on a silica PLC platform [52]. The FP-LD was a spot-size converter integrated laser diode having low coupling loss. The front and rear facets of the FP-LD were coated with anti-reflection film and high reflection film, respectively [14, 40, 54]. The length of the FP-LD was 300μm. The center wavelength of the FP-LD was about 1290nm, and the longitudinal mode spacing was 0.8nm. The sampled Bragg grating was written by exposing the PLC waveguide to a KrF excimer laser beam (λ= 248nm) through a combination of a phase mask and an amplitude mask [52, 53].

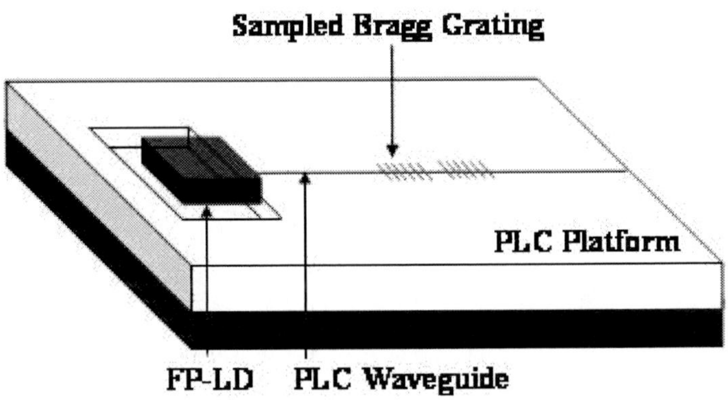

Figure 20. Configuration of a dual-wavelength external cavity laser with a sampled Bragg grating formed in a silica PLC waveguide [52]

The PLC sampled grating was designed such that two strong reflection peaks occur within the gain bandwidth of the FP-LD for the operation of the dual-wavelength laser [50]. The separation of the two lasing wavelengths is given by $\Delta\lambda=\lambda_B^2/2n_{eff}\Lambda_s$, where λ_B is the Bragg wavelength, Λ_s is the sampling period of the PLC sampled grating and n_{eff} is the effective refractive index of the silica PLC waveguide. Thus, for generation of 1THz beating signal, the required sampling period Λ_s for n_{eff}=1.45 is ~103μm. The length of the sampled grating was adjusted to ~10mm with a shutter placed between the KrF excimer laser and the phase mask.

When the gain experienced by the two frequencies fed back in the laser by the sampled grating is approximately equal, and the two modes are perfectly mutually coherent, the output intensity I(t) is modulated as [52]

$$I(t) = I_1 + I_2 + 2\sqrt{I_1 I_2}\cos(2\pi f_b t), \tag{9}$$

Where f_b is the beat frequency of the two wavelengths, I_1 and I_2 are the laser mode intensities for the two different wavelengths. In reality, the lasing modes are only partially mutually coherent, and the amplitude of the modulation term decreases according to equation (9). Thus, a good dual-wavelength design must exhibit high mutual coherence between the lasing modes in order to achieve strong THz radiation generation [52].

4.3.2. Results and Discussion

The PLC gratings with a pitch Λ of 450.91nm were grown after 900s-irradiation [52] with the fluence of 680mJ/cm^2 at 5Hz repetition rate. The transmission spectra of the PLC sampled gratings were shown in Fig.21 for two different sampling periods Λ_S of ~ 100μm and ~ 150μm. The two reflection peaks indicate that the PLC gratings act as an external cavity mirror feeding back two laser wavelengths. The two peak wavelengths of the PLC gratings with Λ_S=~100μm are about 1312.02nm and 1318.86nm, as shown in Fig. 21(a). Those of the PLC gratings with Λ_S of ~ 150μm are 1313.08nm and 1317.02nm as shown in Fig. 21(b). The reflectivity of the reflection peaks ranges between 80% and 94%. The wavelength spacing between the two peaks for Λ_S=~100μm is ~5.84nm, while the spacing for Λ_S= ~150μm is ~3.94nm [52].

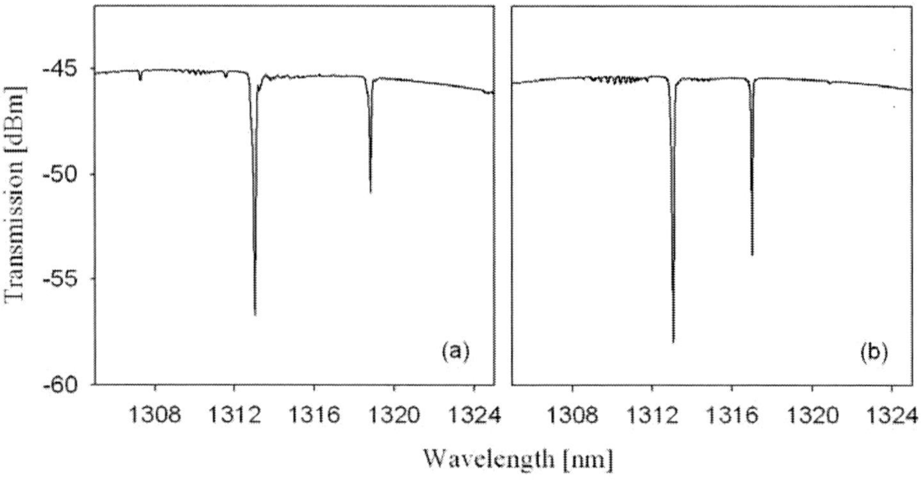

Figure 21. Transmission spectra of the PLC sampled gratings used in the dual-wavelength external cavity laser with two different sampling periods, (a) Λ_s= ~100μm and (b) Λ_s= ~150μm [52].

The optical spectra of the dual-wavelength external cavity lasers with the PLC sampled Bragg gratings corresponding to the Fig. 21(a) and Fig. 21(b) are shown in Fig. 22(a) and Fig. 22(b), respectively. Note that the FP-LD, which runs multimode without the external cavity feedback, becomes stabilized at the two laser modes with SMSR >32dB after the hybrid integration on the silica PLC platform. Fig. 22(a) indicates that the dual-wavelength ECL (with the grating with Λ_S=~100μm) emits laser outputs with equal intensity (i.e., $I_1=I_2$) at 1313.04nm and 1318.74nm. Fig. 22(b) shows that the dual-wavelength ECL with Λ_S=~150μm oscillates with unequal intensity at 1313.08nm and 1317.14nm. This says that the laser output of the first ECL (Λ_S=~100μm) results in a beat signal with f_b~1 THz and that of the second ECL (Λ_S=~150μm) yields to the beat signal with f_b~0.7 THz. However, the dual-wavelength laser with equal intensity is desirable for the Terahertz beat signal generation because it gives the maximum modulation depth.

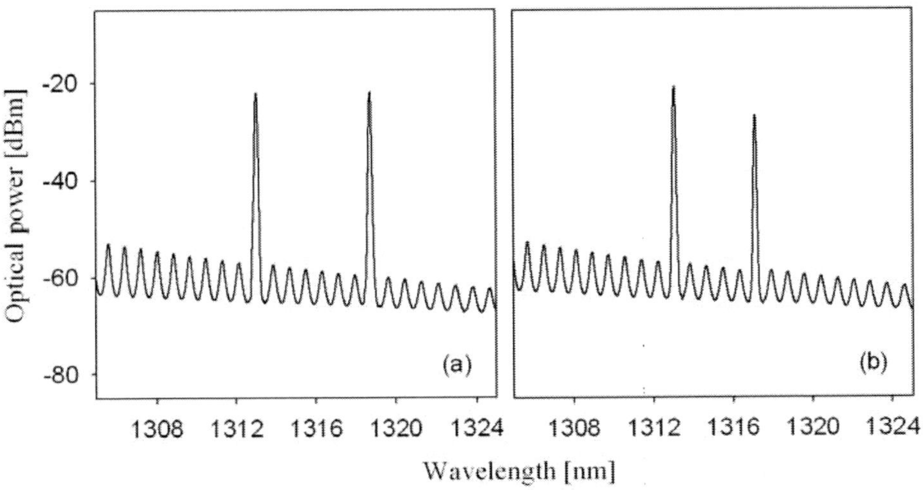

Figure 22. Optical spectra of the dual-wavelength external cavity laser with different grating sampling periods for two different bias conditions, (a) Λ_s=~100μm and (b) Λ_s=~150μm [52].

The oscillation wavelengths of the dual-wavelength ECL are stable and consistent with the Bragg wavelengths of the PLC gratings. This is because the thermal stability of the silica gratings is excellent and the PLC Bragg gratings act as the external mirror of the PLC-based dual-wavelength ECLs. Also, the two modes are oscillating in the same fundamental spatial mode such that the mode overlapping is excellent. To demonstrate the Terahertz beat signal generation from the hybrid-integrated dual-wavelength ECL, computer simulation can be conducted by the simulation technology [52], with the core of the model based on a broad-band laser model capable of resolving semiconductor laser dynamics on very fast time scales while properly capturing all properties of the active structure [55]. The model consists essentially of two parts. The first is the waveguide with the sampled grating, and the second is a Fabry-Perot semiconductor laser. The laser cavity and feedback properties such as the mode spacing of the laser and the both facet reflectivities were taken directly from the experiment [52].

Fig. 23(a) shows the simulated spectrum in the two-color operation regime [52]. The two peaks correspond to two frequencies enforced by the sampled-grating feedback. The almost equal height of the two peaks means roughly equal output powers of the two lasing modes. Simulation shows that their relative intensities fluctuate only slightly on a nano-second time scale. The mutual coherence of the two lasing modes can be seen as a beat signal in the output intensity shown in Fig. 23(b). The depth of the modulation is almost 100% indicating both equalized power and high mutual coherence of the lasing modes. Long-time simulation showed that the mutual coherence is stable and exhibits only small fluctuations [52], indicating that the simulation results indicate that the two-color operation is stable and produces mutually highly coherent modes suitable for THz beat signal generation.

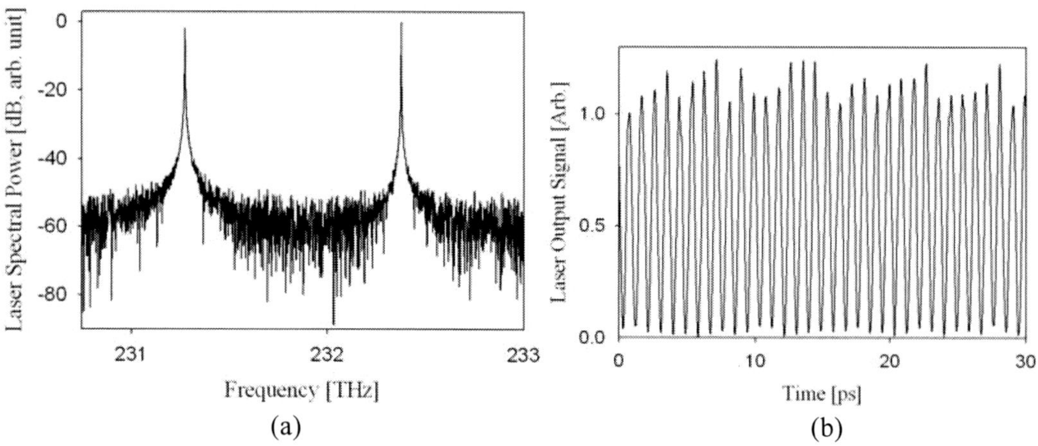

Figure 23. Simulation results of the dual-wavelength ECL with a sampled grating with Λ_S= ~100μm. (a) Optical spectrum and (b) Laser output power exhibiting deep modulation (f_b=~1THz) [52].

4.4. Hybrid Integrated External Cavity Lasers without Mode Hopping

WDM passive optical networks (PON) are a critical solution for ultra speed access networks because of a number of advantages such as large capacity, network security, protocol transparency and upgradeability. However, the WDM-PON is expensive to implement yet because simply the WDM sources are costly. To build up a cost-effective multichannel WDM system, a hybrid integrated multi-wavelength external cavity laser (MWECL) has recently been proposed [56, 57]. Compared with other types of MWECL, the hybrid integrated MWECL has high wavelength stability because the temperature dependence of a Bragg grating written in the PLC is small. Fig. 24 shows a schematic configuration of a typical hybrid integrated MWECL [56, 57]. This device consists of a tuning heater, high-order reflection gratings, AWG and multichannel ECLs with hybrid integrated superluminescent diodes (SLDs) on the PLC platform.

The high order PLC grating was formed with the wavelength separation corresponding to that of the multichannel AWG [56, 57]. Although the gratings can be written by the phase mask method, they were patterned by the photolithography process. The depth and the length of the grating were 1μm and 4mm, respectively. Each SLD was precisely mounted on the terraced silicon by flip-chip bond and the passive alignment technique using marks formed on both the SLD and the terraced silicon platform. The length of the cavity including the SLDs and grating region for the device [57] was 6.345mm and the longitudinal mode spacing was 0.11nm. A tuning heater was formed on the specific grating area to eliminate the mode hopping and to tune the wavelength of each channel. The experimental study on the device shows that one can not only eliminate the mode hopping, which often occurs during the modulation, but also control the lasing wavelengths to the ITU grid within ±0.02nm accurately by tuning the heater [57].

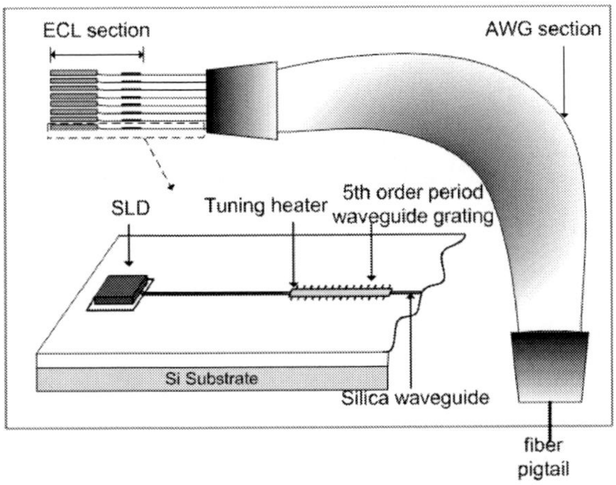

Figure 24. Schematic configuration of a hybrid integrated multichannel ECL [56, 57].

Since the mode hopping makes it difficult to fine-tune the lasing wavelength and improve the bit error rate during transmission, the mode hopping problem has been studied by Tanaka et al [37]. The mode hopping is caused by the difference between the thermooptic coefficients of the LD and the PLC waveguide. Therefore, they proposed a method of eliminating the temperature dependent mode hopping by compensating for the difference between the thermooptic coefficients with silicone. The configuration of the hybrid integrated ECL without mode hopping is shown in Fig. 25.

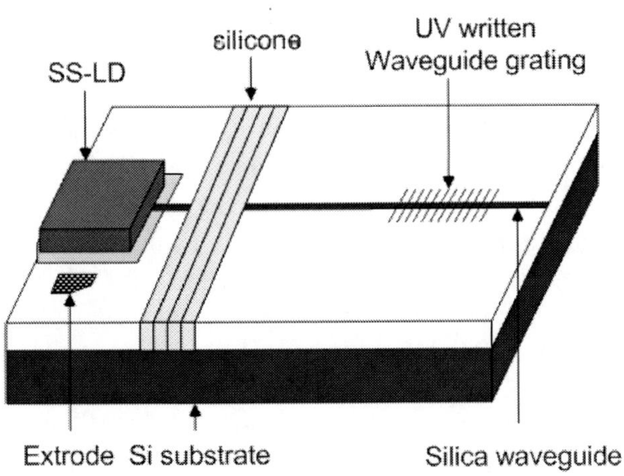

Figure 25. Configuration of the hybrid integrated ECL without mode hopping [37].

To fabricate the ECL, a Si substrate was first etched, except for the area designated as the Si terrace. Then, an undercladding layer, a core layer and an overcladding layer were deposited by the flame hydrolysis deposition. Next, a silica layer was etched by RIE to obtain the Si terrace and grooves for the silicone. The waveguide grating was written by illuminating an ArF excimer laser (λ=193nm) through a phase mask. Finally, a SSC-LD on the Si terrace was aligned with the silica waveguide, and the liquid state silicone was poured into the grooves to finish the device [37]. This hybrid integrated ECL exhibited no mode hopping from 18 °C to 56 °C and controlled the optical power from 18 °C to 70 °C with APC by changing the temperature.

5. Highly Polarized External Cavity Lasers Hybrid Integrated on PLC Platform

Polarization becomes increasingly important in many optical systems. They include optical communication, optical sensing and measurement systems and optical encryption system. In many cases, the light sources highly polarized to certain state of polarization play a critical role for the success of the systems. But, the majority of the polarized light sources developed to date are bulky and costly. In the following sections, we will see some hybrid integrated highly polarized ECLs [54, 58] that are compact and cost-effective.

5.1. Highly Linear-Polarized External Cavity Lasers

A stable linear-polarized output beam is required in many fields including optical communication, optical sensing, spectroscopy and second-harmonic generation. Usually, highly linear-polarized laser diodes have been realized by placing a polarizer after the light source. However, this has made the laser systems costly and complex. For the reason, cost-effective, compact and highly polarized ECLs (hybrid integrated on PLC) have been developed [54, 58], which will be discussed below.

5.1.1. Design and Fabrication

An in-fiber polarizer [59] has often been formed in silica fiber by simply writing a tilted fiber Bragg grating [60], whose grating plane tilted at ~45°. Similarly, one can form an in-PLC polarizer [54] by writing a polarization sensitive tilted grating in the PLC at $\theta_B = tan^{-1}(n_2 / n_1)$ as shown in Fig. 26, where n_1 is the refractive index of the PLC waveguide and n_2 is the refractive index increased during the grating growth process. The principle of the in-PLC polarizer can be explained by the polarization-sensitive mode coupling between core mode and radiation modes in waveguide [59, 60]. Also, one can explain it simply by viewing the tilted Bragg grating as a stack of thin Brewster windows. Fig. 26(a) displays the principle of the operation of an hybrid integrated ECL consisting of an FP-LD, a Bragg grating as an external mirror and a tilted grating as the in-PLC polarizer formed (by a series of grating planes orientated at $\theta_B \approx 45$) in the laser cavity [54]. During the multiple round trip of light between the FP-LD and the Bragg grating, a part of the TM mode reflects from every tilted grating plane and loses some of its energy, whereas the TE mode passes through the tilted grating with no loss. Therefore, by integrating a tilted Bragg grating within the laser cavity, one can achieve a highly linear-polarized light. The degree of linear polarization of the hybrid integrated ECL can be well described by the polarization extinction ratio defined as $PER = 10 \cdot log(P_{max} / P_{min})$, where, P_{max} and P_{min} are the maximum and minimum powers transmitted through a linear polarizer as it is rotated all around the polarization axes. Fig. 26(b) shows the proposed configuration of a highly linear-polarized ECL integrated on PLC, in which a FP-LD, a Bragg grating and an in-PLC polarizer are hybrid integrated on a PLC platform [54]. The PLC used for the ECL had a refractive index difference of 0.75% between the core (refractive index is 1.456) and cladding, the core size was 6.5×6.5μm^2, and the thickness of the PLC cladding was 40μm [54]. The chip size was 21×3mm^2. The Ge concentrations of the core and cladding were 13.5 wt% and 2.1 wt%, respectively, as mentioned in section 3.1.

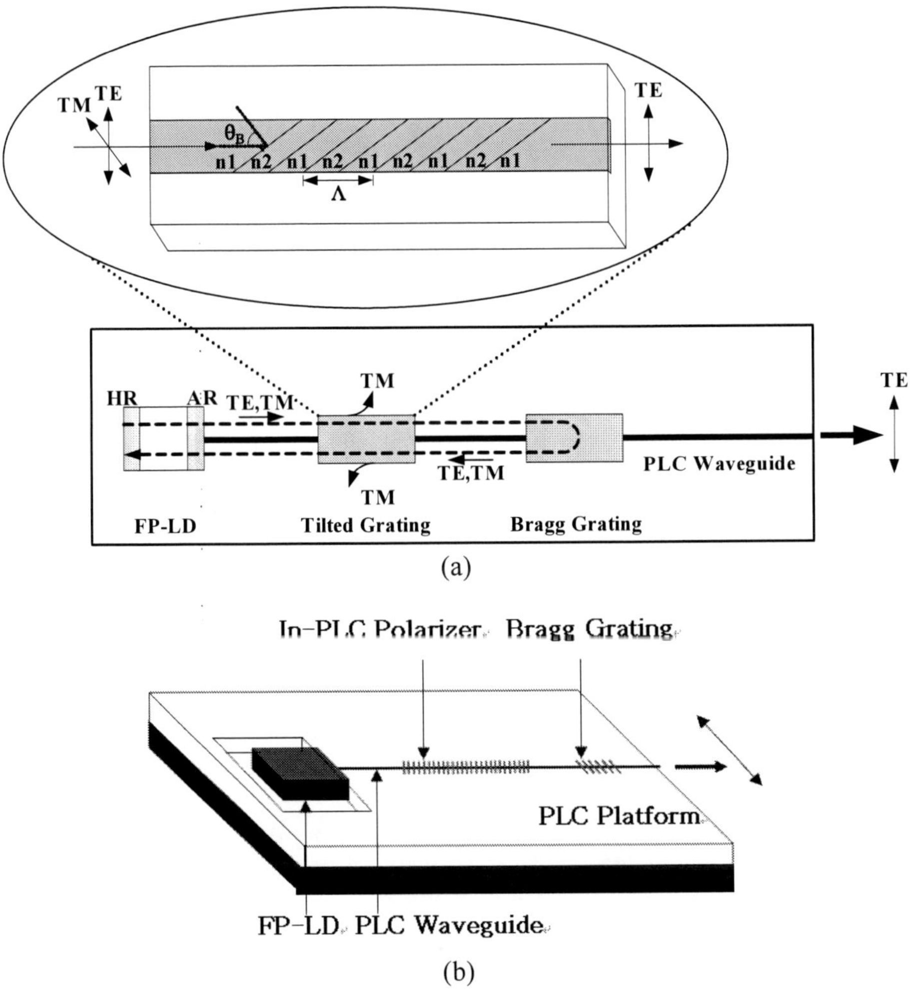

Figure 26. (a) Principle diagram of a hybrid integrated highly linear-polarized ECL with a tilted grating as an in-PLC polarizer (b) Configuration of the hybrid integrated highly polarized ECL consisting of a FP-LD, an in-PLC polarizer and a Bragg grating [54].

The FP-LD was the SSC-LD used in references [14, 40, 61]. The length of the FP-LD was 600μm. The Bragg grating was formed by exposing the PLC waveguide to a KrF excimer laser beam (λ = 248nm) through a cylindrical lens and a phase mask. The Ge-doped silica waveguide was hydrogenated at room temperature to enhance the photosensitivity as described in section 2.2. A 2mm-long grating with a pitch Λ of 450.91nm was written by the laser beam of fluence of 680mJ/cm^2 at the repetition rate of 5Hz [54]. The in-PLC polarizer was also fabricated within the cavity between the FP-LD and the Bragg grating to improve the PER of the PLC-based ECL by writing a grating tilted at ~45° in the core of the PLC waveguide. The phase mask with a period of 1066.3nm and a size of 10×10mm was rotated such that the tilt angle of the tilted grating is aligned to θ_B. The PER of the tilted grating depends on the grating length (*i.e.*, the number of grating planes) and the grating strength [62]. The length of the tilted grating was controlled by using a shutter placed between the KrF excimer laser and the phase mask. The strength of the tilted grating can be adjusted by controlling the exposure time of the excimer laser beam, because the strength is proportional to the UV exposure time [14]. The in-PLC polarizers with different lengths were integrated with the laser beam of fluence of 680mJ/cm^2 under two different exposure times (230s and 900s) to develop the hybrid integrated highly linear-polarized ECLs.

5.1.2. Experimental Results

Fig. 27 displays the oscillation spectrum of the hybrid integrated ECL. This ECL consists of a uniform Bragg grating as an external reflector, a polarization-sensitive tilted grating as an in-line polarizer and a FP-LD exhibiting multimode oscillation with mode spacing of 0.56nm and center wavelength of 1305nm. The transmission spectrum of the uniform Bragg grating after a 90s-irradiation was obtained as shown in the inset in Fig. 27. The center wavelength of the Bragg grating was ~1310nm, the reflectivity was 60% and the FWHM was 0.44nm. The FP-LD oscillating at multimodes becomes stabilized at a single-longitudinal mode with SMSR of 50dB after its hybrid integration on the PLC platform. The oscillation wavelength was 1310nm, which was consistent with the Bragg wavelength of the Bragg grating.

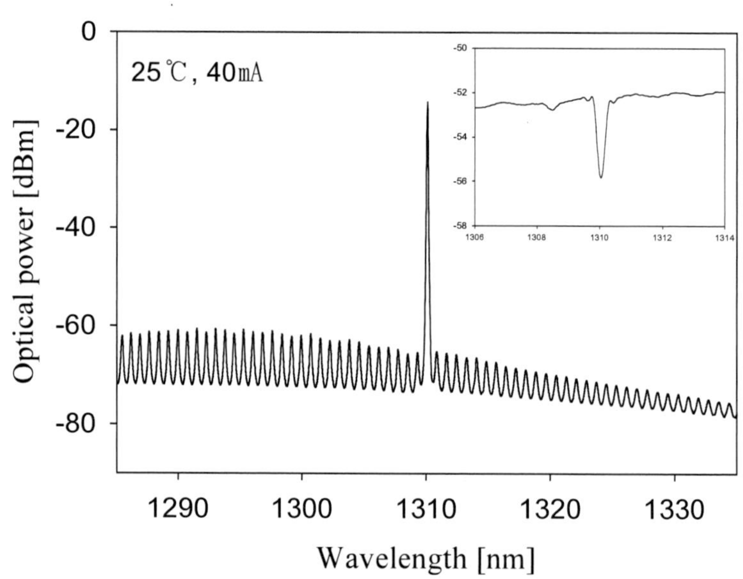

Figure 27. Oscillation spectrum of the highly linear-polarized ECL measured at 40mA. Inset shows the transmission spectrum of the PLC Bragg grating as an external mirror [54].

The PERs of the hybrid integrated ECL are plotted in Fig. 28 as a function of the length of the 45°-tilted grating (*i.e.*, in-PLC polarizer) for two different writing conditions. The tilted Bragg gratings were written during the UV exposure of excimer laser beam with the fluence of 680mJ/cm^2 at the repetition rate of 5Hz. The open and closed circles in Fig. 28 represent the PER of the hybrid integrated ECL for the UV exposure times of 230s and 900s, respectively. The arrow indicates the level of the PER of the ECL with no in-PLC polarizer, which is 21.5dB. The PER increased from 21.5dB to 23.48dB and to 24.98dB by writing 7mm-long and 14mm-long tilted gratings, respectively, in the PLC waveguide with the UV exposure of 230s [54]. The PER further increased from 21.5dB to 25.9dB and to 30dB with 7mm-long and 14mm-long tilted gratings, respectively, for the UV exposure of 900s. The solid and short-dashed lines are the linear lines fitted to the PER values for the exposure time of 230s and 900s, respectively. Fig. 28 shows that the PER increases roughly linearly to the tilted grating length and to the exposure time. This indicates that one can improve the degree of polarization of the PLC-based ECL by increasing the polarizer length and the UV exposure time. The hybrid integrated ECL exhibited highly linear polarization with a PER value as high as 30dB, with a 14mm-long in-PLC polarizer and an UV exposure time of 900s [54].

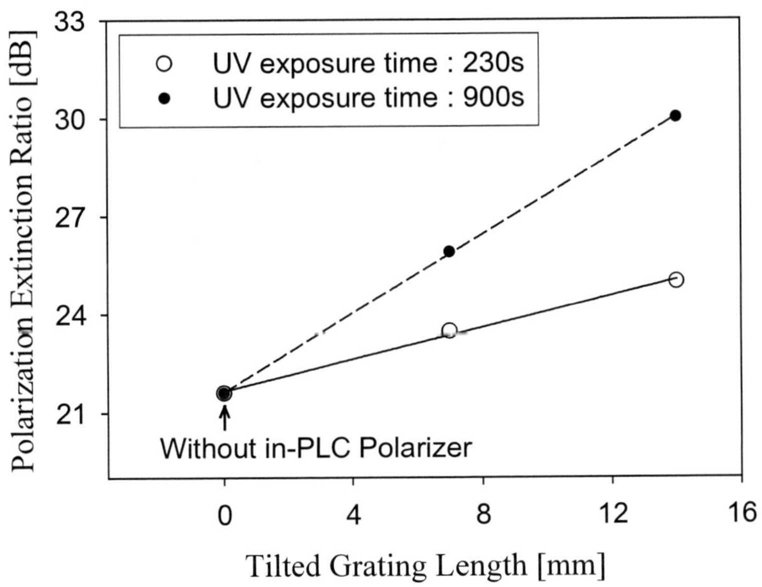

Figure 28. Polarization extinction ratios of the highly linear-polarized ECL as a function of the length of the tilted grating for two different exposure times [54].

5.2. Circularly Polarized External Cavity Laser Hybrid Integrated on PLC Platform

In some optical systems such as optical sensing, optical measurement and optical encryption systems [63], a circularly polarized laser source is very important. One of the most promising candidates for the circularly polarized source is the ECL hybrid integrated on PLC platform because it is compact, cost-effective and exhibits high SMSR. The circularly polarized ECL is hybrid integrated on the PLC platform with a FP-LD, a polarization sensitive filter, Bragg grating and a wave plate as shown in Fig. 29.

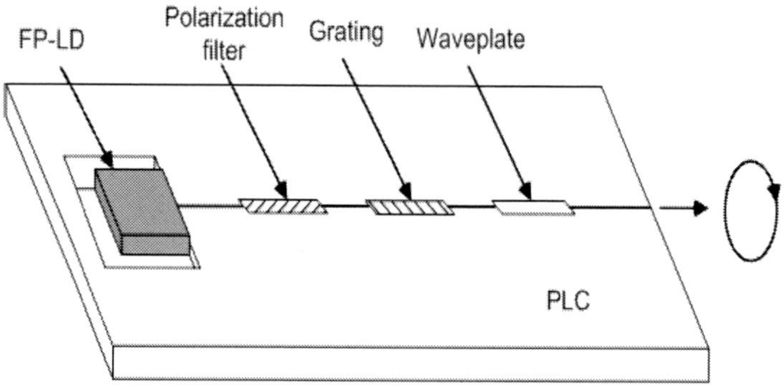

Figure 29. Configuration of a hybrid integrated polarized ECL.

The polarization sensitive filter as the in-PLC polarizer and the Bragg grating in Fig. 29 can be formed as described in section 5.1. Meanwhile, the waveplate can be formed by creating a photoinduced birefringence [64, 65] in PLC waveguide, just as the fiber waveplate can be fabricated by exposing it to 242nm radiation [59]. Although it has advantage to integrate the waveplate in PLC waveguide, it makes the device longer and complex. For the reason, a polyimide quarter-wave plate can be instead hybrid integrated to complete the hybrid integrated circularly polarized ECL [58]. We will discuss the circularly polarized ECL hybrid integrated with a polyimide quarter-wave plate on PLC in this section.

5.2.1. Design and Fabrication

The waveplate should be aligned such that its axes are oriented at ±45° with respect to the polarization direction of the input light. If the waveplate is a quarter waveplate, the state of polarization (SOP) of the laser output becomes a circular-polarized state. However, when the axes of the waveplate are not oriented at ±45°, the laser combined with the wave plate produces an elliptically polarized light. Fig. 30 shows the configuration of a circular-polarized ECL hybrid integrated on the PLC platform, on which a FP-LD, a Bragg grating as an external reflector and a polyimide quarter-wave plate are hybrid integrated [58]. The FP-LD is predominantly linear-polarized [54] as seen in section 5.1. Thus, the hybrid integrated ECL should emit a highly circular-polarized laser beam, provided that the wave plate is rotated properly around the plane perpendicular to the direction of the laser beam such that the principal axis is oriented at ±45°.

The PLC waveguide used for this circularly polarized hybrid integrated ECL is the same PLC waveguide used for the linear-polarized ECL discussed in section 5.1. Thus, it has the refractive index difference of 0.75%, and the Ge concentrations of the core and cladding are 13.5 and 2.1wt%, respectively. The size of the PLC platform is $21\times3mm^2$. The core size is $6.5\times6.5\mu m^2$ and the thickness of the PLC cladding is 40μm. The birefringence of planar waveguide is about 1×10^{-5}. The FP-LD is also the SSC-LD and exhibits multiple oscillations with mode spacing of 0.6nm and a center wavelength of 1550nm. The Bragg grating was also written by exposing the hydrogenated planar waveguide to a KrF excimer laser beam (λ = 248nm) through a phase mask. The UV light beam from a KrF excimer laser was focused with a cylindrical lens through a phase mask to the PLC waveguide. A 2-mm-long grating with a pitch Λ of 533.15nm was written by the laser beam of fluence of 680mJ/cm^2 during a 90-second irradiation [58].

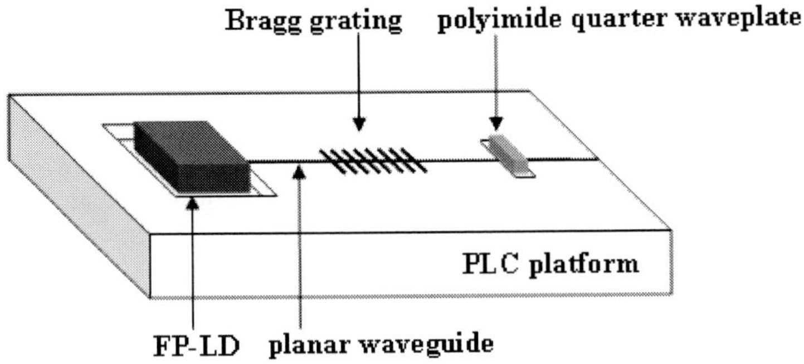

Figure 30. Configuration of a circularly polarized ECL hybrid integrated on the PLC platform [58].

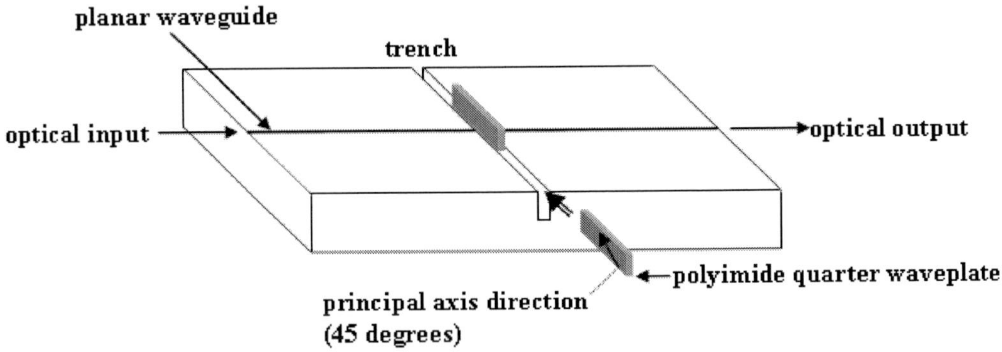

(a)

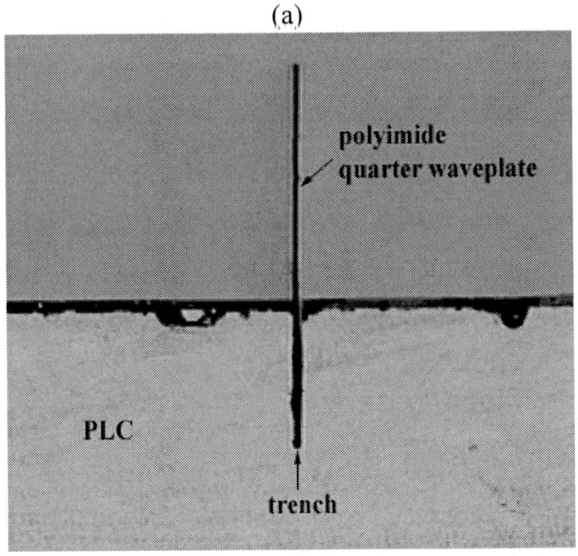

(b)

Figure 31. Polyimide quarter-wave plate part in the circularly polarized ECL: (a) schematic view of wave plate integration, and (b) cross sectional photograph [58].

The polyimide quarter-wave plate from the NTT-AT Corporation was specially designed for an easy insertion with low insertion loss in the PLC during the hybrid-integration process. The thickness and size of the wave plate were 8.8μm and $5\times1mm^2$, respectively. Fig. 31(a) and (b) shows the schematic view of integration and a cross sectional photograph of the polyimide wave plate, respectively [58]. The trench for inserting the polyimide wave plate was formed by a sawing machine with a diamond blade. Then, the polyimide wave plate was inserted in the trench with a width of 20μm and a depth of 300μm. One of the principal axes of the wave plate was set at θ=45° with respect to the polarization direction of an incident light, and then fixed in the trench with a UV-curable index-matching adhesive. The insertion loss of the trench was ~0.5dB. The cross sectional photograph of an integrated wave plate taken by an optical microscope is displayed in Fig. 31(b).

5.2.2. Experimental Results and Discussion

The oscillation spectrum of the circularly polarized ECL is shown in Fig. 32. It was measured at 25°C and 25mA with an optical spectrum analyzer with a resolution of 0.05nm. The center wavelength and spectral width of the ECL are 1550.04nm and ~0.05nm, respectively. Note that the FP-LD exhibiting multi-longitudinal modes become stabilized at a single-longitudinal mode with high SMSR of ~51dB after its hybrid integration on the PLC platform [58]. The oscillation wavelength was consistent with the Bragg wavelength of the grating. The center wavelength of the Bragg grating was 1550.08nm as shown in the inset in Fig. 32. The reflectivity of the Bragg grating was greater than 90%.

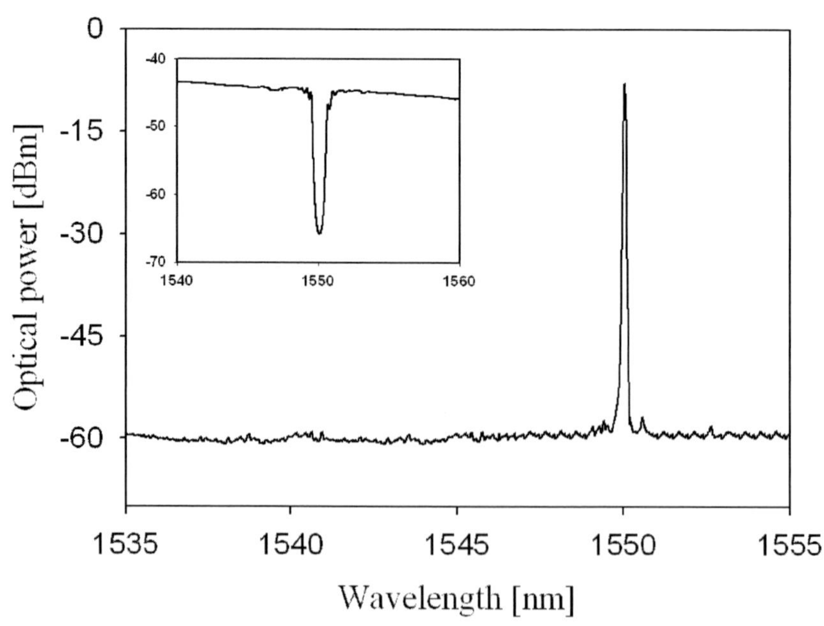

Figure 32. Oscillation spectrum of the circularly polarized ECL. Inset shows the transmission spectrum of the Bragg grating [58].

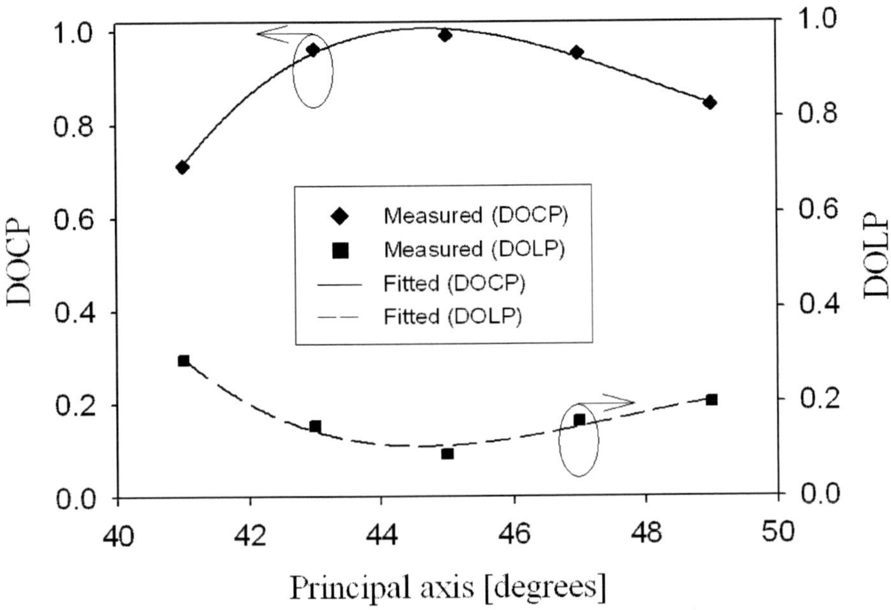

Figure 33. DOCP and DOLP of the ECLs as a function of the angle θ [58].

The polarization state of the light coming out of a light source can be described by the Stokes parameters (S_0, S_1, S_2, and S_3). The Stokes vector S can be measured by a polarization analyzer. The degree of polarization (DOP) is then calculated by DOP=$(S_1^2+ S_2^2+ S_3^2)^{1/2}/S_0$. The degree of circular polarization (DOCP) and the degree of linear polarization (DOLP) can also be determined by DOCP=$|S_3|/S_0$ and DOLP=$(S_1^2+S_2^2)^{1/2}/S_0$. Fig. 33 shows the measured DOCP and DOLP of the ECLs as a function of θ, the direction of the principal axis of the quarter-wave plate with respect to the polarization direction of an incident light, along with the fitted curves [64]. The diamond- and square-marks indicate the values of the DOCP and DOLP measured at the different angles θ. The angles θ were varied from 41° to 49° to measure the values of the DOCP and DOLP. When the polyimide wave plate was rotated such that θ=45°, the value of DOCP reached to maximum (~0.992) meanwhile that of the DOLP became minimum (~0.09). Note that the DOCP of the circularly polarized ECL drops as θ deviates from the 45°. This verifies that the ECL hybrid integrated with a polyimide quarter-wave plate with θ=45° becomes highly circular-polarized. The DOCP and DOLP of the ECL without the polyimide wave plate were measured to be 0.14 and 0.97, respectively. This result indicates that the hybrid integration of the wave plate along with the grating in the PLC waveguide improves substantially the DOCP of the ECL. The length of the waveguide after the quarter-wave plate was 2mm. The corresponding phase retardation over that piece in the waveguide was about 4.645°. However, the computed value of the DOCP for the circularly polarized light after experiencing an additional phase retardation of 4.645° remained to be good (~0.996). But one has to keep in mind in designing this type of hybrid integrated circular-polarized ECL such that the birefringence of the waveguide is small and the length of the waveguide after the quarter-wave plate is short [58]. The emitted beam of the hybrid integrated ECL with the polyimide wave plate at ±45° was highly circular-polarized with a DOCP of 0.992. The circularly polarized ECL also exhibited high SMSR >50dB at ~1550nm.

6. Conclusion

We have reviewed the well-developed grating technology and the hybrid integration technology. Then, we have discussed some types of gratings grown in the silica PLC waveguide by the grating technology. We also have presented several types of ECLs integrated on the PLC-platform using the hybrid integration technology in this chapter. Although some hybrid integrated ECLs have already been demonstrated by a few researchers as discussed in this chapter, numerous other types of the hybrid integrated ECLs need to be more explored. This is because it is compact, cost-effective, simple and flexible to select the oscillation wavelength and polarization properties of the ECL. It also exhibits high SMSR. We strongly encourage researchers to try many other kinds of gratings to integrate in the PLC waveguide for the demonstration of noble hybrid integrated ECLs.

Acknowledgements

The authors thank Dr. Ryun K. Kim and Mr. Byung C. Min of the Sungkyunkwan University for assisting the preparation of this chapter.

References

[1] K. O. Hill, Y. Fujii, D. C. Johnson, and B. S. Kawasaki, "Photosensitivity in optical waveguides: application to reflection filter fabrication," *Appl. Phys. Lett.* vol.32, pp.647-649, 1978.

[2] K. O. Hill, and G. Meltz, "Fiber Bragg grating technology fundamental and overview," *J. Lightw. Technol.*, vol.15, pp.1263-1276, 1997.

[3] R. Kashyap, *Fiber Bragg gratings* (Academic press, San Diego, 1999).

[4] Othonos, "Fiber Bragg gratings," *Rev. Sci. Instrum.* vol.68, No.12, pp.4309-4341, 1997.

[5] Othonos, and K. Kalli, *Fiber Bragg gratings: fundamentals and applications in telecommunications and sensing*, (Artech House, 1999).

[6] T. Erdogan, "Fiber grating spectra," *J. Lightw. Technol.*, vol.15, pp.1277-1294, 1997.

[7] G. Meltz, W. W. Morey, and W. H. Glenn, "Formation of Bragg gratings in optical fiber by a transverse holographic method," *Optics Lett.*, vol.14, pp.823-825, 1989.

[8] K. O. Hill, B. Malo, F. Bilodeau, D. C. Johnson, and J. Albert, "Bragg gratings fabricated in monomode photosensitive optical fiber by UV exposure through a phase mask," *Appl. Phys. Lett.*, vol.62, pp.1035-1037, 1993.

[9] T. A. Strasser and T. Erdogan, "Fiber grating devices in high-performance optical communications systems," *Optical fiber telecommunications IV A components*, Ed. I. P. Kaminow, and T. Li, Academic Press, 2002.

[10] P. J. Lemaire, R. M. Atkins, V. Mizrahi, and W. A. Reed, "High pressure H_2 loading as a technique for achieving ultrahigh UV photosensitivity and thermal sensitivity in GeO_2 doped optical fibers," *Electron. Lett.*, vol. 29, pp.1191-1193, 1993.

[11] J. Crank, *The mathematics of diffusion*, (Oxford University Press, 1975).

[12] J. H. Song, K. S. Lee, and J. W. Lee, "Investigation of enhanced photosensitivity of silica based fibers by hydrogenation," *Proc. OECC2001*, pp.419-420, Sydney, Australia, 2001.

[13] J. H. Song and K. S. Lee, "The hydrogen loading method for FBG fabrication," *J. of the Korea Institute of Electronic Engineers*, vol.37-D, no.7, pp.541-547, 2000, in Korean.

[14] J. H. Lim, G. Lim, K. S. Lee, J. H. Song, Y. Oh, S. T. Jung, and T. I. Kim, "Investigation of grating growth characteristics in planar lightwave circuits for external cavity laser," *Fiber and Integrated Optics*, vol.24, pp.73-82, 2005.

[15] H. Patrick, and S. L. Gilbert, "Growth of Bragg gratings produced by continuous-wave ultraviolet light in optical fiber," *Opt. Lett.* vol.18, pp.1484–1486, 1993.

[16] D. Z. Anderson, V. Mizrahi, T. Erdogan, and A. E. White, "Production of in-fibre gratings using a diffractive optical element," *Electron. Lett.* vol. 29, pp.566–568, 1993.

[17] Y. P. Li, and C. H. Henry, "Silica-base optical integrated circuits," *IEE Proc. Optoelectron.*, vol.143, pp.263-280, 1996.

[18] K. Kato, and Y. Tohmori, "PLC hybrid integration technology and its application to photonic components," *IEEE J. Select. Topics Quantum Electron.*, vol.6, pp.4-13, 2000.

[19] Himeno, K. Kato and T. Miya, "Silica-based planar lightwave circuits," *IEEE J. Sel. Topics Quantum Electron.*, vol.4, no.6, pp. 270–276, Jun. 1996.

[20] K. Okamoto, *Fundamentals of optical waveguides* (Academic press, San Diego, 2000).

[21] Mukherjee, "WDM optical communication networks: progress and challenges," *IEEE J. Select. Areas in Commun.*, vol.18, pp.810-1834, 2000.

[22] T. Koonen, "Fiber to the home/fiber to the premises: what, where, and when?" *Proceedings of the IEEE,* Vol.94, No.5, pp.911-934, 2006

[23] S.-J. Park, C.-H. Lee, K.-T. Jeong, H.-J. Park, J.-G. Ahn, and K.-H. Song, "Fiber-to-the-home services based on wavelength-division-multiplexing passive optical network," *J. Lightw. Technol.*, vol.22, pp.2582-2591, 2004.

[24] J. H. Song, K.-Y. Kim, J. Cho, D. Han, J. Lee, Y. S. Lee, S. Jung, Y. Oh, D.-H. Jang, and K. S. Lee, "Thin film filter-embedded triplexing-filters based on directional couplers for FTTH networks," *IEEE Photon. Technol. Lett.*, vol.17, pp.1668-1670, 2005.

[25] J. H. Song, J. H. Lim, R. K. Kim, K. S. Lee, K.-Y. Kim, J. Cho, D. Han, S. Jung, Y. Oh, and D.-H. Jang, "Bragg grating-assisted WDM filter for integrated optical triplexer transceivers," *IEEE Photon. Technol. Lett.*, vol.17, no.12, pp.2607-2609, Dec. 2005.

[26] J. H. Song, K. S. Lee, and Y. Oh, "Triple wavelength demultiplexers for low-cost optical triplexer transceivers," *J. Lightw. Technol.*, vol. 25, pp. 350-358, 2007.

[27] J. Cho, D. Han, J. H. Song, and S. Jung, "Crosstalk enhancement of AWG fabricated by flame hydrolysis deposition method," *IEEE Photon. Technol. Lett.*, vol.17, no.11, pp.2328-2390, 2005.

[28] L. Wosinski, "Technology for photonic components in silica/silicon material structure," Ph.D thesis, Royal Institute of Technology (KTH), Stockholm, Sweden, 2003.

[29] Y. Yamada, A. Takagi, I. Ogawa, M. Kawachi and M. Kobayashi, "Silica-based optical waveguide on terraced silicon substrate as hybrid integration platform," *Electron. Lett.*, vol.29, pp.444-446, 1993.

[30] J. H. Song, "Optical WDM filters for hybrid integrated triplexer transceivers," Ph.D thesis, Sungkyunkwan Univ. Suwon, Korea, 2006.

[31] Y. Nakasuga, T. Hashimoto, Y. Yamada, H. Terui, M. Yanagisawa, K. Moriwaki, Y. Akahori, Y. Tohmori, K. Kato, S. Sekine, and M. Horiguchi, "Multi-chip hybrid integration on PLC platform using alignment technique," *Proc. IEEE Electronic Components and Technology Conference 1996*, pp.20-25, 1996.

[32] S. Yang, H. Kang, B. Jeon, D. Rhee,Y. Kim, E. Lee, A. Choo, J. Burm, and T. Kim, "A reflecting mirror facet (RMF) photodiode suitable for 2-dimensional optical package," *Proc. ECOC2003*, Rimini, Italy, 2003.

[33] R. K. Kim, "Hybrid integrated external cavity lasers on planer lightwave circuits," Ph.D thesis, Sungkyunkwan Univ. Suwon, Korea, 2008.

[34] Moeyersoon, G. Morthier, and M. Zhao, "Degradation of the mode suppression in single-mode laser diodes due to integrated optical amplifiers," *IEEE J. Quantum Electron.*, vol. 40, no. 3, pp. 241-244, 2004.

[35] L. P. Barry, and P. Anandarajah, "Cross-channel interference due to mode partition noise in WDM optical systems using self-seeded gain-switched pulse sources," *IEEE Photon. Technol. Lett.*, vol. 13, no. 3, pp. 242-244, 2001.

[36] H. Bissessur, C. Caraglia, B. Thedrez, J.–M. Rainsant, and I. Riant, "Wavelength-versatile external fiber grating lasers for 2.5-Gb/s WDM networks," *IEEE Photon. Technol. Lett.*, vol. 11, no. 10, pp. 1304-1306, 1999.

[37] T. Tanaka, Y. Hibino, T. Hashimoto, R. Kasahara, M. Abe, and Y. Tohmori, "Hybrid-integrated external-cavity laser without temperature-dependent mode hopping," *J. Lightw. Technol.*, vol. 20, no. 9, pp. 1730-1739, 2002.

[38] T. Sato, F. Yamamoto, K. Tsuji, H. Takesue, and T. Horiguchi, "An uncooled external cavity diode laser for coarse-WDM access network systems," *IEEE Photon. Technol. Lett.*, vol. 14, no. 7, pp. 1001-1003, 2002.

[39] R. C. Alferness, U. Koren, L. L. Buhl, B.I. Miller, M.G. Young, T. L. Koch, G. Raybon, and C. A. Burrus, "Broadly tunable InGaAsP/InP laser based on a vertical coupler filter with 57-nm tuning range." *Appl. Phys. Lett.*, vol. 60, pp.3209-3211, 1992.

[40] J. H. Lim, J. H. Song, R. K. Kim, K. S. Lee, and J. R. Kim, "External cavity laser with high-sidemode suppression ratio using grating-assisted directional coupler," *IEEE Photon. Technol. Lett.*, vol. 17, no. 11, pp.2430-2432, 2005.

[41] S. Li, K. S. Chiang, W. A. Gambiling, Y. Liu, L. Zhang, and I. Bennion, "Self-seeding of Fabry-Perot laser diode for generating wavelength-tunable chirp-compensated single-mode pulses with high-sidemode suppression Ratio," *IEEE Photon. Technol. Lett.*, vol. 12, no. 11, pp. 1441-1443, 2000.

[42] S. S. Orlov, A. Yariv, and S. V. Essen, "Coupled-mode analysis of fiber-optic add-drop filters for dense wavelength-division multiplexing," *Opt. Lett.*, vol. 22, no. 10, pp. 688-690, 1997.

[43] Riziotis, and M. N. Zervas, "Novel full-cycle-coupler-based optical add-drop multiplexer and performance characteristics at 40-Gb/s WDM networks," *J. Lightw. Technol.*, vol. 21, no. 8, pp. 1828-1837, 2003.

[44] M. Matus, M. Kolesik, J. Moloney, M. Hofmann, S. Koch, "Dynamics of two-color laser systems with spectrally filtered feedback," *J. Opt. Soc. of Am. B,* vol.21, pp.1758-1771, 2004.

[45] C.-L. Wang and C.-L. Pan, "Tunable multiterahertz beat signal generation from a two-wavelength laser-diode array," *Opt. Lett.* vol.20, pp.1292–1294, 1995.

[46] M. Tani, P. Gu, M. Hyodo, K. Sakai, T. Hidaka, "Generation of coherent terahertz radiation by photomixing of dual-mode lasers," *Opt. and Quantum Electron.* vol.32, pp.503-520, 2000.

[47] P. Pellandini, R. Stanley, R. Houdre, U. Oesterle, M. Ilegems, C. Weisbuch, "Dual-wavelength laser emission from a coupled semiconductor microcavity," *Appl. Phys. Lett.* vol.71, pp.864-866, 1997.

[48] M. Brunner, K. Gulden, R. Hovel, M. Moser, J. Carlin, R. Stanley, M. Ilegens, "Continuous-wave dual-wavelength lasing in a two-section vertical-cavity laser," *IEEE Photon. Technol. Lett.* vol.12, pp.1316-1318, 2000.

[49] W. Wang, M. Cada, J. Seregelyi, S. Paquet, S. Mihailov, P. Lu, "A beat-frequency tunable dual-mode fiber-Bragg-grating external-cavity laser," *IEEE Photon. Technol. Lett.* vol.17, pp.2436-2438, 2005.

[50] Tilford, "Analytical procedure for determining lengths from fractional fringes," *Appl. Opt*. vol.16, pp.1857- , 1977.

[51] Hu, M. Nuss, "Imaging with terahertz waves," *Optics Lett*. vol.20, pp.1716-1718, 1995.

[52] Kyung S. Lee, C. S. Kim, R. K. Kim, G. Patterson, M. Kolesik, J. Moloney, N. Peyghambarian, "Dual-wavelength external cavity laser with a sampled grating formed in a silica PLC waveguide for Terahertz beat signal generation," *Appl. Phys. B*, vol.87, pp. 293-296, 2007.

[53] H. Helmers, O. Durand, G.. Duan, E. Gohin, J. Landreau, J. Jacquet, I. Riant, "45nm tunability in C-band obtained with external cavity laser including sampled fiber Bragg grating," *Electron. Lett.*, vol.38, no.24, pp1535-1536, 2002.

[54] R. K. Kim, J. H. Lim, J. H. Song, K. S. Lee, "Highly linear-polarized external cavity lasers hybrid integrated on planar lightwave circuit platform," *IEEE Photon. Technol. Lett.*, vol.18, pp.580-582, 2006.

[55] M. Kolesik, J. V. Moloney, "A spatial digital filter method for broadband simulation of semiconductor lasers," *IEEE J. Quantum Electron.* vol.37, pp.936-944, 2001.

[56] S. Oh, J. Shin, Y. Park, S. Kim, S. Park, H. Sung, Y. Baek and K. Oh, "Multiwavelength lasers for WDM-PON optical line terminal source by silica planar lightwave circuit hybrid integration," *IEEE Photon. Technol. Lett.*, vol.19, pp.1622-1624, 2007.

[57] S. Oh, J. Shin, Y. Park, S. Park, K. Kim, S. Kim, H. Sung, Y. Baek and K. Oh, "Wavelength Tuning of Hybrid Integrated Wavelength Lasers Using a Heat," *IEEE Photon. Technol. Lett.*, vol. 20, pp.422-424, 2008

[58] R. K. Kim, J. H. Song, Y. Oh, D. H. Jang, J. R. Kim, Kyung S. Lee, "Circularly polarized external cavity lasers hybrid integrated with a polymide quarter waveplate on planar lightwave circuit," *IEEE Photon. Technol. Lett.*, vol.19, no.14, 2007.

[59] P. Westbrook, T. Strasser and T. Erdogan, "In-line polarizer using blazed fiber gratings," *IEEE Photon. Technol. Lett.*, vol. 12, no.10, 2000.

[60] T. Erdogan, and J. Sipe, "Tilted fiber phase gratings," *J. Opt. Soc. Am. A*, vol. 13, no.2, pp. 296-313, 1996.

[61] M. Itoh, T. Saida, Y. Hida, M. Ishii, Y. Inoue, Y. Hibino, and A. Sugita, "Large reduction of singlemode-fibre coupling loss in 1.5% Δ planar lightwave circuits using spot-size converters," *Electron. Lett.*, vol. 38, no. 2, pp. 72-74, 2002.

[62] Y. Li, M. Froggatt, and T. Erdogan, "Volume current method for analysis of tilted fiber gratings," J. Lightwave Technol., vol. 19, no. 10, pp. 1580-1591, 2001.

[63] K.-B. Cho, C.-M. Shin, D.-H. Seo, S.-G. Park, Y.-H. Doh, and S.-J. Kim, "Optical encryption system based on circular polarization and interferometer," *The 5th Pacific Rim Conference on Lasers and Electro-Optics, CLEO/Pacific Rim*, vol.2, pp.425, 2003.

[64] Ouellette, D. Gagnon and M. Poirier, "Permanent photoinduced birefringence in a Ge-doped fiber," *Appl. Phys. Lett.*, vol.58, no.17, pp.1813-1815, 1991.

[65] T. Erdogan and V. Mizrahi, "Characterization of UV-induced birefringence in photosensitive Ge-doped silica optical fibers," *J. Opt. Soc. Am. B*, vol.11, no.10, pp.2100-2105, 1994.

In: Horizons in World Physics, Volume 268
Editors: M. Everett and L. Pedroza, pp. 181-209 ISBN 978-1-60692-861-5

Chapter 5

QUANTUM THEORY ON A GALOIS FIELD: MOTIVATION AND FIRST RESULTS

Felix M. Lev*
Artwork Conversion Software Inc., 1201 Morningside Drive,
Manhattan Beach, CA 90266, USA

Abstract

Systems of free particles in a quantum theory based on a Galois field (GFQT) are discussed in detail. In this approach infinities cannot exist, the cosmological constant problem does not arise and one irreducible representation of the symmetry algebra necessarily describes a particle and its antiparticle simultaneously. As a consequence, well known results of the standard theory (spin-statistics theorem; a particle and its antiparticle have the same masses and spins but opposite charges etc.) can be proved without involving local covariant equations. The spin-statistics theorem is simply a requirement that quantum theory should be based on complex numbers. Some new features of GFQT are as follows: i) elementary particles cannot be neutral; ii) the Dirac vacuum energy problem has a natural solution and the vacuum energy (which in the standard theory is infinite and negative) equals zero as it should be. In the AdS version of the theory there exists a dilemma that either the notion of particles and antiparticles is absolute and then only particles with a half-integer spin can be elementary or the notion is valid only when energies are not asymptotically large and then supersymmetry is possible.

1. Introduction

The goal of the present paper is to describe main ideas of a quantum theory on a Galois field (GFQT) in a simplest possible way. An attempt is made to give clear and simple arguments which hopefully might convince physicists that GFQT is a more natural quantum theory than the standard one. Our observations show that physicists are reluctant to accept GFQT for several reasons and the most significant ones are the following two.

*E-mail address: felixlev@hotmail.com

- Theoretical physicists are typically using a sophisticated mathematics involving integral and differential equations, path integrals, Lie groups and algebras, their representations in Hilbert spaces etc. At the same time, the absolute majority of physicists is not familiar even with the basic notions of Galois fields. Meanwhile, mathematics of Galois fields is not only very elegant but also much simpler than the standard one.

- The modern quantum theory has achieved very impressive success in describing verious experimental data, predicted new particles and even new interactions. The majority of physicists believes that the agreement with the data is much more important than the fact that the theory is not based on a solid mathematical basis. Physicists typically believe that, although future quantum theory might involve essentially new ideas, there is no need to drastically change its mathematical basis. They believe that Galois fields is an exotics which has nothing to do with physics.

We understand that the best way to achieve the above goal is to find phenomena which can be explained by GFQT and cannot be explained in the standard approach. In the present paper we discuss possible candidates for such phenomena. At the same time we believe that nature should be described by an elegant and simple mathematics and that logical arguments are of importance by their own.

The paper is organized as follows. In Sects. 2.-5. we discuss why in our opinion the future quantum theory will be based on the operator formulation involving Galois fields. The discussion involves mainly logical arguments and only very simple mathematical facts are mentioned. The subsequent presentation in Sects. 6.-10. however, is impossible without mentioning specific mathematical results. Although these results require rather extensive calculations, they involve only finite sums in Galois field. For this reason all the results can be reproduced even by readers who previously did not have practice in calculations with Galois fields. We do not describe the intermediate calculations but only final results. The reader interested in technical details can found them in Ref. [1].

2. Galois fields vs. 'Infinite' Mathematics

The standard mathematics used in quantum physics is essentially based on the notion of infinity. Although any realistic calculation involves only a finite number of numbers, one usually believes that *in principle* any calculation can be performed with arbitrary large numbers and with any desired accuracy.

Suppose we wish to experimentally verify whether the addition is commutative, i.e. whether the relation $a + b = b + a$ is true for any numbers a and b. It is obvious that if our Universe is finite and N is the maximum possible number of elementary particles in the Universe then the above relation cannot be verified even in principle if $a + b > N$.

This example shows that if the Universe is finite then the validity of the standard mathematics cannot be verified in principle for sufficiently large numbers. What conclusion should be drawn from this observation? We essentially have the following dilemma. The first possibility is to accept that standard mathematics is nevertheless suitable for describing phenomena with any numbers but not all of the phenomena can occur in our Universe. Another possibility is to assume that there exists a number $\mathcal{N}$ such that no physical quantity can have a value greater than $\mathcal{N}$. In that case mathematics describing physics should obviously involve only numbers not greater than $\mathcal{N}$; in particular such a mathematics does not contain the actual infinity at all.

The above dilemma has a well known historical analogy. A hundred years ago nobody believed that there exists an absolute limit of speed. People did not see any reason which in principle does not allow any particle to have an arbitrary speed. The special theory of relativity did not show that the Newtonian mechanics was wrong: it was correct but only for phenomena where velocities are much less than the velocity of light.

Our second example is as follows. The current world record for computing the number π belongs to Y. Kanada et. al. who in 2002 computed 1.24 trillion digits of π at Tokyo University. This result has been achieved by using the HITACHI supercomputer working for 400 hours. Suppose we wish to know N_1 decimal places of π. One might believe that for any N_1 it is possible to find combinations (N_2, T) such that it is possible in principle to build a computer operating with N_2 bits and it will be capable to compute N_1 digits of π for the time T (e.g. it is possible to build a computer as big as the Solar system which will compute π for thousands years). However, it is clear again that if the Universe is finite, this goal cannot be achieved in principle (if N is the number of elementary particles in the Universe we will have no room for writing down $N + 1$ decimal digits of π).

One might think that division is a natural operation since we know from everyday experience that any macroscopic object can be divided by two, three and even a million parts. But is it possible to divide by two or three the electron or neutrino? It is obvious that if elementary particles exist, then division has only a limited sense. Indeed, let us consider, for example, the gram-molecule of water having the mass 18 grams. It contains the Avogadro number of molecules $6 \cdot 10^{23}$. We can divide this gram-molecule by ten, million, billion, but when we begin to divide by numbers greater than the Avogadro one, the division operation loses its sense. The obvious conclusion is that that the notion of division has a sense only within some limits. The well known fact is that more than two thousand years ago ancient Greeks believed that division should have a limit, and the word 'atom' in Greek means indivisible.

It is well known that mathematics starts from natural numbers (recall the famous Kronecker expression that 'God made the natural numbers, all else is the work of man') which have a clear physical meaning. However only two operations are always possible in the set of natural numbers: addition and multiplication.

In order to make addition reversible, we introduce negative integers -1, -2 etc. Then, instead of the set of natural numbers we can work with the ring of integers where three operations are always possible: addition, subtraction and multiplication. However, the negative numbers do not have a direct physical meaning (we cannot say, for example, 'I have minus two apples'). Their only role is to make addition reversible.

The next step is the transition to the field of rational numbers in which all four operations excepting division by zero are possible. However, as noted above, division has only a limited sense.

In mathematics the notion of linear space is widely used, and such important notions as the basis and dimension are meaningful only if the space is considered over a field or body. Therefore if we start from natural numbers and wish to have a field, then we have to introduce negative and rational numbers. However, if, instead of all natural numbers, we consider only p numbers 0, 1, 2, ... $p-1$ where p is prime (we treat zero as a natural number), then we can easily construct a field without adding any new elements. This construction, called Galois field, contains nothing that could prevent its understanding even by pupils of elementary schools.

Let us denote the set of numbers 0, 1, 2,...$p-1$ as $GF(p)$. Define addition and multiplication as usual but take the final result modulo p. For simplicity, let us consider the case $p=5$. Then F_5 is the set 0, 1, 2, 3, 4. Then $1+2=3$ and $1+3=4$ as usual, but $2+3=0$, $3+4=2$ etc. Analogously, $1\cdot 2=2$, $2\cdot 2=4$, but $2\cdot 3=1$, $3\cdot 4=2$ etc. By definition, the element $y\in GF(p)$ is called opposite to $x\in GF(p)$ and is denoted as $-x$ if $x+y=0$ in GF_p. For example, in $GF(5)$ we have -2=3, -4=1 etc. Analogously $y\in GF(p)$ is called inverse to $x\in GF(p)$ and is denoted as $1/x$ if $xy=1$ in $GF(p)$. For example, in $GF(5)$ we have 1/2=3, 1/4=4 etc. It is easy to see that addition is reversible for any natural $p>0$ but for making multiplication reversible we should choose p to be a prime. Otherwise the product of two nonzero elements may be zero modulo p. If p is chosen to be a prime then indeed $GF(p)$ becomes a field without introducing any new objects (like negative numbers or fractions). For example, in this field each element can obviously be treated as positive and negative **simultaneously**!

One might say: well, this is beautiful but impractical since in physics and everyday life 2+3 is always 5 but not 0. Let us suppose, however that fundamental physics is described not by 'usual mathematics' but by 'mathematics modulo p' where p is a very large number. Then, operating with numbers much smaller than p we shall not notice this p, at least if we only add and multiply. We will feel a difference between 'usual mathematics' and 'mathematics modulo p' only while operating with numbers comparable to p.

We can easily extend the correspondence between $GF(p)$ and the ring of integers Z in such a way that subtraction will also be included. To make it clearer we note the following. Since the field $GF(p)$ is cyclic (adding 1 successively, we will obtain 0 eventually), it is convenient to visually depict its elements by the points of a circle of the radius $p/2\pi$ on the plane (x,y). In Fig. 1 only a part of the circle near the origin is depicted. Then the

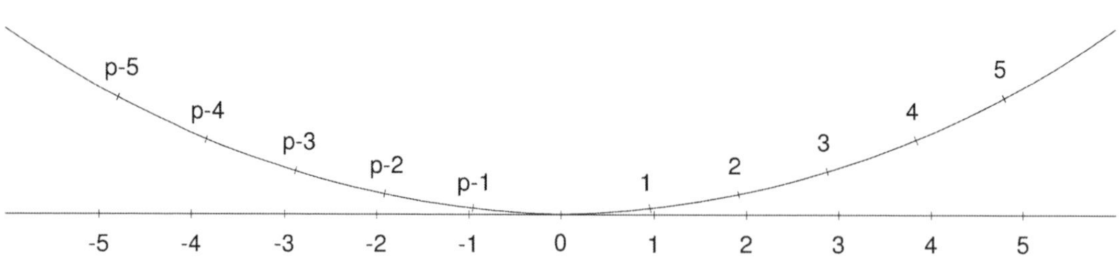

Figure 1. Relation between $GF(p)$ and the ring of integers.

distance between neighboring elements of the field is equal to unity, and the elements 0, 1, 2,... are situated on the circle counterclockwise. At the same time we depict the elements of Z as usual such that each element $z\in Z$ is depicted by a point with the coordinates $(z,0)$. We can denote the elements of $GF(p)$ not only as 0, 1,... $p-1$ but also as 0, ± 1, ± 2,,...$\pm(p-1)/2$, and such a set is called the set of minimal residues. Let f be a map from $GF(p)$ to Z, such that the element $f(a)\in Z$ corresponding to the minimal residue a has the same notation as a but is considered as the element of Z. Denote $C(p)=p^{1/(\ln p)^{1/2}}$ and let U_0 be the set of elements $a\in GF(p)$ such that $|f(a)|<C(p)$. Then if $a_1, a_2, ...a_n \in U_0$

and n_1, n_2 are such natural numbers that

$$n_1 < (p-1)/2C(p), \;\; n_2 < ln((p-1)/2)/(lnp)^{1/2} \tag{1}$$

then

$$f(a_1 \pm a_2 \pm ...a_n) = f(a_1) \pm f(a_2) \pm ...f(a_n)$$

if $n \leq n_1$ and

$$f(a_1 a_2 ... a_n) = f(a_1) f(a_2) ... f(a_n)$$

if $n \leq n_2$.

The meaning of the above relations is simple: although f is not a homomorphism of rings $GF(p)$ and Z, but if p is sufficiently large, then for a sufficiently large number of elements of U_0 the addition, subtraction and multiplication are performed according to the same rules as for elements $z \in Z$ such that $|z| < C(p)$. Therefore f can be treated as a local isomorphism of rings $GF(p)$ and Z.

The above discussion has a well known historical analogy. For many years people believed that our Earth was flat and infinite, and only after a long period of time they realized that it was finite and had a curvature. It is difficult to notice the curvature when we deal only with distances much less than the radius of the curvature R. Analogously one might think that the set of numbers describing physics has a curvature defined by a very large number p but we do not notice it when we deal only with numbers much less than p.

Let us note that even for elements from U_0 the result of division in the field $GF(p)$ differs generally speaking, from the corresponding result in the field of rational number Q. For example the element 1/2 in $GF(p)$ is a very large number $(p+1)/2$. For this reason one might think that physics based on Galois fields has nothing to with the reality. We will see in the subsequent section that this is not so since the spaces describing quantum systems are projective.

Since the standard quantum theory is based on complex numbers, the question arises whether it is possible to construct a finite analog of such numbers. By analogy with the field of complex numbers, we can consider a set $GF(p^2)$ of p^2 elements $a+bi$ where $a, b \in GF(p)$ and i is a formal element such that $i^2 = 1$. The question arises whether $GF(p^2)$ is a field, i.e. we can define all the four operations excepting division by zero. The definition of addition, subtraction and multiplication in $GF(p^2)$ is obvious and, by analogy with the field of complex numbers, one could define division as $1/(a+bi) = a/(a^2+b^2) - ib/(a^2+b^2)$ if a and b are not equal to zero simultaneously. This definition can be meaningful only if $a^2 + b^2 \neq 0$ in $GF(p)$ for any $a, b \in GF(p)$ i.e. $a^2 + b^2$ is not divisible by p. Therefore the definition is meaningful only if p *cannot* be represented as a sum of two squares and is meaningless otherwise. We will not consider the case $p = 2$ and therefore p is necessarily odd. Then we have two possibilities: the value of $p\,(mod\,4)$ is either 1 or 3. The well known result of number theory (see e.g. the textbooks [3]) is that a prime number p can be represented as a sum of two squares only in the former case and cannot in the latter one. Therefore the above construction of the field $GF(p^2)$ is correct if $p = 3\,(mod\,4)$. In that case the above local homomorphism of the rings Z and $GF(p)$ can be extended to the homomorphism between the rings $Z + iZ$ and $GF(p^2)$ if we consider a set U such that $a + bi \in U$ if $a \in U_0$ and $b \in U_0$.

The first impression is that if Galois fields can somehow replace the conventional field of complex numbers then this can be done only for p satisfying $p = 3\,(mod\,4)$ and therefore the case $p = 1\,(mod\,4)$ is of no interest for this purpose. It can be shown however, [4, 1] that correspondence between complex numbers and Galois fields containing p^2 elements can also be established if $p = 1\,(mod\,4)$. Nevertheless, arguments given in Refs. [4, 1] indicate that if quantum theory is based on a Galois field then p is probably such that $p = 3\,(mod\,4)$ rather than $p = 1\,(mod\,4)$. In general, it is well known (see e.g. Ref. [3]) that any Galois field consists of p^n elements where p is prime and $n > 0$ is natural. The numbers p and n define the field F_{p^n} uniquely up to isomorphism and p is called the characteristic of the Galois field.

The idea to replace the field of complex numbers in quantum theory by something else is well known. There exists a wide literature where quantum theory is based on quaternions, p-adic numbers or other constructions. However, as noted above, if we accept that the future quantum physics should not contain the actual infinity at all then the only possible choice is a Galois field.

3. Spacetime and Operator Formalism

If one accepts the arguments given in the preceding section then the problem arises how to construct a quantum theory based on a Galois field rather than the field of complex numbers.

A well known historical fact is that originally quantum theory has been proposed in two formalisms which seemed to be essentially different: the Schroedinger wave formalism and the Heisenberg operator (matrix) formalism. It has been shown later by Born, von Neumann and others that the both formalisms are equivalent and, in addition, the path integral formalism has been developed. From time to time physicists change their preferences and at present the wave approach prevails. In the spirit of the wave or path integral approach one might try to replace classical spacetime by a finite lattice which may even be not a field. In that case the problem arises what is the natural 'quantum of spacetime' and some physical quantities should necessarily have the field structure. A detailed discussion can be found in Ref. [5] and references therein. On the other hand, GFQT is a direct generalization of the operator formalism when the field of complex numbers is replaced by a Galois field. However, as it will be clear in the subsequent section, such a generalization is meaningful only if the symmetry algebra is of the de Sitter type (e.g. so(2,3) or so(1,4)) but not Poincare one.

Although the existing versions of the standard quantum theory are equivalent, the philosophies behind them are essentially different. One of the main problems is how quantum theory should treat the notion of spacetime. One of the main postulate of the standard quantum theory is that any observable physical quantity is represented by a selfadjoint operator in a Hilbert space. Then the first question which immediately arises is that, even in nonrelativistic quantum mechanics, there is no operator corresponding to time [6]. In particular, we cannot prepare a state which is the eigenstate of the time operator with the eigenvalue 5000BC or 2005AD.

It is also well known that, when quantum mechanics is combined with relativity, there is no operator satisfying all the properties of the spatial position operator (see e.g. Ref. [7]).

For these reasons the quantity x in the Lagrangian density $L(x)$ is only a parameter which becomes the coordinate in the classical limit.

These facts were well known already in 30th of the last century and became very popular in 60th (recall the famous Heisenberg S-matrix program). In the first section of the well-known textbook [8] it is claimed that spacetime and local quantum fields are rudimentary notions which will disappear in the ultimate quantum theory. Since that time, no arguments questioning those ideas have been given, but in view of the great success of gauge theories in 70th and 80th, such ideas became almost forgotten.

In the standard approach to elementary particle theory it is assumed from the beginning that there exists a background spacetime (e.g. Minkowski or de Sitter spacetime), and the system under consideration is described by local quantum fields defined on that spacetime. Then by using Lagrangian formalism and Noether theorem, one can (at least in principle) construct global quantized operators (e.g. the four-momentum operator) for the system as a whole. It is interesting to note that after this stage has been implemented, one can safely forget about spacetime and concentrate his or her efforts on calculating S-matrix and other physical observables.

The problem of whether the empty classical spacetime has a physical meaning, has been discussed for a long time. In particular, according to the famous Mach's principle, the properties of space at a given point depend on the distribution of masses in the whole Universe. As described in a wide literature (see e.g. Refs. [9, 10, 11, 12] and references therein), Mach's principle was a guiding one for Einstein in developing general relativity (GR), but when it has been constructed, it has been realized that it does not contain Mach's principle at all! As noted in Refs. [9, 10, 11], this problem is not closed.

Consider now how one should define the notion of elementary particles. Although particles are observable and fields are not, in the spirit of local quantum field theory (LQFT), fields are more fundamental than particles, and a possible definition is as follows [13]: 'It is simply a particle whose field appears in the Lagrangian. It does not matter if it's stable, unstable, heavy, light — if its field appears in the Lagrangian then it's elementary, otherwise it's composite'.

Another approach has been developed by Wigner in his investigations of unitary irreducible representations (IRs) of the Poincare group [14]. In view of this approach, one might postulate that a particle is elementary if the set of its wave functions is the space of a unitary IR of the symmetry group in the given theory.

Although in standard well-known theories (QED, electroweak theory and QCD) the above approaches are equivalent, the following problem arises. The symmetry group is usually chosen as a group of motions of some classical manifold. How does this agree with the above discussion that quantum theory in the operator formulation should not contain spacetime? A possible answer is as follows. One can notice that for calculating observables (e.g. the spectrum of the Hamiltonian) we need in fact not a representation of the group but a representation of its Lie algebra by Hermitian operators. After such a representation has been constructed, we have only operators acting in the Hilbert space and this is all we need in the operator approach. The representation operators of the group are needed only if it is necessary to calculate some macroscopic transformation, e.g. time evolution. In the approximation when classical time is a good approximate parameter, one can calculate evolution, but nothing guarantees that this is always the case (e.g. at the very early stage of the Universe). Let us also note that in the stationary formulation of scattering theory, the S-matrix can be defined without any mentioning of time (see e.g. Ref. [15]). For these reasons we can assume that on quantum level the symmetry algebra is more fundamental than the symmetry group.

In other words, instead of saying that some operators satisfy commutation relations of a Lie algebra A because spacetime X has a group of motions G such that A is the Lie algebra of G, we say that there exist operators satisfying commutation relations of the Lie algebra A such that: for some operator functions $\{O\}$ of them the classical limit is a good approximation, a set X of the eigenvalues of the operators $\{O\}$ represents a classical manifold with the group of motions G and its Lie algebra is A. This is not of course in the spirit of famous Klein's Erlangen program or LQFT.

Consider for illustration the well-known example of nonrelativistic quantum mechanics. Usually the existence of the Galilei spacetime is assumed from the beginning. Let $(\mathbf{r}, t)$ be the spacetime coordinates of a particle in that spacetime. Then the particle momentum operator is $-i\partial/\partial\mathbf{r}$ and the Hamiltonian describes evolution by the Schroedinger equation. In our approach one starts from an IR of the Galilei algebra. The momentum operator and the Hamiltonian represent four of ten generators of such a representation. If it is implemented in a space of functions $\psi(\mathbf{p})$ then the momentum operator is simply the operator of multiplication by $\mathbf{p}$. Then the position operator can be *defined* as $i\partial/\partial\mathbf{p}$ and time can be *defined* as an evolution parameter such that evolution is described by the Schroedinger equation with the given Hamiltonian. Mathematically the both approaches are equivalent since they are related to each other by the Fourier transform. However, the philosophies behind them are essentially different. In the second approach there is no empty spacetime (in the spirit of Mach's principle) and the spacetime coordinates have a physical meaning only if there are particles for which the coordinates can be measured.

Summarizing our discussion, we assume that, *by definition*, on quantum level a Lie algebra is the symmetry algebra if there exist physical observables such that their operators satisfy the commutation relations characterizing the algebra. Then, a particle is called elementary if the set of its wave functions is a space of IR of this algebra by Hermitian operators. Such an approach is in the spirit of that considered by Dirac in Ref. [16]. By using the abbreviation 'IR' we will always mean an irreducible representation by Hermitian operators.

One can now define GFQT as a theory where *quantum states are represented by elements of a linear projective space over a field $GF(p^n)$ and physical quantities are represented by linear operators in that space.* Then a Lie algebra $\mathcal{A}$ over $GF(p)$ is called the symmetry algebra if the operators in $GF(p^n)$ representing the observables belong to a representation of $\mathcal{A}$ in $GF(p^n)$. If this representation is irreducible then the system is called elementary particle.

The problem now arises what values of p and n should be used in GFQT. As noted above, if the principal number field in GFQT is $GF(p^2)$ where p is very large and $p = 3\ (mod\,4)$ then there exists the correspondence between GFQT and the standard theory. However, since we treat GFQT as a more general theory than the standard one, it is desirable not to postulate that GFQT is based on $GF(p^2)$ (with $p = 3\ (mod\,4)$) because the standard theory is based on complex numbers but vice versa, explain the fact that the standard theory is based on complex numbers since GFQT is based on $GF(p^2)$. Therefore we should find a motivation for the choice of $GF(p^2)$ in GFQT. A possible motivation is discussed in Ref. [4] and another motivation will be given in the present paper. We will not assume from the beginning that $n = 2$ and $p = 3\ (mod\, 4)$. Therefore we should discuss the meaning of the scalar product and Hermiticity for any n.

It is well known [3] that the field $GF(p^n)$ has $n - 1$ nontrivial automorphisms. When $n = 2$ and $p = 3\,(mod\, 4)$, there exists only one automorphism coinciding with the complex conjugation $z \to \bar{z}$. In the general case we use $\bar{z}$ to denote the element obtained from $z \in GF(p^n)$ by some chosen automorphism of $GF(p^n)$. Suppose that the representation space over $GF(p^n)$ is supplied by a scalar product $(x, y) \in GF(p^n)$ such that

$$(x,y) = \overline{(y,x)}, \quad (ax,y) = \bar{a}(x,y), \quad (x,ay) = a(x,y) \tag{2}$$

Then the operator B is called adjoint to A if $(Bx, y) = (y, Ax)$ for any x, y from the representation space. As usual, in this case one can write $A^* = B$. If $A^* = A$, the operator is called selfadjoined or Hermitian (in finite dimensional spaces over Galois fields these properties are equivalent).

Representations in spaces over a field of nonzero characteristics are called modular representations. There exists a wide literature devoted to such representations (see e.g. Ref. [17] and references therein). In particular, it has been shown by Zassenhaus [18] that all modular IRs are finite-dimensional and in numerous papers the maximum dimension of such representations is considered.

4. PCT, Spin-Statistics and All That

The title of this section is borrowed from that in the well known book [19] where the famous results of the particle theory are derived from LQFT. In this theory each elementary particle is described in two ways: i) by using an IR of the Poincare algebra; ii) by using a local Poincare covariant equation. For each values of the mass and spin, there exist two IRs - with positive and negative energies, respectively. However, the negative energy IRs are not used since particles and its antiparticles are described only by positive energy IRs. It is usually assumed that a particle and its antiparticle are described by their own equivalent positive energy IRs. In Poincare invariant theories there exists a possibility that a massless particle and its antiparticle are described by different IRs. For example, the massles neutrino can be described by an IR with the left-handed helicity while the antineutrino — by an IR with the right-handed helicity. At the same time, in the AdS case there exists a nonzero probability for transitions between the left-handed and right-handed massless states (see e.g. Ref. [20]).

A description in terms of a covariant equation seems to be more fundamental since one equation describes a particle and its antiparticle simultaneously. Namely, the negative energy solutions of the covariant equation are associated with antiparticles by means of quantization such that the creation and annihilation operators for the antiparticle have the usual meaning but they enter the quantum Lagrangian with the coefficients representing the negative energy solutions. On the other hand, covariant equations describe functions defined on a classical spacetime and for this reason their applicability is limited to local theories only.

The necessity to have negative energy solutions is related to the implementation of the idea that the creation or annihilation of an antiparticle can be treated, respectively, as the annihilation or creation of the corresponding particle with the negative energy. However, since negative energies have no direct physical meaning in the standard theory, this idea is implemented implicitly rather than explicitly.

The fact that on the level of Hilbert spaces a particle and its antiparticle are treated independently, poses a problem why they have equal masses, spins and lifetimes. The usual explanation (see e.g. [21, 8, 22, 19]) is that this is a consequence of CPT invariance. Therefore if it appears that the masses of a particle and its antiparticle were not equal, this would indicate the violation of CPT invariance. In turn, as shown in well-known works [23, 19], any local Poincare invariant quantum theory is automatically CPT invariant.

Such an explanation seems to be not quite convincing. Although at present there are no theories which explain the existing data better than the standard model based on LQFT, there is no guarantee that the ultimate quantum theory will be necessarily local. The modern theories aiming to unify all the known interactions (loop quantum gravity, noncommutative quantum theory, string theory etc.) do not adopt the exact locality. Therefore each of those theories should give its own explanation.

Consider a model example when isotopic invariance is exact (i.e. electromagnetic and weak interactions are absent). Then the proton and the neutron have equal masses and spins as a consequence of the fact that they belong to the same IR of the isotopic algebra. In this example the proton and the neutron are simply different states of the same object - the nucleon, and the problem of why they have equal masses and spins has a natural explanation.

It is clear from this example that in theories where a particle and its antiparticle are described by the same IR of the symmetry algebra, the fact that they have equal masses and spins has a natural explanation. In such theories the very existence of antiparticles is inevitable (i.e. the existence of a particle without its antiparticle is impossible) and a particle and its antiparticle are different states of the same object. We will see in Sect. 6. that in GFQT such a situation indeed takes place.

Another famous result of LQFT is the Pauli spin-statistics theorem [24]. After the original Pauli proof, many authors investigated more general approaches to the theorem (see e.g. Ref. [25] and references therein) but in all the approaches the locality was used by assuming that particles are described by covariant equations. Meanwhile, if, for simplicity, we consider only free particles then all the information about them is known from the corresponding IR while local covariant wave functions are needed only for constructing interaction Lagrangian in LQFT. From this point view the problem arises whether, at least in the case of free particles, the spin-statistics theorem can be proved by using only the properties of the particle IRs.

5. Poincare Invariance vs. de Sitter Invariance

As follows from our definition of symmetry on quantum level, the standard theory is Poincare invariant if the representation operators for the system under consideration satisfy the well-known commutation relations

$$[P^\mu, P^\nu] = 0, \quad [M^{\mu\nu}, P^\rho] = -2i(g^{\mu\rho}P^\nu - g^{\nu\rho}P^\mu),$$
$$[M^{\mu\nu}, M^{\rho\sigma}] = -2i(g^{\mu\rho}M^{\nu\sigma} + g^{\nu\sigma}M^{\mu\rho} - g^{\mu\sigma}M^{\nu\rho} - g^{\nu\rho}M^{\mu\sigma}) \tag{3}$$

where $\mu, \nu, \rho, \sigma = 0, 1, 2, 3$, P^μ are the four-momentum operators, $M^{\mu\nu}$ are the angular momentum operators, the metric tensor in Minkowski space has the nonzero components $g^{00} = -g^{11} = -g^{22} = -g^{33} = 1$, and for further convenience we use the system of units with $\hbar/2 = c = 1$. The operators $M^{\mu\nu}$ are antisymmetric: $M^{\mu\nu} = -M^{\nu\mu}$ and therefore there are only six independent angular momentum operators.

The question arises whether Poincare invariant quantum theory can be a starting point for its generalization to GFQT. The answer is probably 'no' and the reason is the following. GFQT is discrete and finite because the only numbers it can contain are elements of a Galois field. By analogy with integers, those numbers can have no dimension and all operators in GFQT cannot have the continuous spectrum. In the Poincare invariant quantum theory the angular momentum operators are dimensionless (if $\hbar/2 = c = 1$) but the momentum operators have the dimension of the inverse length. In addition, the momentum operators and the generators of the Lorentz boosts M^{0i} $(i = 1, 2, 3)$ contain the continuous spectrum.

This observation might prompt a skeptical reader to immediately conclude that no GFQT can describe the nature. However, a simple way out of this situation is as follows.

First we recall the well-known fact that conventional Poincare invariant theory is a special case of de Sitter invariant one. The symmetry algebra of the de Sitter invariant quantum theory can be either so(2,3) or so(1,4). The algebra so(2,3) is the Lie algebra of symmetry group of the four-dimensional manifold in the five-dimensional space, defined by the equation

$$x_5^2 + x_0^2 - x_1^2 - x_2^2 - x_3^2 = R^2 \tag{4}$$

where a constant R has the dimension of length. We use x_0 to denote the conventional time coordinate and x_5 to denote the fifth coordinate. The notation x_5 rather than x_4 is used since in the literature the latter is sometimes used to denote ix_0. Analogously, so(1,4) is the Lie algebra of the symmetry group of the four-dimensional manifold in the five-dimensional space, defined by the equation

$$x_0^2 - x_1^2 - x_2^2 - x_3^2 - x_5^2 = -R^2 \tag{5}$$

The quantity R^2 is often written as $R^2 = 3/\Lambda$ where Λ is the cosmological constant. The existing astronomical data show that it is very small. The nomenclature is such that $\Lambda < 0$ corresponds to the case of Eq. (4) while $\Lambda > 0$ - to the case of Eq. (5). In the literature the latter is often called the de Sitter (dS) space while the former is called the anti de Sitter (AdS) one. Analogously, some authors prefer to call only so(1,4) as the dS algebra while so(2,3) is called the AdS one. In view of recent cosmological investigations it is now believed that $\Lambda > 0$ (see e.g. Ref. [26]).

The both de Sitter algebras are ten-parametric, as well as the Poincare algebra. However, in contrast to the Poincare algebra, all the representation operators of the de Sitter algebras are angular momenta, and in the units $\hbar/2 = c = 1$ they are dimensionless. The commutation relations now can be written in the form of one tensor equation

$$[M^{ab}, M^{cd}] = -2i(g^{ac}M^{bd} + g^{bd}M^{cd} - g^{ad}M^{bc} - g^{bc}M^{ad}) \tag{6}$$

where a, b, c, d take the values 0,1,2,3,5 and the operators M^{ab} are antisymmetric. The diagonal metric tensor has the components $g^{00} = -g^{11} = -g^{22} = -g^{33} = 1$ as usual, while $g^{55} = 1$ for the algebra so(2,3) and $g^{55} = -1$ for the algebra so(1,4).

When R is very large, the transition from the de Sitter symmetry to Poincare one (this procedure is called contraction [27]) is performed as follows. We define the operators $P^\mu = M^{\mu 5}/2R$. Then, when $M^{\mu 5} \to \infty$, $R \to \infty$, but their ratio is finite, Eq. (6) splits into the set of expressions given by Eq. (3).

Note that our definition of the de Sitter symmetry on quantum level does not involve the cosmological constant at all. It appears only if one is interested in interpreting results in terms of the de Sitter spacetime or in the Poincare limit. Since all the operators M^{ab} are dimensionless in units $\hbar/2 = c = 1$, the de Sitter invariant quantum theories can be formulated only in terms of dimensionless variables. In particular one might expect that the gravitational and cosmological constants are not fundamental in the framework of such theories. Mirmovich has proposed a hypothesis [28] that only quantities having the dimension of the angular momentum can be fundamental.

If one assumes that spacetime is fundamental then in the spirit of GR it is natural to think that the empty space is flat, i.e. that the cosmological constant is equal to zero. This was the subject of the well-known dispute between Einstein and de Sitter described in a wide literature (see e.g. Refs. [10, 29] and references therein). In the LQFT the cosmological constant is given by a contribution of vacuum diagrams, and the problem is to explain why it is so small. On the other hand, if we assume that symmetry on quantum level in our formulation is more fundamental then the cosmological constant problem does not arise at all. Instead we have a problem of why nowadays Poincare symmetry is so good approximate symmetry. It seems natural to involve the anthropic principle for the explanation of this phenomenon (see e.g. Ref. [30] and references therein).

Let us note that there is no continuous transition from de Sitter invariant to Poincare invariant theories. If R is finite then we have de Sitter invariance even if R is very large. Therefore if we accept the de Sitter symmetry from the beginning, we should accept that energy and momentum are no longer fundamental physical quantities. We can use them at certain conditions but only as approximations.

In local Poincare invariant theories, the energy and momentum can be written as integrals from energy-momentum tensor over a space-like hypersurface, while the angular momentum operators can be written as integrals from the angular momentum tensor over the same hypersurface. In local de Sitter invariant theories all the representation generators are angular momenta. Therefore the energy-momentum tensor in that case is not needed at all and all the representation generators can be written as integrals from a five-dimensional angular momentum tensor over some hypersurface. Meanwhile, in GR and quantum theories in curved spaces the energy-momentum tensor is used in situations where spacetime is much more complicated than the dS or AdS space-time. It is well-known that Einstein was not satisfied by the presence of this tensor in his theory, and this was the subject of numerous discussions in the literature.

Finally we discuss the differences between the so(2,3) and so(1,4) invariant theories. There exists a wide literature devoted to the both of them.

The former has many features analogous to those in Poincare invariant theories. There exist IRs of the AdS algebra where the operator M^{05}, which is the de Sitter analog of the energy operator, is bounded below by some positive value which can be treated as the AdS analog of the mass. With such an interpretation, a system of free particles with the AdS masses $m_1, m_2 ... m_n$ has the minimum energy equal to $m_1 + m_2 + ... m_n$. There also exist IRs with negative energies where the energy operator is bounded above. The so(2,3) invariant theories can be easily generalized if supersymmetry is required while supersymmetric generalizations of the so(1,4) invariant theories is problematic.

On the contrary, the latter has many unusual features. For example, even in IRs, the operator M_{05} (which is the SO(1,4) analog of the energy operator) has the spectrum in the interval $(-\infty, +\infty)$. The dS mass operator of the system of free particles with the dS masses $m_1, m_2 ... m_n$ is not bounded below by the value of $m_1 + m_2 + ... m_n$ and also has the spectrum in the interval $(-\infty, +\infty)$ [31, 32].

For these reasons there existed an opinion that de Sitter invariant quantum theories can be based only on the algebra so(2,3) and not so(1,4). However, as noted above, in view of recent cosmological observations it is now believed that $\Lambda > 0$, and the opinion has been changed. As noted in our works [31, 32, 33, 34] (done before these developments), so(1,4) invariant theories have many interesting properties, and the fact that they are unusual does not mean that they contradict experiment. For example, a particle has only an exponentially small probability to have the energy less than its mass (see e.g. Ref. [32]). The fact that the mass operator is not bounded below by the value of $m_1 + m_2 + ... m_n$ poses an interesting question whether some interactions (even gravity) can be a direct consequence of the dS symmetry [32, 33]. A very interesting property of the so(1,4) invariant theories is that they do not contain bound states at all and, as a consequence, the free and interacting operators are unitarily equivalent [33]. This poses the problem whether the notion of interactions is needed at all. Finally, as shown in Ref. [34], each IR of the so(1,4) algebra cannot be interpreted in a standard way but describe a particle and its antiparticle simultaneously.

The main goal of the present paper is to convince the reader that GFQT is a more natural quantum theory than the standard one based on complex numbers. For this reason (in view of the above remarks) we consider below only the so(2,3) version of GFQT. The problem of the Galois field generalization of so(1,4) invariant theories has been discussed in Ref. [32, 35].

6. Modular IRs of the so(2,3) Algebra

The above discussion indicates that the first problem in constructing the so(2,3) version of GFQT is to consider modular IRs of this algebra. In the standard theory the representation operators of the so(2,3) algebra in units $\hbar/2 = c = 1$ are given by Eq. (6). If a modular IR is considered in a linear space over $GF(p^2)$ with $p = 3\ (mod\ 4)$ then Eq. (6) is also valid. However, as noted in Sect. 3., we consider modular IRs in linear spaces over $GF(p^k)$ where k is arbitrary. In this case it is convenient to work with another set of ten operators. Let $(a'_j, a_j", h_j)$ $(j = 1, 2)$ be two independent sets of operators satisfying the commutation relations for the sp(2) algebra

$$[h_j, a'_j] = -2a'_j \quad [h_j, a_j"] = 2a_j" \quad [a'_j, a_j"] = h_j \tag{7}$$

The sets are independent in the sense that for different j they mutually commute with each other. We denote additional four operators as $b', b", L_+, L_-$. The meaning of L_+, L_- is as follows. The operators $L_3 = h_1 - h_2, L_+, L_-$ satisfy the commutation relations of the su(2) algebra

$$[L_3, L_+] = 2L_+ \quad [L_3, L_-] = -2L_- \quad [L_+, L_-] = L_3 \tag{8}$$

while the other commutation relations are as follows

$$\begin{gathered}[a_1', b'] = [a_2', b'] = [a_1", b"] = [a_2", b"] = \\ [a_1', L_-] = [a_1", L_+] = [a_2', L_+] = [a_2", L_-] = 0 \\ [h_j, b'] = -b' \quad [h_j, b"] = b" \quad [h_1, L_\pm] = \pm L_\pm, \\ [h_2, L_\pm] = \mp L_\pm \quad [b', b"] = h_1 + h_2 \\ [b', L_-] = 2a_1' \quad [b', L_+] = 2a_2' \quad [b", L_-] = -2a_2" \\ [b", L_+] = -2a_1", \quad [a_1', b"] = [b', a_2"] = L_- \\ [a_2', b"] = [b', a_1"] = L_+, \quad [a_1', L_+] = [a_2', L_-] = b' \\ [a_2", L_+] = [a_1", L_-] = -b"\end{gathered} \tag{9}$$

At first glance these relations might seem rather chaotic but in fact they are very natural in the Weyl basis of the so(2,3) algebra.

In spaces over $GF(p^2)$ with $p = 3\ (mod\ 4)$ the relation between the above sets of ten operators is

$$\begin{gathered}M_{10} = i(a_1" - a_1' - a_2" + a_2') \quad M_{15} = a_2" + a_2' - a_1" - a_1' \\ M_{20} = a_1" + a_2" + a_1' + a_2' \quad M_{25} = i(a_1" + a_2" - a_1' - a_2') \\ M_{12} = L_3 \quad M_{23} = L_+ + L_- \quad M_{31} = -i(L_+ - L_-) \\ M_{05} = h_1 + h_2 \quad M_{35} = b' + b" \quad M_{30} = -i(b" - b')\end{gathered} \tag{10}$$

and therefore the sets are equivalent. However, the relations (7-9) are more general since they can be used when the representation space is a space over $GF(p^k)$ with an arbitrary k.

By analogy with Refs. [36, 37], we require the existence of the vector e_0 satisfying the conditions

$$a_j' e_0 = b' e_0 = L_+ e_0 = 0 \quad h_j e_0 = q_j e_0 \quad (j = 1, 2) \tag{11}$$

where $q_j \in F_p$, $|f(q_j)| \ll p$, $f(q_j) > 0$ for $j = 1, 2$ and $f(q_1 - q_2) \geq 0$. It is well known (see e.g. Ref. [1]) that $M^{05} = h_1 + h_2$ is the AdS analog of the energy operator. As follows from Eqs. (7) and (9), the operators (a_1', a_2', b') reduce the AdS energy by two units. Therefore e_0 is an analog the state with the minimum energy which can be called the rest state, and the spin in our units is equal to the maximum value of the operator $L_3 = h_1 - h_2$ in that state. For these reasons we use s to denote $q_1 - q_2$ and m to denote $q_1 + q_2$.

Let us denote

$$\begin{gathered}A^{++} = b"(h_1 - 1)(h_2 - 1) - a_1" L_-(h_2 - 1) - a_2" L_+(h_1 - 1) + \\ a_1" a_2" b' \quad A^{-+} = L_-(h_2 - 1) - a_2" b' \\ e(n_1 n_2 n k) = (a_1")^{n_1} (a_2")^{n_2} (A^{++})^n (A^{-+})^k e_0\end{gathered} \tag{12}$$

Then it can be shown [1] that the elements $e(n_1n_2nk)$ form a basis of IR. The quantity k is in the range $0, 1, ..., s$ as well as in the standard theory. However, in contrast with the standard theory where the quantities n_1, n_2, n can be any natural numbers, in the modular case

$$\begin{aligned} n_1 = 0, 1, ...N_1(n,k) \quad n_2 = 0, 1, ...N_2(n,k) \\ N_1(n,k) = p - q_1 - n + k \quad N_2(n,k) = p - q_2 - n - k \\ n = 0, 1, ...n_{max}(m,s) \end{aligned} \tag{13}$$

where $n_{max}(m, s)$ is different for different types of IR. For example, in the massive case $n_{max}(m, s) = p + 2 - m$. Therefore the dimension of the IR is always finite in agreement with the Zassenhaus theorem [18].

We will use the notation

$$Norm(n_1n_2nk) = (e(n_1n_2nk), e(n_1n_2nk)) \tag{14}$$

This quantity can be represented as [1]

$$Norm(n_1n_2nk) = F(n_1n_2nk)G(nk) \tag{15}$$

where

$$\begin{aligned} F(n_1n_2nk) = n_1!(Q_1(n,k) + n_1 - 1)!n_2!(Q_2(n,k) + n_2 - 1)! \\ G(nk) = \{(q_2 + k - 2)!n!(m + n - 3)!(q_1 + n - 1)! \\ (q_2 + n - 2)!k!s!\}\{(q_1 - k - 2)![(q_2 - 2)!]^3(q_1 - 1)! \\ (m - 3)!(s - k)![Q_1(n,k) - 1][Q_2(n,k) - 1]\}^{-1} \end{aligned} \tag{16}$$

In standard Poincare and AdS theories there also exist IRs with negative energies. They can be constructed by analogy with positive energy IRs. Instead of Eq. (11) one can require the existence of the vector e_0' such that

$$\begin{aligned} a_j"e_0' = b"e_0' = L_-e_0' = 0 \quad h_je_0' = -q_je_0' \\ (e_0', e_0') \neq 0 \quad (j = 1, 2) \end{aligned} \tag{17}$$

where the quantities q_1, q_2 are the same as for positive energy IRs. It is obvious that positive and negative energy IRs are fully independent since the spectrum of the operator M^{05} for such IRs is positive and negative, respectively. However, *the modular analog of a positive energy IR characterized by q_1, q_2 in Eq. (11), and the modular analog of a negative energy IR characterized by the same values of q_1, q_2 in Eq. (17) represent the same modular IR.* This is the crucial difference between the standard quantum theory and GFQT, and a proof is given in Ref. [1].

7. Quantization and AB Symmetry

Let us first recall how the Fock space is defined in the standard theory. Let $a(n_1n_2nk)$ be the operator of particle annihilation in the state described by the vector $e(n_1n_2nk)$. Then the adjoint operator $a(n_1n_2nk)^*$ has the meaning of particle creation in that state. Since we do not normalize the states $e(n_1n_2nk)$ to one, we require that the operators $a(n_1n_2nk)$ and $a(n_1n_2nk)^*$ should satisfy either the anticommutation relations

$$\{a(n_1n_2nk), a(n_1'n_2'n'k')^*\} = \\ Norm(n_1n_2nk)\delta_{n_1n_1'}\delta_{n_2n_2'}\delta_{nn'}\delta_{kk'} \tag{18}$$

or the commutation relations

$$[a(n_1n_2nk), a(n_1'n_2'n'k')^*] = \\ Norm(n_1n_2nk)\delta_{n_1n_1'}\delta_{n_2n_2'}\delta_{nn'}\delta_{kk'} \tag{19}$$

In the standard theory the representation describing a particle and its antiparticle are fully independent and therefore quantization of antiparticles should be described by other operators. If $b(n_1n_2nk)$ and $b(n_1n_2nk)^*$ are operators of the antiparticle annihilation and creation in the state $e(n_1n_2nk)$ then by analogy with Eqs. (18) and (19)

$$\{b(n_1n_2nk), b(n_1'n_2'n'k')^*\} = \\ Norm(n_1n_2nk)\delta_{n_1n_1'}\delta_{n_2n_2'}\delta_{nn'}\delta_{kk'} \tag{20}$$

in the case of anticommutation relations and

$$[b(n_1n_2nk), b(n_1'n_2'n'k')^*] = \\ Norm(n_1n_2nk)\delta_{n_1n_1'}\delta_{n_2n_2'}\delta_{nn'}\delta_{kk'} \tag{21}$$

in the case of commutation relations. It is assumed additionally that in the case of anticommutation relations all the operators (a, a^*) anticommute with all the operators (b, b^*) while in the case of commutation relations they commute with each other. It is also assumed that the Fock space contains the vacuum vector Φ_0 such that

$$a(n_1n_2nk)\Phi_0 = b(n_1n_2nk)\Phi_0 = 0 \quad \forall\ n_1, n_2, n, k \tag{22}$$

The Fock space in the standard theory can now be defined as a linear combination of all elements obtained by the action of the operators (a^*, b^*) on the vacuum vector, and the problem of second quantization of representation operators can now be formulated as follows. Let $(A_1, A_2....A_n)$ be representation operators describing IR of the AdS algebra. One should replace them by operators acting in the Fock space such that the commutation relations between their images in the Fock space are the same as for original operators (in other words, we should have a homomorphism of Lie algebras of operators acting in the space of IR and in the Fock space). We can also require that our map should be compatible with the Hermitian conjugation in both spaces. It is easy to verify that a possible solution satisfying all the requirements is as follows. Taking into account the fact that the matrix elements satisfy the proper commutation relations, the operators A_i in the quantized form

$$A_i = \sum A_i(n_1'n_2'n'k', n_1n_2nk)[a(n_1'n_2'n'k')^*a(n_1n_2nk) + \\ b(n_1'n_2'n'k')^*b(n_1n_2nk)]/Norm(n_1n_2nk) \tag{23}$$

satisfy the commutation relations (7-9). We will not use special notations for operators in the Fock space since in each case it will be clear whether the operator in question acts in the space of IR or in the Fock space.

A well known problem in the standard theory is that the quantization procedure does not define the order of the annihilation and creation operators uniquely. For example, another possible solution is

$$A_i = \mp \sum A_i(n_1'n_2'n'k', n_1n_2nk)[a(n_1n_2nk)a(n_1'n_2'n'k')^* + b(n_1n_2nk)b(n_1'n_2'n'k')^*]/Norm(n_1n_2nk) \tag{24}$$

for the cases of anticommutation and commutation relations, respectively. It is clear that the solutions (23) and (24) are different since the energy operators M^{05} in these expressions differ by an infinite constant. In the standard theory the solution (23) is selected by imposing an additional requirement that all operators should be written in the normal form where annihilation operators should always precede creation ones. Then the vacuum has zero energy and Eq. (24) should be rejected. Such a requirement does not follow from the theory. Ideally there should be a procedure which correctly defines the order of operators from first principles.

In the standard theory there also exist neutral particles. In that case there is no need to have two independent sets of operators (a, a^*) and (b, b^*), and Eq. (23) should be written without the (b, b^*) operators. The problem of neutral particles in GFQT is discussed in Sect. 10..

We now proceed to quantization in the modular case. As noted in the preceding section, one modular IR corresponds to two standard IRs with the positive and negative energies, respectively. This indicates to a possibility that one modular IR describes a particle and its antiparticle simultaneously. However, we don't know yet what should be treated as a particle and its antiparticle in the modular case. We have a description of an object such that (n_1n_2nk) is the full set of its quantum numbers, and these numbers take the values described in the preceding section.

We now assume that $a(n_1n_2nk)$ in GFQT is the operator describing annihilation of the object with the quantum numbers (n_1n_2nk) regardless of whether the numbers are physical or nonphysical. Analogously $a(n_1n_2nk)^*$ describes creation of the object with the quantum numbers (n_1n_2nk). If these operators anticommute then they satisfy Eq. (18) while if they commute then they satisfy Eq. (19). Then, by analogy with the standard case, the operators

$$A_i = \sum A_i(n_1'n_2'n'k', n_1n_2nk) a(n_1'n_2'n'k')^* a(n_1n_2nk)/Norm(n_1n_2nk) \tag{25}$$

satisfy the commutation relations (7-9). In this expression the sum is taken over all possible values of the quantum numbers in the modular case.

In the modular case the solution can be taken not only as in Eq. (25) but also as

$$A_i = \mp \sum A_i(n_1'n_2'n'k', n_1n_2nk) a(n_1n_2nk)a(n_1'n_2'n'k')^*/Norm(n_1n_2nk) \tag{26}$$

for the cases of anticommutators and commutators, respectively. An essential difference between the standard approach and GFQT is that in the latter the trace of any operator is zero while in the former this is obviously not the case since even the AdS energy operator can be either positive definite or negative definite (see Ref. [1] for a detailed discussion). Therefore, as follows from Eqs. (18) and (19), the solutions (25) and (26) are the same and in the modular case there is no need to impose an artificial requirement that all operators should be written in the normal form.

The problem with the treatment of the (a, a^*) operators is as follows. When the values of $(n_1 n_2 n)$ are much less than p, the modular IR corresponds to the standard one and therefore the (a, a^*) operator can be treated as those describing the particle annihilation and creation, respectively. However, when the AdS energy is negative, the operators $a(n_1 n_2 nk)$ and $a(n_1 n_2 nk)^*$ become unphysical since they describe annihilation and creation, respectively, in the unphysical region of negative energies.

One can show [1] that at any fixed values of n and k, the quantities n_1 and n_2 can take only the values described in Eq. (13) and the eigenvalues of the operators h_1 and h_2 are given by $Q_1(n,k) + 2n_1$ and $Q_2(n,k) + 2n_2$, respectively, where $Q_1(n,k) = q_1 + n - k$ and $Q_2(n,k) = q_2 + n + k$. One can also show [1] that the first IR of the sp(2) algebra has the dimension $N_1(n,k) + 1$ and the second IR has the dimension $N_2(n,k) + 1$. Therefore if $n_1 = N_1(n,k)$ then the first eigenvalue is equal to $-Q_1(n,k)$ in $GF(p)$, and if $n_2 = N_2(n,k)$ then the second eigenvalue is equal to $-Q_2(n,k)$ in $GF(p)$. We use $\tilde{n}_1$ to denote $N_1(n,k) - n_1$ and $\tilde{n}_2$ to denote $N_2(n,k) - n_2$. Then $e(\tilde{n}_1 \tilde{n}_2 nk)$ is the eigenvector of the operator h_1 with the eigenvalue $-(Q_1(n,k) + 2n_1)$ and the eigenvector of the operator h_2 with the eigenvalue $-(Q_2(n,k) + 2n_2)$.

The standard theory implicitly involves the idea that creation of the antiparticle with the positive energy can be treated as annihilation of the corresponding particle with the negative energy and annihilation of the antiparticle with the positive energy can be treated as creation of the corresponding particle with the negative energy. In GFQT we can implement this idea explicitly. Namely, we can define the operators $b(n_1 n_2 nk)$ and $b(n_1 n_2 nk)^*$ in such a way that they will replace the (a, a^*) operators if the quantum numbers are unphysical. In addition, if the values of $(n_1 n_2 n)$ are much less than p, the operators $b(n_1 n_2 nk)$ and $b(n_1 n_2 nk)^*$ should be interpreted as physical operators describing annihilation and creation of antiparticles, respectively.

In GFQT the (b, b^*) operators cannot be independent of the (a, a^*) operators since the latter are defined for all possible quantum numbers. Therefore the (b, b^*) operators should be expressed in terms of the (a, a^*) ones. We can implement the above idea if the operator $b(n_1 n_2 nk)$ is defined in such a way that it is proportional to $a(\tilde{n}_1, \tilde{n}_2, n, k)^*$ and hence $b(n_1 n_2 nk)^*$ is proportional to $a(\tilde{n}_1, \tilde{n}_2, n, k)$.

Since Eq. (16) should now be considered in F_p, it follows from the well known Wilson theorem $(p-1)! = -1$ in F_p (see e.g. [3]) that

$$F(n_1 n_2 nk) F(\tilde{n}_1 \tilde{n}_2 nk) = (-1)^s \tag{27}$$

We now define the b-operators as

$$a(n_1 n_2 nk)^* = \eta(n_1 n_2 nk) b(\tilde{n}_1 \tilde{n}_2 nk) / F(\tilde{n}_1 \tilde{n}_2 nk) \tag{28}$$

where $\eta(n_1n_2nk)$ is some function. As a consequence of this definition,

$$\begin{aligned} a(n_1n_2nk) &= \bar{\eta}(n_1n_2nk)b(\tilde{n}_1\tilde{n}_2nk)^*/F(\tilde{n}_1\tilde{n}_2nk) \\ b(n_1n_2nk)^* &= a(\tilde{n}_1\tilde{n}_2nk)F(n_1n_2nk)/\bar{\eta}(\tilde{n}_1\tilde{n}_2nk) \\ b(n_1n_2nk) &= a(\tilde{n}_1\tilde{n}_2nk)^*F(n_1n_2nk)/\eta(\tilde{n}_1\tilde{n}_2nk) \end{aligned} \tag{29}$$

Eqs. (28) and (29) define a relation between the sets (a, a^*) and (b, b^*). Although our motivation was to replace the (a, a^*) operators by the (b, b^*) ones only for the nonphysical values of the quantum numbers, we can consider this definition for all the values of (n_1n_2nk).

We have not discussed yet what exact definition of the physical and nonphysical quantum numbers should be. This problem will be discussed in the subsequent section. However, one might accept

Physical-nonphysical states assumption: Each set of quantum numbers (n_1n_2nk) is either physical or unphysical. If it is physical then the set $(\tilde{n}_1\tilde{n}_2nk)$ is unphysical and vice versa.

With this assumption we can conclude from Eqs. (28) and (29) that if some operator a is physical then the corresponding operator b^* is unphysical and vice versa while if some operator a^* is physical then the corresponding operator b is unphysical and vice versa.

We have no ground to think that the set of the (a, a^*) operators is more fundamental than the set of the (b, b^*) operators and vice versa. Therefore the question arises whether the (b, b^*) operators satisfy the relations (19) or (20) in the case of anticommutation or commutation relations, respectively and whether the operators A_i (see Eq. (25)) have the same form in terms of the (a, a^*) and (b, b^*) operators. In other words, if the (a, a^*) operators in Eq. (25) are expressed in terms of the (b, b^*) ones then the problem arises whether

$$\begin{aligned} A_i = \sum A_i(n_1'n_2'n'k', n_1n_2nk) \\ b(n_1'n_2'n'k')^*b(n_1n_2nk)/Norm(n_1n_2nk) \end{aligned} \tag{30}$$

is valid. It is natural to accept the following

Definition of the AB symmetry: If the (b, b^*) operators satisfy Eq. (20) in the case of anticommutators or Eq. (21) in the case of commutators and all the representation operators (25) in terms of the (b, b^*) operators have the form (30) then it is said that the AB symmetry is satisfied.

We will first investigate whether Eqs. (20) and (21) follow from Eqs. (18) and (19), respectively. As follows from Eqs. (27-29), Eq. (20) follows from Eq. (18) if

$$\eta(n_1n_2nk)\bar{\eta}(n_1, n_2, nk) = (-1)^s \tag{31}$$

while Eq. (21) follows from Eq. (19) if

$$\eta(n_1n_2nk)\bar{\eta}(n_1, n_2, nk) = (-1)^{s+1} \tag{32}$$

We now represent $\eta(n_1n_2nk)$ in the form

$$\eta(n_1n_2nk) = \alpha f(n_1n_2nk) \tag{33}$$

where $f(n_1n_2nk)$ should satisfy the condition

$$f(n_1n_2nk)\bar{f}(n_1,n_2,nk) = 1 \tag{34}$$

Then α should be such that

$$\alpha\bar{\alpha} = \pm(-1)^s \tag{35}$$

where the plus sign refers to anticommutators and the minus sign to commutators, respectively. If the normal spin-statistics connection is valid, i.e. we have anticommutators for odd values of s and commutators for even ones then the r.h.s. of Eq. (35) equals -1 while in the opposite case it equals 1. In Sect. 10. Eq. (35) is discussed in detail and for now we assume that solutions of this relation exist.

A direct calculation [1] shows that if $\eta(n_1n_2nk)$ is given by Eq. (33) and

$$f(n_1n_2nk) = (-1)^{n_1+n_2+n} \tag{36}$$

then the AB symmetry is valid regardless of whether the normal spin-statistics connection is valid or not.

8. Physical and Nonphysical States

Although we have called the sets (a, a^*) and (b, b^*) annihilation and creation operators for particles and antiparticles, respectively, it is not clear yet whether these operators indeed can be treated in such a way. The operator $a(n_1n_2nk)$ can be called the annihilation one only if it annihilates the vacuum vector Φ_0. Then if the operators $a(n_1n_2nk)$ and $a(n_1n_2nk)^*$ satisfy the relations (18) or (19), the vector $a(n_1n_2nk)^*\Phi_0$ has the meaning of the one-particle state. The same can be said about the operators $b(n_1n_2nk)$ and $b(n_1n_2nk)^*$. For these reasons in the standard theory it is required that the vacuum vector should satisfy the conditions (22). Then the elements

$$\Phi_+(n_1n_2nk) = a(n_1n_2nk)^*\Phi_0 \quad \Phi_-(n_1n_2nk) = b(n_1n_2nk)^*\Phi_0 \tag{37}$$

have the meaning of one-particle states for particles and antiparticles, respectively.

However, if one requires the condition (22) in GFQT then it is obvious from Eqs. (28) and Eq. (29), that the elements defined by Eq. (37) are null vectors. Note that in the standard approach the AdS energy is always greater than m while in the GFQT the AdS energy is not positive definite. We can therefore try to modify Eq. (22) as follows. Suppose that 'Physical-nonphysical states assumption' (see Sect. 7.) can be substantiated. Then we can break the set of elements (n_1n_2nk) into two equal nonintersecting parts, S_+ and S_-, such that if $(n_1n_2nk) \in S_+$ then $(\tilde{n}_1\tilde{n}_2nk) \in S_-$ and vice versa. Then, instead of the condition (22) we require

$$a(n_1n_2nk)\Phi_0 = b(n_1n_2nk)\Phi_0 = 0 \quad \forall\ (n_1,n_2,n,k) \in S_+ \tag{38}$$

In that case the elements defined by Eq. (37) will indeed have the meaning of one-particle states for $(n_1n_2nk) \in S_+$.

It is clear that if we wish to work with the full set of elements (n_1n_2nk) then, as follows from Eqs. (28) and (29), the operators (b, b^*) are redundant and we can work only with the operators (a, a^*). However, if one works with the both sets, (a, a^*) and (b, b^*) then such operators can be independent of each other only for a half of the elements $(n_1 n_2 n k)$.

Regardless of how the sets S_+ and S_- are defined, the 'Physical-nonphysical states assumption' cannot be consistent if there exist quantum numbers $(n_1 n_2 n k)$ such that $n_1 = \tilde{n}_1$ and $n_2 = \tilde{n}_2$. Indeed, in that case the sets $(n_1 n_2 n k)$ and $(\tilde{n}_1 \tilde{n}_2 n k)$ are the same what contradicts the assumption that each set $(n_1 n_2 n k)$ belongs either to S_+ or S_-.

Since the replacements $n_1 \to \tilde{n}_1$ and $n_2 \to \tilde{n}_2$ change the signs of the eigenvalues of the h_1 and h_2 operators (see Sect. 7.), the condition that that $n_1 = \tilde{n}_1$ and $n_2 = \tilde{n}_2$ should be valid simultaneously implies that the eigenvalues of the operators h_1 and h_2 should be equal to zero simultaneously. If one considers an IR of the sp(2) algebra and treats the eigenvalues of the diagonal operator h not as elements of $GF(p)$ but as integers, then they take the values of $q_0, q_0 + 2, ...2p - q_0 - 2, 2p - q_0$ (see the discussion in Ref. [1]). It is clear that the eigenvalue is equal to zero in $GF(p)$ only if it is equal to p when considered as an integer. Since the AdS energy is defined as $E = h_1 + h_2$, it is now clear that the above situation can take place only if the energy considered as an integer is equal to 2p. It now follows from Eq. (13) that the energy can be equal to $2p$ only if m is even. Since $m = q_1 + q_2$ and $s = q_1 - q_2$ we conclude that m can be even if and only if s is even. In that case we will necessarily have quantum numbers $(n_1 n_2 n k)$ such that the sets $(n_1 n_2 n k)$ and $(\tilde{n}_1 \tilde{n}_2 n k)$ are the same and therefore the 'Physical-nonphysical states assumption' is not valid. On the other hand, if s is odd (i.e. half-integer in the usual units) then there are no quantum numbers $(n_1 n_2 n k)$ such that the sets $(n_1 n_2 n k)$ and $(\tilde{n}_1 \tilde{n}_2 n k)$ are the same.

Our conclusion is as follows: *If the separation of states should be valid for any quantum numbers then the spin s should be necessarily odd.* In other words, if the notion of particle and antiparticle state is absolute then elementary particles can have only a half-integer spin in the usual units.

In view of the above observations it seems natural to implement the 'Physical-nonphysical states assumption' as follows. *If the quantum numbers $(n_1 n_2 n k)$ are such that $m + 2(n_1 + n_2 + n) < 2p$ then the corresponding state is physical and belongs to S_+, otherwise the state is unphysical and belongs to S_-.*

9. Dirac Vacuum Energy Problem

The Dirac vacuum energy problem is discussed in practically every textbook on LQFT. In its simplified form it can be described as follows. Suppose that the energy spectrum is discrete and n is the quantum number enumerating the states. Let $E(n)$ be the energy in the state n. Consider the electron-positron field. As a result of quantization one gets for the energy operator

$$E = \sum_n E(n)[a(n)^* a(n) - b(n) b(n)^*] \tag{39}$$

where $a(n)$ is the operator of electron annihilation in the state n, $a(n)^*$ is the operator of electron creation in the state n, $b(n)$ is the operator of positron annihilation in the state n and $b(n)^*$ is the operator of positron creation in the state n. It follows from this expression that only anticommutation relations are possible since otherwise the energy of positrons will be negative. However, if anticommutation relations are assumed, it follows from Eq.

(39) that

$$E = \{\sum_n E(n)[a(n)^*a(n) + b(n)^*b(n)]\} + E_0 \tag{40}$$

where E_0 is some infinite negative constant. Its presence was a motivation for developing Dirac's hole theory. In the modern approach it is usually required that the vacuum energy should be zero. This can be obtained by assuming that all operators should be written in the normal form. However, this requirement is not quite consistent since the result of quantization is Eq. (39) where the positron operators are not written in that form (see also the discussion in Sect. 7.).

Consider now the AdS energy operator $M^{05} = h_1 + h_2$ in GFQT. As follows from Eqs. (10) and (25)

$$\begin{aligned} M^{05} = &\sum [m + 2(n_1 + n_2 + n)]a(n_1n_2nk)^* \\ &a(n_1n_2nk)/Norm(n_1n_2nk) \end{aligned} \tag{41}$$

where the sum is taken over all possible quantum numbers (n_1n_2nk). We now wish to replace only the nonphysical (a, a^*) operators by the physical (b, b^*) ones and represent M^{05} in terms of only physical operators. As follows from Eqs. (27-29) and (33-35)

$$\begin{aligned} M^{05} = &\{\textstyle\sum_{S_+} [m + 2(n_1 + n_2 + n)][a(n_1n_2nk)^*a(n_1n_2nk) + \\ &b(n_1n_2nk)^*b(n_1n_2nk)]/Norm(n_1n_2nk)\} + E_{vac} \end{aligned} \tag{42}$$

where

$$E_{vac} = \mp \sum_{S_+} [m + 2(n_1 + n_2 + n)] \tag{43}$$

Here $\mp$ refers to the cases when the (b, b^*) operators anticommute and commute, respectively. The quantity E_{vac} has the meaning of the vacuum energy in GFQT.

The explicit calculation carried out in Ref. [1] gives that the result for E_{vac} in the massive case is

$$E_{vac} = \pm(m - 3)(s - 1)(s + 1)^2(s + 3)/96 \tag{44}$$

Therefore *if s is odd and the separation of states into physical and nonphysical ones is accomplished as in Sect. 8. then $E_{vac} = 0$ only if $s = 1$ (i.e. $s = 1/2$ in the usual units).*

10. Neutral Particles and Spin-Statistics Theorem

The nonexistence of neutral elementary particles in GFQT is one of the most striking differences between GFQT and the standard theory. For this reason we discuss this problem in detail. One could give the following definition of neutral particle:

- i) it is a particle coinciding with its antiparticle
- ii) it is a particle which does not coincide with its antiparticle but they have the same properties

In the standard theory only i) is meaningful since neutral particles are described by real (not complex) fields and this condition is required by Hermiticity. One might think that the definition ii) is only academic since if a particle and its antiparticle have the same properties then they are indistinguishable and can be treated as the same. However, the cases i) and ii) are essentially different from the operator point of view. In the case i) only the (a, a^*) operators are sufficient for describing the operators (23). This is the reflection of the fact that the real field has the number of degrees of freedom twice as less as the complex field. On the other hand, in the case ii) both (a, a^*) and (b, b^*) operators are required, i.e. in the standard theory such a situation is described by a complex field. Nevertheless, the case ii) seems to be rather odd: it implies that there exists a quantum number distinguishing a particle from its antiparticle but this number is not manifested experimentally. We now consider whether the conditions i) or ii) can be implemented in GFQT.

Since each operator a is proportional to some operator b^* and vice versa (see Eqs. (28) and (29)), it is clear that if the particles described by the operators (a, a^*) have some nonzero charge then the particles described by the operators (b, b^*) have the opposite charge and the number of operators cannot be reduced. However, if all possible charges are zero, one could try to implement i) by requiring that each $b(n_1n_2nk)$ should be proportional to $a(n_1n_2nk)$ and then $a(n_1n_2nk)$ will be proportional to $a(\tilde{n}_1, \tilde{n}_2, nk)^*$. In this case the operators (b, b^*) will not be needed at all.

Suppose, for example, that the operators (a, a^*) satisfy the commutation relations (19). In that case the operators $a(n_1n_2nk)$ and $a(n_1'n_2'n'k')$ should commute if the sets (n_1n_2nk) and $(n_1'n_2'n'k')$ are not the same. In particular, one should have $[a(n_1n_2nk), a(\tilde{n}_1\tilde{n}_2nk)] = 0$ if either $n_1 \neq \tilde{n}_1$ or $n_2 \neq \tilde{n}_2$. On the other hand, if $a(\tilde{n}_1\tilde{n}_2nk)$ is proportional to $a(n_1n_2nk)^*$, it follows from Eq. (19) that the commutator cannot be zero. Analogously one can consider the case of anticommutators.

The fact that the number of operators cannot be reduced is also clear from the observation that the (a, a^*) or (b, b^*) operators describe an irreducible representation in which the number of states (by definition) cannot be reduced. Our conclusion is that in GFQT the definition of neutral particle according to i) is fully unacceptable.

Consider now whether it is possible to implement the definition ii) in GFQT. Recall that we started from the operators (a, a^*) and defined the operators (b, b^*) by means of Eq. (28). Then the latter satisfy the same commutation or anticommutation relations as the former and the AB symmetry is valid. Does it mean that the particles described by the operators (b, b^*) are the same as the ones described by the operators (a, a^*)? If one starts from the operators (b, b^*) then, by analogy with Eq. (28), the operators (a, a^*) can be defined as

$$b(n_1n_2nk)^* = \eta'(n_1n_2nk)a(\tilde{n}_1\tilde{n}_2nk)/F(\tilde{n}_1\tilde{n}_2nk) \tag{45}$$

where $\eta'(n_1n_2nk)$ is some function. By analogy with the consideration in Sect. 7. one can show that

$$\eta'(n_1n_2nk) = \beta(-1)^{n_1+n_2+n} \quad \beta\bar{\beta} = \mp 1 \tag{46}$$

where the minus sign refers to the normal spin-statistics connection and the plus sign — to the broken one.

As follows from Eqs. (28), (31-34), (45), (46) and the definition of the quantities $\tilde{n}_1$ and $\tilde{n}_2$ in Sect. 7., the relation between the quantities α and β is $\alpha\bar{\beta} = 1$. Therefore, as follows from Eq. (46), there exist only two possibilities, $\beta = \mp\alpha$, depending on whether the normal spin-statistics connection is valid or not. We conclude that the broken spin-statistics connection implies that $\alpha\bar{\alpha} = \beta\bar{\beta} = 1$ and $\beta = \alpha$ while the normal spin-statistics connection implies that $\alpha\bar{\alpha} = \beta\bar{\beta} = -1$ and $\beta = -\alpha$. In the first case solutions for α and β obviously exist (e.g. $\alpha = \beta = 1$) and the particle and its antiparticle can be treated as neutral in the sense of the definition ii). Since such a situation is clearly unphysical, one might treat the spin-statistics theorem as a requirement excluding neutral particles in the sense ii).

We now consider another possible treatment of the spin-statistics theorem, which seems to be much more interesting. In the case of normal spin-statistics connection we have that

$$\alpha\bar{\alpha} = -1 \tag{47}$$

and the problem arises whether solutions of this relation exist. Such a relation is obviously impossible in the standard theory.

As noted in Sect. 2., -1 is a quadratic residue in $GF(p)$ if $p = 1\ (mod\ 4)$ and a quadratic nonresidue in $GF(p)$ if $p = 3\ (mod\ 4)$. For example, -1 is a quadratic residue in $GF(5)$ since $2^2 = -1\ (mod\ 5)$ but in $GF(7)$ there is no element a such that $a^2 = -1\ (mod\ 7)$. We conclude that if $p = 1\ (mod\ 4)$ then Eq. (47) has solutions in $GF(p)$ and in that case the theory can be constructed without any extention of $GF(p)$.

Consider now the case $p = 3\ (mod\ 4)$. Then Eq. (47) has no solutions in $GF(p)$ and it is necessary to consider this equation in an extention of $GF(p)$ (i.e. there is no real version of GFQT). The minimum extention is obviously $GF(p^2)$ and therefore the problem arises whether Eq. (47) has solutions in $GF(p^2)$.

It is well known [3] that any Galois field without its zero element is a cyclic multiplicative group. Let r be a primitive root, i.e. the element such that any nonzero element of $GF(p^2)$ can be represented as r^k $(k = 1, 2, ..., p^2 - 1)$. It is also well known that the only nontrivial automorphism of $GF(p^2)$ is $\alpha \to \bar{\alpha} = \alpha^p$. Therefore if $\alpha = r^k$ then $\alpha\bar{\alpha} = r^{(p+1)k}$. On the other hand, since $r^{(p^2-1)} = 1$, $r^{(p^2-1)/2} = -1$. Therefore there exists at least a solution with $k = (p-1)/2$.

Our conclusion is that *if* $p = 3\ (mod\ 4)$ *then the spin-statistics theorem implies that the field* $GF(p)$ *should necessarily be extended and the minimum possible extention is* $GF(p^2)$. Therefore the spin-statistics theorem can be treated as a requirement that GFQT should be based on $GF(p^2)$ and the standard theory should be based on complex numbers.

Let us now discuss a different approach to the AB symmetry. A desire to have operators which can be interpreted as those relating separately to particles and antiparticles is natural in view of our experience in the standard approach. However, one might think that in the spirit of GFQT there is no need to have separate operators for particles and antiparticles since they are different states of the same object. We can therefore reformulate the AB symmetry in terms of only (a, a^*) operators as follows. Instead of Eqs. (28) and (29), we consider a *transformation* defined as

$$\begin{aligned} a(n_1n_2nk)^* &\to \eta(n_1n_2nk)a(\tilde{n}_1\tilde{n}_2nk)/F(\tilde{n}_1\tilde{n}_2nk) \\ a(n_1n_2nk) &\to \bar{\eta}(n_1n_2nk)a(\tilde{n}_1\tilde{n}_2nk)^*/F(\tilde{n}_1\tilde{n}_2nk) \end{aligned} \tag{48}$$

Then the AB symmetry can be formulated as a requirement that physical results should be invariant under this transformation.

Let us now apply the AB transformation twice. Then, by analogy with the derivation of Eq. (35), we get

$$a(n_1n_2nk)^* \to \mp a(n_1n_2nk)^* \quad a(n_1n_2nk) \to \mp a(n_1n_2nk) \tag{49}$$

for the normal and broken spin-statistic connections, respectively. Therefore, as a consequence of the spin-statistics theorem, any particle (with the integer or half-integer spin) has the AB^2 parity equal to -1. Therefore in GFQT any interaction can involve only an even number of creation and annihilation operators. In particular, this is additional demonstration of the fact that in GFQT the existence of neutral elementary particles is incompatible with the spin-statistics theorem.

11. Discussion

In the present paper we have discussed in detail the description of free elementary particles in a quantum theory based on a Galois field (GFQT). As noted in Sect. 1., GFQT does not contain infinities at all and all operators are automatically well defined. In my discussions with physicists, some of them commented this fact as follows. This is the approach where a cutoff (the characteristic p of the Galois field) is introduced from the beginning and for this reason there is nothing strange in the fact that the theory does not have infinities. It has a large number p instead and this number can be practically treated as infinite.

However, the difference between Galois fields and usual numbers is not only that the former are finite and the latter are infinite. If the set of usual numbers is visualized as a straight line from $-\infty$ to $+\infty$ then the simplest Galois field can be visualized not as a segment of this line but as a circle (see Fig. 1 in Sect. 2.). This reflects the fact that in Galois fields the rules of arythmetic are different and, as a result, GFQT has many unusual features which have no analog in the standard theory.

The original motivation for investigating GFQT was as follows. Let us take the standard QED in dS or AdS space, write the Hamiltonian and other operators in angular momentum representation and replace standard irreducible representations (IRs) for the electron, positron and photon by corresponding modular IRs. Then we will have a theory with a natural cutoff p and all renormalizations will be well defined. In other words, instead of the standard approach, which, according to Polchinski's joke [38], is essentially based on the formula '$\infty - \infty = physics$', we will have a well defined scheme. One might treat this motivation as an attempt to substantiate standard momentum regularizations (e.g. the Pauli-Villars regularization) at momenta p/R (where R is the radius of the Universe). In other terms this might be treated as introducing fundamental length of order R/p. We now discuss reasons explaining why this naive attempt fails.

Consider first the construction of modular IR for the electron. We start from the state with the minimum energy (where energy=mass) and gradually construct states with higher and higher energies. In such a way we are moving counterclockwise along the circle on Fig. 1 in Sect. 2.. Then sooner or later we will arrive at the left half of the circle, where the energy is negative, and finally we will arrive at the point where energy=-mass (in fact, as noted in Sect. 8., almost four full rotations are needed for that). In other words, instead of the modular analog of IR describing only the electron, we obtain an IR describing the electron and positron simultaneously.

In the standard theory a particle and its antiparticle are described by different IRs but they are combined together by a local covariant equation (in the given case this is the Dirac equation). We see that in GFQT the idea of the Dirac equation is implemented without assuming locality but already at the level of IRs. This automatically explains the existence of antiparticles, shows that a particle cannot exist by itself without its antiparticle and that a particle and its antiparticle are necessarily *different* states of the same object. In particular, there are no elementary particles which in the standard theory are called neutral.

One might immediately conclude that since in GFQT the photon cannot be elementary, this theory cannot be realistic and does not deserve attention. We believe however, that the nonexistence of neutral elementary particles in GFQT shows that the photon, the graviton and other neutral particles should be considered on a deeper level. For example, several authors considered a model where the photon is a composite state of Dirac singletons [39]. They have several unusual properties and probably the importance of singleton IRs is not realized yet. It is interesting to note that Dirac called his paper [40] 'A remarkable representation of the 3 + 2 de Sitter group'.

The nonexistence of neutral elementary particles in GFQT is discussed in detail in Sect. 10.. A possible elementary explanation is as follows. In the standard theory, elementary particles having antiparticles are described by complex fields while neutral elementary particles — by real ones. In GFQT there is no possibility to choose between real and complex fields. As noted in Sect. 10., a possible treatment of the spin-statistics theorem is simply that this is a requirement that quantum theory should be based on complex numbers, and this requirement excludes the existence of neutral elementary particles.

The Dirac vacuum energy problem discussed in Sect. 9. also is a good illustration of the fact that replacement of usual numbers by a Galois field results in a qualitatively new picture. Indeed, in the standard theory the vacuum energy is infinite and, if GFQT is treated simply as a theory with a cutoff p, one would expect that the vacuum energy will be of order p. However, since the rules of arythmetic in Galois fields are different from the standard ones, the result for the vacuum energy is exactly zero. The consideration of the vacuum energy also poses the following very interesting problem. The result is based on the prescription of Sect. 8. for separating physical and nonphysical states. With such a prescription the vacuum energy is zero only for particles with the spin 1/2. Is this an indication that only such particles can be elementary or the prescription (although it seems very natural) should be changed?

In general, the Dirac vacuum energy problem is only one of the examples demonstrating that the standard quantization procedure does not define the order of annihilation and creation operators uniquely and for this reason one has to require additionally that the operators should be written in the normal form (see the discussion in Sect. 7.). However, as noted in Sects. 7., in GFQT this problem has a natural solution and such a requirement is not needed.

The results of Sect. 8. show that in GFQT there exists the following dilemma. If the notion of particles and antiparticles is valid at all energies then only particles with the half-integer spin (in the usual units) can be elementary. Therefore supersymmetry is possible only if the notion of particles and particles is not valid at asymptotically large energies. This problem requires further study.

Since in GFQT one IR necessarily describes a particle and its antiparticle simultaneously, the problem arises whether it is possible to implement this requirement in the standard theory. This problem has been discussed in our recent work [34]. It has been shown that with such a modification of the standard theory, among the Poincare, AdS and dS groups only the latter can be the symmetry group and then only fermions can be elementary. As discussed above, in GFQT there exist much stronger limitations but such a conclusion can be drown only if the above dilemma has the first solution. The difference between GFQT and the standard theory is roughly as follows. If $a = \bar{\eta}b^*$ then necessarily $a^* = \eta b$, $\{a, a^*\} = \eta\bar{\eta}\{b, b^*\}$ and $[a, a^*] = -\eta\bar{\eta}[b, b^*]$. It is easy to satisfy the condition $\eta\bar{\eta} = 1$ but in the field of complex numbers it is impossible to satisfy the condition $\eta\bar{\eta} = -1$. On the other hand, in GFQT this condition is possible and moreover, as shown in Sect. 10., it is the key condition in the spin-statistics theorem.

The above discussion shows that GFQT has very interesting features which might be very important for constructing new quantum theory (which, according to Weinberg [11], 'may be centuries away'). We believe, however, that not only this makes GFQT very attractive.

For centuries, scientists and philosophers have been trying to understand why mathematics is so successful in explaining physical phenomena (see e.g. Ref. [41]). However, such a branch of mathematics as number theory and, in particular, Galois fields, have practically no implications in particle physics. Historically, every new physical theory usually involved more complicated mathematics. The standard mathematical tools in modern quantum theory are differential and integral equations, distributions, analytical functions, representations of Lie algebras in Hilbert spaces etc. At the same time, very impressive results of number theory about properties of natural numbers (e.g. the Wilson theorem) and even the notion of primes are not used at all! The reader can easily notice that GFQT involves only arithmetic of Galois fields (which are even simpler than the set of natural numbers). The very possibility that the future quantum theory could be formulated in such a way, is of indubitable interest.

Acknowledgements: The author is grateful to H. Doughty and C. Hayzelden for numerous useful discussions.

References

[1] F. Lev, hep-th/0403231.

[2] http://pw1.netcom.com/ hjsmith/Pi/Rec1240.html.

[3] B.L. Van der Waerden, Algebra I, Springer-Verlag, Berlin - Heidelberg - New York, 1967; K. Ireland and M. Rosen, A Classical Introduction to Modern Number Theory, *Graduate Texts in Mathematics*-**87**, New York - Heidelberg - Berlin, Springer, 1987.

[4] F. Lev, hep-th/0309003.

[5] H.R. Coish, *Phys. Rev.* **114**, 383 (1959); I.S. Shapiro, *Nucl. Phys.* **21**, 474 (1960); O. Martin, A.M. Odlyzko and S. Wolfram, *Comm. Math. Phys.* **93**, 219 (1984); B. Grossman, *Phys. Lett.* **B197**, 101 (1987); Y. Nambu, in *Quantum Field Theory and Quantum Statistics*, I.A. Batalin et. al. eds. (Bristol, Adam Hilger , 1987); A. Vourdas,

Rep. Progr. Phys. **67**, 267 (2004); H. Doughty, private communication of June 24th, 2004.

[6] W. Pauli, Handbuch der Physik, vol. V/1 (Berlin, 1958); Y. Aharonov and D. Bohm, *Phys. Rev.* **122**, 1649 (1961), 134, 1417 (1964); V.A. Fock, *ZhETF* **42**, 1135 (1962); B.A. Lippman, *Phys. Rev.* **151**, 1023 (1966).

[7] T.D. Newton and E.P. Wigner, *Rev. Mod. Phys*. **21**, 400 (1949).

[8] V.B. Berestetsky, E.M. Lifshits and L.P. Pitaevsky, *Relativistic Quantum Theory*, Vol. IV, Part 1 (Nauka, Moscow 1968).

[9] C. Brans and R.H. Dicke, *Phys. Rev.* **124**, 925 (1961).

[10] S. Weinberg, *Gravitation and Cosmology* (John Wiley & Sons Inc., 1972),

[11] S. Weinberg, *Dreams of a Final Theory* (A Division of Random House Inc., New York, 1992).

[12] H.I. Hartman and C. Nissin-Sabat, *Am. J. Phys.* **21**, 1163, (2003).

[13] S. Weinberg, hep-th/9702027.

[14] E.P. Wigner, *Ann. Math.* **40**, 149 (1939).

[15] T. Kato, *Perturbation Theory for Linear Operators* (Springer-Verlag, Berlin - Heidelberg - New York, 1966).

[16] P.A.M. Dirac, *Rev. Mod. Phys.* **21**, 392 (1949).

[17] J.C. Jantzen, *Representation of Lie algebras in prime characteristics*, NATO Asi series, series C, *Math. Phys. Sci.* **514** "Representation theories and algebraic geometry", 185-235, Kluwer Acad. Publ. Dordrecht, 1998.

[18] H. Zassenhaus, The representations of Lie algebras of prime characteristics, *Proc. Glasgow Math. Assoc.* **2** (1954) 1-36.

[19] R.F. Streater and A.S. Wightman, PCT, Spin, *Statistics and All That* (W.A.Benjamin Inc. New York-Amsterdam, 1964).

[20] F.M. Lev, *Theor. Math. Phys.* **138**, 208 (2004).

[21] A.I.Akhiezer and V.B.Berestetsky, *Quantum Electrodynamics* (Nauka, Moscow, 1969); Itzykson, C., and Zuber, J.-B., *Quantum Field Theory* (McGraw-Hill Book Company, New York, 1982).

[22] S. Weinberg, *The Quantum Theory of Fields* (Cambridge, Cambridge University Press, 1995).

[23] W. Pauli, in "N. Bohr and the Development of Physics" (Pergamon Press, London, 1955); G. Gravert, G. Luders and G. Rollnik, *Fortschr. Phys.* **7**, 291 (1959); R. Jost, *Helv. Phys. Acta* **30**, 409 (1957); J. Schwinger, *Proc. Nat. Acad. Sci.* **44**, 223 (1958); O.W. Greenberg, hep-ph/0309309.

[24] W. Pauli, *Phys. Rev.* **58**, 116 (1940).

[25] R. Verch, *Commun. Math. Phys.* **223**, 261 (2001); B. Kuckert, *Phys. Lett.* **A322**, 47 (2004).

[26] S. Perlmutter et. al. *Astrophys. J.* **517**, 565 (1999); A. Melchiori et. al. *Astrophys. J.*, **536**, L63 (2000).

[27] E. Inonu and E.P. Wigner, *Nuovo Cimento*, IX, 705 (1952).

[28] F.M. Lev and E.G. Mirmovich, VINITI No 6099 Dep.

[29] P.J.E. Peebles and B. Ratra, *Rev. Mod. Phys.* **75**, 559 (2003).

[30] A. Linde, *Particle Physics and Inflationary Cosmology* (Harwood Academic Publishers, Paris, 1990); R. Kallosh and A. Linde, *Phys. Rev.* **D67**, 023510 (2003); A. Linde, hep-th/0211048.

[31] F.M. Lev, *J. Phys.* **A21**, 599 (1988).

[32] F.M. Lev, *J. Math. Phys.*, **34**, 490 (1993).

[33] F.M. Lev, *J. Phys.* **A32**, 1225 (1999).

[34] F.M. Lev, *J. Phys.* **A37**, 3285 (2004).

[35] F.M. Lev, hep-th/0207087.

[36] N.T. Evans, *J. Math. Phys.* **8**, 170 (1967).

[37] B. Braden, *Bull. Amer. Math. Soc.* **73**, 482 (1967).

[38] J. Polchinski, hep-th/0209105.

[39] M. Flato and C. Fronsdal, *Lett. Math. Phys.* **2**, 421 (1978); L. Castell and W. Heidenreich, *Phys. Rev.* **D24**, 371 (1981); C. Fronsdal, *Phys. Rev.* **D26**, 1988 (1982).

[40] P.A.M. Dirac, *J. Math. Phys.* **4**, 901 (1963).

[41] E.P. Wigner, in 'The World Treasury of Physics, Astronomy and Mathematics', p. 526 (Timothy Ferris ed., Little Brown and Company, Boston-New York-London, 1991).

In: Horizons in World Physics, Volume 268
Editors: M. Everett and L. Pedroza, pp. 211-246
ISBN 978-1-60692-861-5

Chapter 6

BLACK HOLE ENTROPY FROM ENTANGLEMENT: A REVIEW

***Saurya Das*[1]*, *S. Shankaranarayanan*[2]† *and Sourav Sur*[1]‡**
[1] Department of Physics, University of Lethbridge,
4401 University Drive, Lethbridge, Alberta, Canada T1K 3M4
[2] Institute of Cosmology and Gravitation,
University of Portsmouth, Portsmouth P01 2EG, U.K

Abstract

We review aspects of the thermodynamics of black holes and in particular take into account the fact that the quantum entanglement between the degrees of freedom of a scalar field, traced inside the event horizon, can be the origin of black hole entropy. The main reason behind such a plausibility is that the well-known Bekenstein-Hawking entropy-area proportionality — the so-called 'area law' of black hole physics — holds for entanglement entropy as well, provided the scalar field is in its ground state, or in other minimum uncertainty states, such as a generic coherent state or squeezed state. However, when the field is either in an excited state or in a state which is a superposition of ground and excited states, a power-law correction to the area law is shown to exist. Such a correction term falls off with increasing area, so that eventually the area law is recovered for large enough horizon area. On ascertaining the location of the microscopic degrees of freedom that lead to the entanglement entropy of black holes, it is found that although the degrees of freedom close to the horizon contribute most to the total entropy, the contributions from those that are far from the horizon are more significant for excited/superposed states than for the ground state. Thus, the deviations from the area law for excited/superposed states may, in a way, be attributed to the far-away degrees of freedom. Finally, taking the scalar field (which is traced over) to be massive, we explore the changes on the area law due to the mass. Although most of our computations are done in flat space-time with a hypothetical spherical region, considered to be the analogue of the horizon, we show that our results hold as well in curved space-times representing static asymptotically flat spherical black holes with single horizon.

*E-mail address: saurya.das@uleth.ca
†E-mail address: shanki.subramaniam@port.ac.uk
‡E-mail address: sourav.sur@uleth.ca

1. Introduction

One of the most remarkable features of black hole physics is the realization that black holes behave as thermodynamic systems and possess entropy and temperature. The pioneering works in the field of black hole thermodynamics started with Bekenstein [1], who argued that the universal applicability of the second law of thermodynamics rests on the fact that a black hole must possess an entropy ($S_{\rm BH}$) proportional to the area ($\mathcal{A}_{\rm H}$) of its horizon. The macroscopic properties of black holes were subsequently formalized by Bardeen, Carter and Hawking [2] as the four laws of black hole mechanics, in analogy with ordinary thermodynamics. They showed that (i) the surface gravity κ, which is the force applied by an observer at spatial infinity to hold a particle of unit mass in place at the location of the horizon, is same everywhere on the horizon for a stationary black hole — the statement of the zeroth law of black hole physics. (ii) The surface gravity κ of the black hole analogically resembles the temperature ($T_{\rm H}$) of the hole, in accordance with the interpretation of the horizon area as the black hole entropy. This may be conceived from the first law, which states that the change in mass (energy) of the black hole is proportional to the surface gravity times the change in horizon area.

Hawking's demonstration of black hole thermal radiation [3] paved the way to understand the physical significance of the temperature $T_{\rm H}$ (and hence the entropy-area proportionality). Hawking showed that quantum effects in the background of a body collapsing to a Schwarzschild black hole leads to the emission of a thermal radiation at a characteristic temperature:

$$T_{\rm H} = \left(\frac{\hbar c}{k_{_B}}\right)\frac{\kappa}{2\pi} = \left(\frac{\hbar c^3}{G k_{_B}}\right)\frac{1}{8\pi M}\,, \tag{1}$$

where G is the Newton's constant in four dimensions, $k_{_B}$ is the Boltzmann constant, and M is the mass of the black hole. The factor of proportionality between temperature and surface gravity (and as such between entropy and area) gets fixed in Hawking's derivation [3], thus leading to the Bekenstein-Hawking area law:

$$S_{\rm BH} = \left(\frac{k_{_B}}{4}\right)\frac{\mathcal{A}_{\rm H}}{\ell_{\rm Pl}^2} \quad , \tag{2}$$

where $\ell_{\rm Pl} = \sqrt{G\hbar/c^3}$ is the four dimensional Planck length.

Black-hole thermodynamics and, in particular, black-hole entropy raises several important questions which can be broadly classified into two categories:

- **Gravitational collapse leading to black-hole formation**
  - (i) What is the dynamical mechanism that makes $S_{\rm BH}$ a universal function, independent of the black-hole's past history and detailed internal condition?
  - (ii) How does a pure state evolve into a mixed (thermal) state? Is there a information loss due to the formation of black-hole and Hawking process? Does the usual quantum mechanics need to be modified in the context of black-holes?
  - (iii) Can quantum theory of gravity remove the formation of space-time singularity due to the gravitational collapse?

- **Near thermodynamical equilibrium**

  (i) Unlike other thermodynamical systems, why is black-hole entropy non-extensive? i.e., why $S_{\rm BH}$ is proportional to area and not volume?

  (ii) Why is the black-hole entropy large?[1]

  (iii) How $S_{\rm BH}$ concords with the standard view of the statistical origin? What are the black-hole microstates?

$$S \stackrel{?}{=} k_B \ln(\# \text{ of microstates})$$

  (iv) Are there corrections to $S_{\rm BH}$? If there are, how generic are they?

  (v) Where are the microscopic degrees of freedom responsible for black-hole entropy located?

These questions often seem related, which a correct theory of quantum gravity is expected to address. In the absence of a workable theory of quantum gravity, there have been several approaches which address one or several of the above questions. Most of the effort in the literature, as in this review, has been to understand the microscopic statistical mechanical origin of $S_{\rm BH}$ assuming that the black-hole is in a (near) thermal equilibrium or not interacting with surroundings.

The various approaches may broadly be classified into two categories [5]: (a) the ones that associate $S_{\rm BH}$ with fundamental states such as strings, D-Branes, spin-networks, etc. [6, 7], and (b) the others that associate $S_{\rm BH}$ with quantum fields in a fixed BH background, like the brick-wall model [8], the quantum entanglement of modes inside and outside of the horizon [9–11] and the Noether charge [12]. As mentioned above, although, none of these approaches can be considered to be complete; all of them — within their domains of applicability — by counting certain microscopic states yield (2). This is in complete contrast to other physical systems, such as ideal gas, where quantum degrees of freedom (DOF) are uniqely identified and lead to the classical thermodynamic entropy.

The above discussion raises three important questions which we try to address in this review:

[1] In order to see that, let us compare $S_{\rm BH}$ with the entropy of the current universe. The entropy of the universe within our horizon today [4] is

$$S_{\rm Univ} \sim 10^{87} \left(\frac{Th^{-1}}{2.75K}\right) \tag{3}$$

where T is the temperature of the universe, h is of the order unity. On the other hand the entropy of a Schwarzschild black-hole is

$$S_{\rm BH} \sim 10^{77} \left(\frac{M}{M_\odot}\right)^2 \tag{4}$$

where $M_\odot$ is the solar mass. Hence, a couple of hundred thousand solar mass black holes can contain as much entropy as is free in the entire universe. There is increasing evidence that super-massive black holes exist at the centers of many galaxies. By now we know that a large fraction of galaxies — of the total (of order) 10^{11} — contain such super-massive black holes with mass range $10^6 < M_{\rm BH}/M_\odot < 10^{10}$. This implies that the entropy of the black holes dominates all other sources of entropy. Hence, understanding the origins of black-hole entropy may help us explain the entropy budget of the universe.

1. *Is it sufficient for an approach to reproduce (2) or need to go beyond $S_{\rm BH}$?*

   As we know, $S_{\rm BH}$ is a semi-classical result and there are strong indications that Eq. (2) is valid for large black holes [i.e. $\mathcal{A}_{\rm H} \gg \ell_{\rm Pl}$]. However, it is not clear, whether this relation will continue to hold for the Planck-size black-holes. Besides, there is no reason to expect that $S_{\rm BH}$ to be the whole answer for a correct theory of quantum gravity. In order to have a better understanding of black-hole entropy, it is imperative for any approach to go beyond $S_{\rm BH}$ and identify the subleading corrections.

2. *Are the quantum DOF that contribute to $S_{\rm BH}$ and its subleading corrections, identical or different?*

   In general, the quantum DOF can be different. However, several approaches in the literature [13] that do lead to subleading corrections either assume that the quantum DOF are identical or do not *disentangle* DOF contribution to $S_{\rm BH}$ and the subleading corrections.

3. *Can one locate the quantum DOF that give rise to $S_{\rm BH}$ and its subleading corrections?*

   More specifically, can we determine to what extent do the quantum DOF close to the horizon or far from the horizon contribute to the $S_{\rm BH}$ and its corrections. Depending on the approach, one either counts certain DOF on the horizon, or abstract DOF related to the black hole, and there does not appear to be a consensus about which DOF are relevant or about their precise location [5].

In this review, using the approach of entanglement of modes across the black-hole horizon, we address the above three issues and show that:

1. The entanglement leads to generic power-law corrections to the Bekenstein-Hawking entropy, of the form [14]:

   $$S = \sigma_0 \left(\frac{\mathcal{A}}{a^2}\right) + \sigma_1 \left(\frac{\mathcal{A}}{a^2}\right)^{-\nu} \quad ; \quad (0 < \nu < 1)\,, \tag{5}$$

   where σ_0, σ_1 are constants and a is the lattice spacing.

2. The quantum degrees of freedom that lead to $S_{\rm BH}$ and subleading corrections are different, and

3. The contribution to $S_{\rm BH}$ comes from the region close to the horizon while the subleading corrections have a larger contribution from the region far from the horizon [15].

This review is organized as follows: in the next section we briefly review the basic features of quantum entanglement and the concept of entanglement entropy. In sec. (3.), we provide a heuristic picture of the link between the entanglement entropy and black-hole entropy. In sec. (4.) we discuss the procedure and assumptions to compute the entanglement entropy of a scalar field in black-hole space-times. In sec. (5.) we review the cases where the scalar field is either in its ground state, or in a generalized coherent state, or in a class of squeezed states — for all these cases the area law is found to hold. In sec. (6.), we show that the for the superposition of ground and 1-particle state, the entanglement entropy has a subleading power-law correction to the AL. In sec. (7.) we study the locations scalar field DOF that are responsible for the entanglement entropy. In sec. (8.) we examine the entanglement entropy due to a massive scalar field and compare it with that obtained for a massless scalar field. We conclude with a summary and open questions in sec. (9.). In Appendix A. we discuss the relevance for considering massless or massive scalar field for computing the entanglement entropy of black holes, from the perspective of gravitational perturbations in static black hole space-times. In Appendix B. we discuss the steps to obtain the Hamiltonian of a scalar field in the static black-hole background, which in Lemaître coordinates, and at a *fixed* Lemaître time, reduces to the scalar field Hamiltonian in flat space-time.

Before we proceed, we outline our conventions and notations: We work in four-dimensions and our signature for the metric is $(-,+,+,+)$. Hereafter, we use units with $k_B = c = \hbar = 1$ and set the Planck mass $M_{\rm Pl}^2 = 1/(16\pi G)$. The quantum field φ is a minimally coupled scalar field.

2. Entanglement Entropy

Let us consider a bipartite quantum mechanical system, i.e., a system which can be decomposed into two subsystems u and v, such that the Hilbert space of the system is a tensor product of the subsystem Hilbert spaces:

$$\mathcal{H} = \mathcal{H}_u \otimes \mathcal{H}_v \,. \tag{6}$$

Let $|u_i\rangle$ and $|v_j\rangle$ are eigen-bases which span the Hilbert spaces $\mathcal{H}_u$ and $\mathcal{H}_v$ respectively. Then $|u_i\rangle \otimes |v_j\rangle$ forms an eigen-basis in $\mathcal{H}$, and in terms of this a generic wave-function $|\Psi\rangle$ in $\mathcal{H}$ can be expanded as:

$$|\Psi\rangle = \sum_{ij} d_{ij} |u_i\rangle \otimes |v_j\rangle \ \in \mathcal{H} \,. \tag{7}$$

The density matrix operator of the whole system defined by:

$$\rho \equiv |\Psi\rangle\langle\Psi| \,, \tag{8}$$

has the following properties: (i) it is non-negative, i.e., for any vector $|\phi\rangle$, $\langle\phi|\rho|\phi\rangle \geq 0$, (ii) it is self-adjoint ($\rho^\dagger = \rho$), (iii) its trace is unity ($\mathrm{Tr}(\rho) = 1$), and (iv) it is idempotent ($\rho^2 = \rho$), which means that it has only two eigenvalues $p_n = 0, 1$.

Now the expectation values of the operators which represent the observables in one subsystem, say u, can be obtained using the *reduced* density matrix operator (ρ_u) for the subsystem u, which is the trace of the full density matrix operator (ρ) over the other subsystem v:

$$\rho_u = \mathrm{Tr}_v(\rho) = \sum_l \langle v_l|\rho|v_l\rangle = \sum_{i,j,k} |u_i\rangle d_{ik} d^\star_{jk} \langle u_j| \,. \tag{9}$$

When the correlation coefficients $d_{ij} = 1$ (for all i, j), the wave-function $|\Psi\rangle$ describing the entire system is just the product of the wave-functions $|\Psi_u\rangle$ and $|\Psi_v\rangle$ which describe the subsystems u and v respectively, and the full density matrix is $\rho = \rho_u \rho_v$. The reduced density matrix for the subsystem u (say) is then simply given by $\rho_u = |\Psi_u\rangle\langle\Psi_u|$, and bears the same properties as those listed above for the full density matrix. In general, however, $|\Psi\rangle \neq |\Psi_u\rangle \otimes |\Psi_v\rangle$, and the two subsystems are said to be in an *entangled* or *EPR* state[2]. For such entanglement the reduced density matrix (ρ_u) for the subsystem u is given by the general expression (9). A similar expression can be obtained for the reduced density matrix (ρ_v) for the subsystem v, by taking the trace of full density matrix (ρ) over the subsystem u. Both ρ_u and ρ_v have the same properties as listed above for ρ, except that the idempotency is lost, i.e., $\rho_u^2 \neq \rho_u, \rho_v^2 \neq \rho_v$. As such the eigenvalues of ρ_u and ρ_v are no longer restricted to 0 and 1, but are in between 0 and 1. However, an important property of reduced density matrices is that irrespective of the tracing over the eigenvalues remain the same, i.e., both ρ_u and ρ_v have the same set of eigenvalues.

The *entanglement entropy* or *Von Neumann entropy* is a manifestation of one's ignorance resulting from tracing over one part of the system, and is defined by

$$S_{\rm ent} \equiv -Tr_u\left(\rho_u \ln \rho_u\right) \equiv -Tr_v\left(\rho_v \ln \rho_v\right) = -\sum_n p_n \ln p_n \ ; \qquad (0 < p_n < 1)\,. \quad (10)$$

Since the eigenvalues p_n of ρ_u and ρ_v are the same, $S_{\rm ent}$ is the same regardless of which part of the system is being traced over.

3. Entanglement Entropy and Black-Hole Entropy — Connection

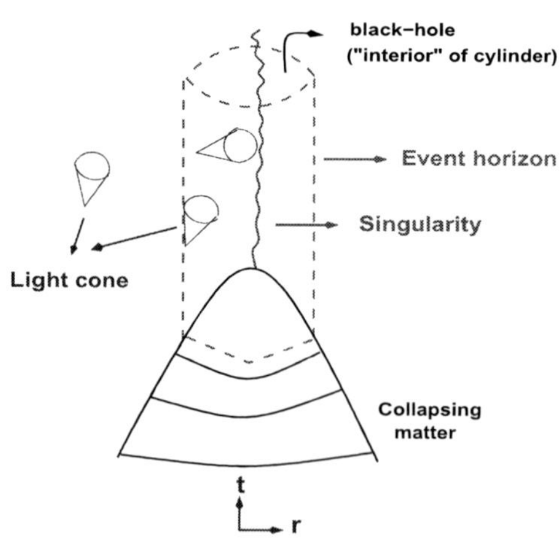

Figure 1. Space-time diagram depicting the collapse of a star to form a black-hole.

Before we proceed with the setup and assumption, in this section, we provide a heuristic link between the entanglement entropy and black-hole entropy.

[2]For details, see the review [17].

To understand this, let us consider a scalar field on a background of a collapsing star. Before the collapse, an outside observer, atleast theoretically, has all the information about the collapsing star. Hence, the entanglement entropy is zero. During the collapse and once the horizon forms, $S_{\rm BH}$ is non-zero. The outside observers at spatial infinity do not have the information about the quantum degrees of freedom inside the horizon. Thus, the entanglement entropy is non-zero. In other words, both the entropies are associated with the existence of horizon[3].

Besides this, $S_{\rm ent}$ and $S_{\rm BH}$ are pure quantum effects with no classical analogue. Based on this picture, we obtain $S_{\rm ent}$ of scalar fields in a fixed background. Although, this analysis is semiclassical, since the entanglement is a quantum effect and should be present in any theory of quantum gravity, the results presented here do have implications beyond the semiclassical regime.

4. Entanglement Entropy of Scalar Fields — Assumptions and Setup

In this section, we consider a free scalar field propagating in spherically symmetric black-hole space-times. The motivation for such scalar field is presented in Appendix A..

We consider a massless scalar field (φ) propagating in an asymptotically flat, four-dimensional black-hole background given by the Lemaître line-element (95) [see Appendix B.]. The Hamiltonian for the scalar field propagating in the above line-element is

$$H_{\rm BH}(\tau) = \sum_{lm} \frac{1}{2}\int_{\tau}^{\infty} d\xi \left[\frac{1}{r^2\sqrt{1-f(r)}}\Pi_{lm}^2 + \frac{r^2}{\sqrt{1-f(r)}}\left(\partial_\xi \varphi_{lm}\right)^2 \right. \\ \left. + \, l(l+1)\sqrt{1-f(r)}\,\varphi_{lm}^2\right] , \tag{11}$$

where φ_{lm} is the spherical decomposed field and Π_{lm} is the canonical conjugate of φ_{lm}, i.e.,

$$\varphi_{lm}(r) = r\int d\Omega\; Z_{lm}(\theta,\phi)\varphi(\vec{r})\;; \qquad \Pi_{lm}(r) = r\int d\Omega\; Z_{lm}(\theta,\phi)\Pi(\vec{r})\,, \tag{12}$$

where $Z_{lm}(\theta, \phi)$ are the real spherical harmonics. [For discussion about the Lemaître co-ordinates and explicit derivation of Eq. (11), see Appendix (B.).] Note that the above Hamiltonian is explicitly time-dependent.

Having obtained the Hamiltonian, the next step is quantization. We use Schrödinger representation since it provides a simple and intuitive description of vacuum states for time-dependent Hamiltonian. Formally, we take the basis vector of the state vector space to be the eigenstate of the field operator $\hat{\varphi}(\tau,\xi)$ on a fixed τ hypersurface, with eigenvalues $\varphi(\xi)$, i.e.,

$$\hat{\varphi}(\tau,\xi)|\,\varphi(\xi),\tau\,\rangle = \varphi(\xi)|\,\varphi(\xi),\tau\,\rangle \tag{13}$$

[3]It is important to understand that it is possible to obtain a non-vanishing entanglement entropy for a scalar field in flat space-time by artificially creating a horizon [10]. However, in the case of black-hole, the event horizon is a physical boundary beyond which the observers do not have access to information.

The quantum states are explicit functions of time and are represented by wave functionals $\Psi[\varphi(\xi), \tau]$ which satisfy the functional Schrödinger equation:

$$i\frac{\partial \Psi}{\partial \tau} = \int_{\tau}^{\infty} d\xi \, H_{\rm BH}(\tau) \, \Psi[\varphi(\xi), \tau]. \tag{14}$$

We now assume that the above Hamiltonian evolves adiabatically. Technically, this implies that the evolution of the late-time modes leading to Hawking particles are negligible. In the Schroedinger formulation, the above assumption translates to $\Psi[\varphi(\xi), \tau]$ being independent of time. At a fixed Lemaître time, the Hamiltonian (11) reduces to the following flat space-time Hamiltonian [see Appendix (B.) for details]

$$H = \sum_{lm} H_{lm} = \sum_{lm} \frac{1}{2} \int_0^{\infty} dr \left\{ \pi_{lm}^2(r) + x^2 \left[\frac{\partial}{\partial r} \left(\frac{\varphi_{lm}(r)}{r} \right) \right]^2 + \frac{l(l+1)}{r^2} \, \varphi_{lm}^2(r) \right\} . \tag{15}$$

The quantum states [defined above in Eq. (13)] of this Hamiltonian is time-independent and $\Psi[\varphi]$ satisfies the time independent Scrödinger equation

$$\int_0^{\infty} dr H \Psi[\varphi(r)] = E \Psi \tag{16}$$

The procedure of finding the entropy involves the following steps:

(i) *Discretize Hamiltonian (15)*: Obtaining an analytic expression for the entropy using the Von Neumann definition (10) is prohibitively difficult. For the field theory, even if we obtain closed-form expression of the density matrix, it is not possible to analytically evaluate the entanglement entropy. Hence, we discretize the Hamiltonian in a spherical lattice of spacing a such that $r \to r_i$; $r_{i+1} - r_i = a$. The ultraviolet cutoff is therefore $M = a^{-1}$. The lattice is of very large but finite size $L = (N+1)a$ $(N \gg 1)$, with a chosen closed spherical region of radius $R(n+1/2)a$ inside of it, as shown in the Fig. 2. It is this closed region, by tracing over the inside or outside of which, one can obtain the reduced density matrix. We demand that the field variables $\varphi_{lm}(r) = 0$ for $r \geq L$ so that the infrared cutoff is $\tilde{M} = L^{-1}$.The ultraviolet cutoff is therefore $M = a^{-1}$.

Now, in discretizing the terms containing the derivatives in the Hamiltonian (15), one usually adopts the middle-point prescription, i.e., the derivative of the form $f(x)d_x[g(x)]$ is replaced by $f_{j+1/2}[g_{j+1} - g_j]/a$. The discretized Hamiltonian is given by

$$\begin{aligned} H_{lm} &= \frac{1}{2a} \sum_{j=1}^{N} \left[\pi_{lm,j}^2 + \left(j + \frac{1}{2} \right)^2 \left(\frac{\varphi_{lm,j}}{j} - \frac{\varphi_{lm,j+1}}{j+1} \right)^2 + \frac{l(l+1)}{j^2} \, \varphi_{lm,j}^2 \right], \\ H &= \sum_{l,m} H_{lm} \, , \end{aligned} \tag{17}$$

where $\varphi_{lm,j} \equiv \varphi_{lm}(r_j)$, $\pi_{lm} \equiv \pi_{lm,j}(r_j)$, which satisfy the canonical commutation relations:

$$[\varphi_{lm,j}, \pi_{l'm',j'}] = i\delta_{ll'}\delta_{mm'}\delta_{jj'} \, . \tag{18}$$

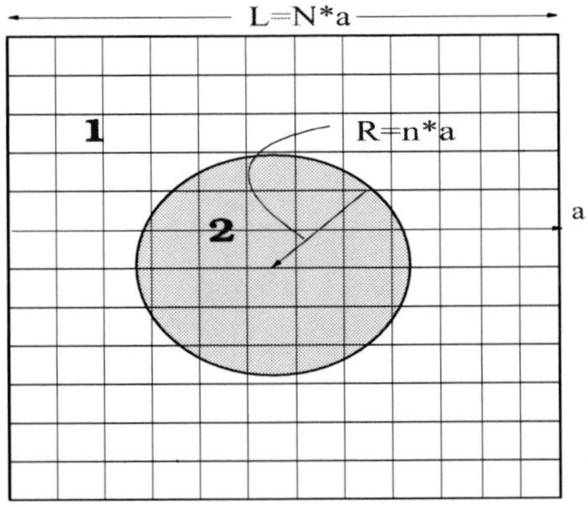

Figure 2. Discretization of the scalar field propagating in the flat space-time. The tracing is done over the shaded region.

Up to an overall factor of a^{-1}, the Hamiltonian H_{lm}, given by Eq. (17), represents the Hamiltonian of $N-$coupled harmonic oscillators (HOs):

$$H_{(N-\text{HO})} = \frac{1}{2}\sum_{i=1}^{N} p_i^2 + \frac{1}{2}\sum_{i,j=1}^{N} x_i K_{ij} x_j \,, \tag{19}$$

where the coordinates x_i replace the field variables φ_{lm}, the momenta p_i replace the conjugate momentum variables π_{lm}, and the $N \times N$ matrix K_{ij} $(i, j = 1, \ldots, N)$ represents the potential energy and interaction between the oscillators:

$$\begin{aligned} K_{ij} &= \frac{1}{i^2}\left[l(l+1)\,\delta_{ij} + \frac{9}{4}\,\delta_{i1}\delta_{j1} + \left(N - \frac{1}{2}\right)^2 \delta_{iN}\delta_{jN} \right. \\ &+ \left. \left\{\left(i + \frac{1}{2}\right)^2 + \left(i - \frac{1}{2}\right)^2\right\} \delta_{i,j(i\neq 1,N)} \right] \\ &- \left[\frac{(j+\frac{1}{2})^2}{j(j+1)}\right] \delta_{i,j+1} - \left[\frac{(i+\frac{1}{2})^2}{i(i+1)}\right] \delta_{i,j-1}. \end{aligned} \tag{20}$$

The last two terms which originate from the derivative term in Eq. (15) denote the nearest-neighbor interactions. Schematically, the matrix K is as shown below, where the off-diagonal terms represent interactions:

$$K = \begin{pmatrix} \times & \times & & & & & \\ \times & \times & \times & & & & \\ & \times & \times & \times & & & \\ & & \times & \times & \times & & \\ & & & \times & \times & \times & \\ & & & & \times & \times & \times \\ & & & & & \times & \times \end{pmatrix}. \tag{21}$$

(ii) *Choose the quantum state of the field*: The most general eigen-state of the Hamiltonian

(19) for the $N-$coupled HOs is given by

$$\psi(x_1,\ldots,x_N) = \prod_{i=1}^{N} \mathcal{N}_i\, \mathcal{H}_{\nu_i}\left(k_{Di}^{1/4}\,\underline{\mathrm{x}}_i\right) \exp\left(-\frac{1}{2}k_{Di}^{1/2}\,\underline{\mathrm{x}}_i^2\right), \tag{22}$$

where $\mathcal{N}_i$'s are the normalization constants given by

$$\mathcal{N}_i = \frac{k_{Di}^{1/4}}{\pi^{1/4}\sqrt{2^{\nu_i}\nu_i!}} \quad , \quad (i = 1, \ldots N), \tag{23}$$

$\underline{\mathrm{x}} = Ux$, $(U^TU = I_N)$, $x^T = (x_1,\ldots,x_N)$, $\underline{\mathrm{x}}^T = (\underline{\mathrm{x}}_1,\ldots,\underline{\mathrm{x}}_N)$, $K_D \equiv UKU^T$ is a diagonal matrix with elements k_{Di}, and $\nu_i\,(i = 1\ldots N)$ are the indices of the Hermite polynomials $(\mathcal{H}_\nu)$. The frequencies are ordered such that $k_{Di} > k_{Dj}$ for $i > j$.

The difficulty however is that it is not possible to work with an arbitrary $N-$particle state as the density matrix (24) cannot be expressed in a closed form. Therefore one has to make some specific choices for the quantum state in order to make the calculations tractable. In the following two sections we will discuss several possible choices.

(iii) *Obtain the reduced density matrix by tracing over certain closed region*: The reduced density matrix is obtained by tracing over the first n of the N oscillators:

$$\rho\left(t;t'\right) = \int \prod_{i=1}^{n} dx_i\, \psi(x_1,\ldots,x_n;t)\, \psi^\star(x_1,\ldots,x_n;t') \tag{24}$$

$$= \int \prod_{i=1}^{n} dx_i \exp\left[-\frac{x^T\Omega x}{2}\right] \prod_{i=1}^{N} \mathcal{N}_i \mathcal{H}_{\nu_i}\left(k_{Di}^{1/4}\underline{\mathrm{x}}_i\right) \exp\left[-\frac{x'^T\Omega x'}{2}\right] \prod_{j=1}^{N} \mathcal{N}_j \mathcal{H}_{\nu_j}\left(k_{Di}^{1/4}\underline{\mathrm{x}}'_i\right)$$

where $\Omega = U^TK_D^{1/2}U$ is an $N\times N$ matrix, such that $|\Omega| = |K_D|^{1/2}$, and we have made the following change in notation: $x^T = (x_1,\ldots,x_n;t_1,\ldots,t_{N-1}) = (x_1,\ldots,x_n;t)$, with $t \equiv t_1,\ldots,t_{N-n}\,; t_j \equiv x_{n+j}$, $j = 1,\ldots,(N-n)$. One may verify that $\rho^2 \neq \rho$, i.e., the state obtained by integrating over n of the HOs is mixed, although the full state is pure.

(iv) *Compute the entropy:* The entanglement entropy can be calculated by substituting the reduced density matrix (24) into the expression (10).

5. Warm up — Entanglement Entropy for (Displaced) Ground State

In this section, we review the procedure for obtaining $S_{\rm ent}$ in the ground state, coherent state (which is a displaced ground state) and a class of squeezed state (which are unitarily related to the ground state). We show that the in these three cases that the entanglement entropy is proportional to area.

Ground State (GS):

When all the HOs are in their GS, then by setting $\nu_i = 0$, for all i, the wave function (22) takes the form

$$\psi_{\rm GS}(x_1,\ldots,x_N) = \prod_{i=1}^{N}\left(\frac{k_{Di}}{\pi}\right)^{1/4} \exp\left(-\frac{1}{2}k_{Di}^{1/2}\underline{\mathrm{x}}_i^2\right) = \left(\frac{|\Omega|}{\pi^N}\right)^{1/4} \exp\left[-\frac{x^T\cdot\Omega\cdot x}{2}\right]. \tag{25}$$

The corresponding density matrix can be evaluated exactly as [10]:

$$\rho_{\rm GS}(t;t') = \sqrt{\frac{|\Omega|}{\pi^{N-n}|A|}}\, \exp\left[-\frac{t^T\gamma t + t'^T\gamma t'}{2} + t^T\beta t'\right], \tag{26}$$

where we have decomposed

$$\Omega \sim K^{1/2} = \begin{pmatrix} A & B \\ B^T & C \end{pmatrix}, \tag{27}$$

and defined

$$\beta = \frac{B^T A^{-1} B}{2} \;;\;\; \gamma = C - \beta\,. \tag{28}$$

A is an $n \times n$ symmetric matrix, B is an $n \times (N-n)$ matrix, and C, β, γ are all $(N-n) \times (N-n)$ symmetric matrices. The matrices B and β are non-zero only when the HOs are interacting.

Performing a series of unitary transformations:

$$\begin{aligned} V\gamma V^T &= \gamma_D = \text{diag}\;,\; \bar\beta \equiv \gamma_D^{-1/2} V\beta V^T \gamma_D^{-1/2}\,, \\ W\bar\beta W^T &= \bar\beta_D = \text{diag}\;,\; v \equiv W^T \gamma_D^{1/2} V\,, \end{aligned} \tag{29}$$

one can reduce $\rho_{\rm GS}(t;t')$ to a product of the reduced density matrices $\rho_{(2-{\rm HO})}(t;t')$ for $(N-n)$ two coupled HOs with one oscillator traced over (i.e., $N=2, n=1$) [10]:

$$\rho_{\rm GS}(t;t') = \prod_{i=1}^{N-n} \rho_{(2-{\rm HO})}(t;t')\,, \quad \rho_{(2-{\rm HO})}(t;t') = \sqrt{\frac{|\Omega|}{\pi^{N-n}|A|}}\, \exp\left[-\frac{v_i^2 + v_i'^2}{2} + \bar\beta_i v_i v_i'\right], \tag{30}$$

where $v_i \in v$ and $\bar\beta_i \in \bar\beta$. Correspondingly, the total entanglement entropy is a sum of $(N-n)$ two-HO entropies $S_i^{(2-{\rm HO})}$, $(i = 1, \ldots, N-n)$ which are obtained using the Von Neumann relation (10) as [10]:

$$S_i^{(2-{\rm HO})} = -\ln[1-\xi_i] - \frac{\xi_i}{1-\xi_i}\ln\xi_i \quad,\quad \xi_i = \frac{\bar\beta_i}{1+\sqrt{1-\bar\beta_i^2}}\,. \tag{31}$$

The total GS entropy for the full Hamiltonian $H = \sum_{lm} H_{lm}$, Eq.(17), is therefore given by

$$S_{\rm GS}(n,N) = \sum_{l=0}^{l_{max}} (2l+1) S_l(n,N)\,, \qquad S_l(n,N) = -\ln[1-\xi_l] - \frac{\xi_l}{1-\xi_l}\ln\xi_l\,, \tag{32}$$

where $(2l+1)$ is the degeneracy factor found by summing over all values of m for the Hamiltonian. Ideally the upper bound l_{max} of the sum should be infinity, however in practice one assigns a very large value of l_{max} to compute the entropy upto a certain precision. The precision limit Pr is set by demanding l_{max} to be such that the percentage change in entropy for a change in l by a step size l_{st}, never exceeds Pr, i.e.,

$$\left|\frac{S(l_{max}) - S(l_{max} - l_{st})}{S(l_{max} - l_{st})}\right| \times 100 \;<\; Pr\,. \tag{33}$$

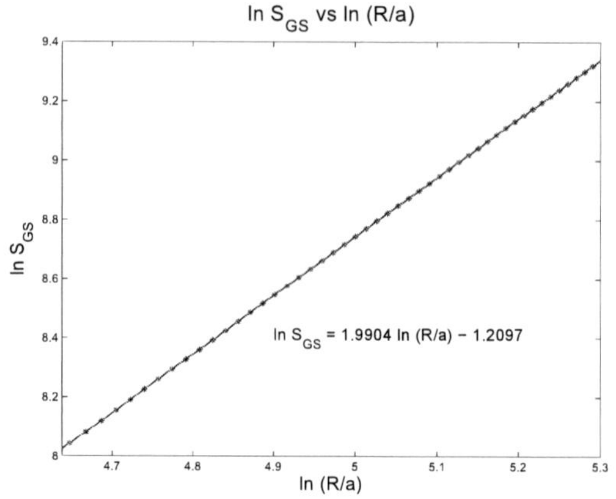

Figure 3. Logarithm of the GS entropy versus $\ln(R/a)$, where $R = (n+1/2)a$ is the radius of the hypothetical sphere (horizon), for $N = 300,\ n = 100 - 200$. The graph is a straight line with slope $\simeq 2$.

Thus $Pr = 0.01$ (say) implies that the numerical error in the computation of the total entropy is less than 0.01%.

It has been shown in ref. [10] that for $N \simeq 60$ and $n \leq N/2$, the GS entropy computed numerically follows the relation:

$$\begin{aligned} S_{\rm GS} &= 0.3\left(n + \frac{1}{2}\right)^2 \\ &= 0.3\left(\frac{R}{a}\right)^2 = \frac{0.3}{4\pi}\left(\frac{\mathcal{A}}{a^2}\right) \end{aligned} \tag{34}$$

where $R = (n + 1/2)a$ is the radius of the hypothetical spherical surface (which is an analogue of the horizon in flat space-time) the DOF inside or outside of which are traced over, and $\mathcal{A} = 4\pi R^2$ is the surface area of this sphere. For a fixed value of n, if N is varied the GS entropy remains the same. This implies that $S_{\rm GS}$ does not depend on the infrared cutoff $\tilde{M} = L^{-1}$, where $L = (N + 1)a$, although it depends on the ultraviolet cutoff $M = a^{-1}$: $S_{\rm GS} = 0.3M^2R^2$ [10].

In Fig. 3 we have plotted $\ln S_{\rm GS}$ vs $R/a = n+1/2$ for the parametric values: $N = 300$ and $n = 100 - 200$. The data fits into a straight line with slope very close to 2 (for a precision limit $Pr = 0.1\%$ in the computation performed using MATLAB has been). The area law ($S_{\rm GS} \propto \mathcal{A}$) is thus found to hold generically for GS.

Generalized Coherent State (GCS):

We now examine the situation where all the oscillators are in GCS, instead of being in GS. Similar to GS, a GCS is a minimum uncertainty state ($\Delta x \Delta p = 1/2$). However unlike GS, the GCS is not an energy eigenstate of HO, it is an eigenstate of the HO annihilation operator with real eigenvalue. The GCS wave-function $\psi_{\rm GCS}$ differs from the GS wave-function, Eq. (25), by constant *shifts* α_i in the coordinates $\underline{x}_i$, as:

$$\begin{aligned}\psi_{\rm GCS}(x_1,\ldots,x_N) &= \prod_{i=1}^{N}\left(\frac{k_{Di}}{\pi}\right)^{1/4}\exp\left[-\frac{1}{2}k_{Di}^{1/2}\left(\underline{x}_i-\alpha_i\right)^2\right]\\ &\sim \prod_{i=1}^{N}\exp\left(-i\underline{p}_i\alpha_i\right)\exp\left(-\frac{1}{2}k_{Di}^{1/2}\underline{x}_i^2\right].\end{aligned} \qquad (35)$$

where $\underline{p}_i = -i\partial/\partial\underline{x}_i$ are the momenta. Physically, the real and imaginary parts of the N complex GCS parameters α_i correspond to the classical position (x_0) and momentum (p_0) of the individual HOs respectively, i.e., $\alpha_i = x_0 - ip_0/k_{Di}$. Defining shifted coordinate variables [11]:

$$\tilde{x} \equiv x - U^{-1}\alpha \quad , \quad d\tilde{x} = dx\,, \qquad (36)$$

one can show that

$$\psi_{\rm GCS}(x_1,\ldots,x_N) = \left(\frac{|\Omega|}{\pi^N}\right)^{1/4}\exp\left[-\frac{\tilde{x}^T\cdot\Omega\cdot\tilde{x}}{2}\right] = \psi_{\rm GS}(\tilde{x}_1,\ldots,\tilde{x}_N)\,. \qquad (37)$$

Thus the GCS wave-function is of the same form as the GS wave-function, albeit in terms of the shifted variables. As such, the reduced density matrix $\rho_{\rm GCS}$ for GCS will also be of the same form as that for GS, Eq.(26), with the variables $(t;t')$ replaced by $(\tilde{t};\tilde{t}')$, where $\tilde{t} \equiv \tilde{t}_1,\ldots,\tilde{t}_{N-n}\,;\tilde{t}_j \equiv \tilde{x}_{n+j},\ j=1,\ldots,(N-n)$:

$$\rho_{\rm GCS}\left(t;t'\right) = \int\prod_{i=1}^{n}dx_i\,\psi_{\rm GCS}(x_1,\ldots,x_n;t)\,\psi^{\star}_{\rm GCS}(x_1,\ldots,x_n;t') = \rho_{\rm GS}\left(\tilde{t};\tilde{t}'\right)\,. \qquad (38)$$

Consequently, the eigenvalues of the density matrix, and hence the entropy for GCS will be the same as those for GS [11]. Thus the area law holds for the case of GCS as well.

Squeezed State (SS):

Let us now consider the case where all the HOs are in a class of SS. The squeezed states are also minimum uncertainty packets ($\Delta x\Delta p = 1/2$), similar to GS and GCS. However, there is a squeezing either in the positions ($\Delta x \ll 1, \Delta p \gg 1$) or in the momenta ($\Delta p \ll 1, \Delta x \gg 1$). We consider a class of SS, which is characterized by a unique squeezing parameter ζ, and described by the wave-function

$$\psi_{\rm SS}(x_1,\ldots,x_N) \sim \zeta^{N/2}\prod_{i=1}^{N}\exp\left[-\zeta\sum_{i=1}^{N}\frac{1}{2}k_{Di}^{1/2}\underline{x}_i^2\right]. \qquad (39)$$

In this case, defining scaled coordinate variables [11]:

$$\tilde{x} \equiv \sqrt{\zeta}\,\underline{x} \quad , \quad d\tilde{x} = \sqrt{\zeta}\,d\underline{x}\,, \qquad (40)$$

one finds that the SS wave-function and the corresponding density matrix reduce to the forms same as those for GS, albeit with the replacement $x \to \tilde{x}$. As such, upto an irrelevant multiplicative factor, the SS entropy turns out to be the same as the GS entropy, Eq. (34), and hence the area law holds.

Thus, in all the above cases where we have the states which form minimum uncertainty packets, the entanglement entropy is proportional to the area. In the next section, we show that if we consider a higher excited state, the area law is not robust. In particular, we show that the superposition of the ground and excited state leads to power-law corrections to area.

6. Power-Law Corrections to the Area-Law

Ideally the most interesting thing would be to find the entanglement entropy for a general eigenstate, given by Eq. (22), which may be the ground state or the first, second, ..., etc. excited states, or a superposition of such states. However, as mentioned earlier, the density matrix cannot be written in a closed form for such a general state. We therefore resort to a simple class of excited states, viz., (i) 1-particle excited state and (ii) superposition of GS and 1-particle excited state. In the case of 1-particle excited state, we show that the entanglement entropy *does not* scale as area while in the superposed state the entanglement entropy has power-law corrections to the area-law.

6.1. 1-Particle Excited State

1-particle excited state is described by a wave-function $\psi_{\rm ES}$ which is a linear superposition of N HO wave functions, each of which has *exactly one* HO in the first ES and the rest $(N-1)$ in their GS [11, 16]. Using Eq. (22), such an 1-particle ES wave-function is expressed as:

$$
\begin{aligned}
\psi_{\rm ES}(x_1,\ldots,x_N) &= \sum_{i=1}^{N}\left(\frac{k_{Di}}{4\pi}\right)^{1/4}\alpha_i\mathcal{H}_1\left(k_{Di}^{1/4}\underline{\mathrm{x}}_i\right)\exp\left(-\frac{1}{2}\sum_j k_{Dj}^{1/2}\,\underline{\mathrm{x}}_j^2\right)\\
&= \sqrt{2}\left(\alpha^T K_D^{1/2}\underline{\mathrm{x}}\right)\psi_{\rm GS}(x_1,\ldots,x_N), \qquad (41)
\end{aligned}
$$

where $\alpha^T = (\alpha_1,\ldots,\alpha_N)$ are the expansion coefficients, and $\alpha^T\alpha = 1$ so that $\psi_{\rm ES}$ is normalized.

From Eq.(24) one finds the reduced density matrix $\rho_{\rm ES}(t;t')$ for ES as:

$$
\begin{aligned}
\rho_{\rm ES}(t;t') &= \int\prod_{i=1}^{n}dx_i\psi_{\rm ES}(x_i;t)\,\psi^\star_{\rm ES}(x_i;t')\\
&= 2\int\prod_{i=1}^{n}dx_i\left[x'^T\Lambda x\right]\psi_{\rm GS}(x_i;t)\,\psi^\star_{\rm GS}(x_i;t'), \qquad (42)
\end{aligned}
$$

where Λ is a $N\times N$ matrix given by

$$
\Lambda = U^T\,K_D^{1/4}\,\alpha\,\alpha^T\,K_D^{1/4}\,U \equiv \begin{pmatrix}\Lambda_A & \Lambda_B\\ \Lambda_B^T & \Lambda_C\end{pmatrix}, \qquad (43)
$$

Λ_A is an $n \times n$ symmetric matrix; Λ_B is an $n \times (N-n)$ matrix; Λ_C is an $(N-n) \times (N-n)$ symmetric matrix. Defining two $(N-n) \times (N-n)$ square matrices Λ_β and Λ_γ such that

$$\begin{aligned} \Lambda_\beta &= \frac{1}{\kappa}\left(2\Lambda_C - \Lambda_B^T A^{-1} B - B^T A^{-1} \Lambda_B + B^T A^{-1} \Lambda_A A^{-1} B\right) \\ \Lambda_\gamma &= \frac{1}{\kappa}\left(2\Lambda_B^T A^{-1} B - B^T A^{-1} \Lambda_A A^{-1} B\right), \end{aligned} \tag{44}$$

one can express the above density matrix for ES in terms of the density matrix for GS as [11]:

$$\rho_{\rm ES}(t;t') = \frac{1}{\kappa}\left[1 - \frac{t^T \Lambda_\gamma t + t'^T \Lambda_\gamma t'}{2} + t^T \Lambda_\beta t'\right] \rho_{\rm GS}(t;t'), \tag{45}$$

where $\kappa = \text{Tr}(\Lambda_A A^{-1})$. In the above, the matrix Λ_β is symmetric, whereas the matrix Λ_γ is not necessarily symmetric due to the presence of the first term in the parentheses on the right hand side.

Eq. (45) is an exact expression for the density matrix for a discretized scalar field with any one HO in the first ES and the rest in the GS. However, unlike the GS density matrix $\rho_{\rm GS}$, given by Eq. (26), the ES density matrix $\rho_{\rm ES}$, Eq. (45), contains non-exponential terms and hence can not be written as a product of $(N-n)$, 2-coupled HO density matrices. Therefore, the ES entropy cannot be expressed as a sum of 2-HO entropies, as in the case of GS. One may however note that the GS density matrix $\rho_{\rm GS}$, Eq. (26), is a Gaussian that attenuates virtually to zero beyond its few sigma limits. Therefore, if

$$\epsilon_1 \equiv t_{max}^T \Lambda_\beta t_{max} \ll 1 \quad , \quad \epsilon_2 \equiv t_{max}^T \Lambda_\gamma t_{max} \ll 1 \tag{46}$$

where

$$t_{max}^T = \left(\frac{3(N-n)}{\sqrt{2\text{Tr}(\gamma-\beta)}}\right)(1,1,\ldots) \tag{47}$$

corresponding to 3σ limits of the Gaussian inside $\rho_{\rm GS}$, then one may approximate

$$1 - \frac{t^T \Lambda_\gamma t + t'^T \Lambda_\gamma t'}{2} + t^T \Lambda_\beta t' \approx \exp\left[-\frac{t^T \Lambda_\gamma t + t'^T \Lambda_\gamma t'}{2} + t^T \Lambda_\beta t'\right]. \tag{48}$$

Consequently, with a shift of parameters: $\beta' \equiv \beta + \Lambda_\beta, \gamma' \equiv \gamma + \Lambda_\gamma$, the ES density matrix above can also be approximated as a Gaussian

$$\rho_{\rm ES}(t;t') \approx \frac{1}{\kappa}\exp\left[-\frac{t^T \gamma' t + t'^T \gamma' t'}{2} + t^T \beta' t'\right]. \tag{49}$$

Factorizing this once again into $N-n$ two-HO density matrices, albeit in terms of the shifted parameters (β', γ'), one can evaluate the ES entanglement entropy $S_{\rm ES}$.

The entropy computation is done numerically (using MATLAB) in [11], with a precision setting $Pr = 0.1\%$ for the set of parametric values: $N = 300$, $n = 100 - 200$, $o = 10 - 50$, where o is the number of last non-vanishing entries in the vector α^T, i.e., $\alpha^T = (1/\sqrt{o})\,(0,\cdots,0;1,\cdots,1)$. The criteria (46) are found to be satisfied for these choices of the parameters. The $\ln S_{\rm ES}$ vs. $\ln(R/a)$ data for different fixed values of the amount of excitation o $(= 20, 30, 40, 50)$ fit approximately to straight lines as shown in Fig.

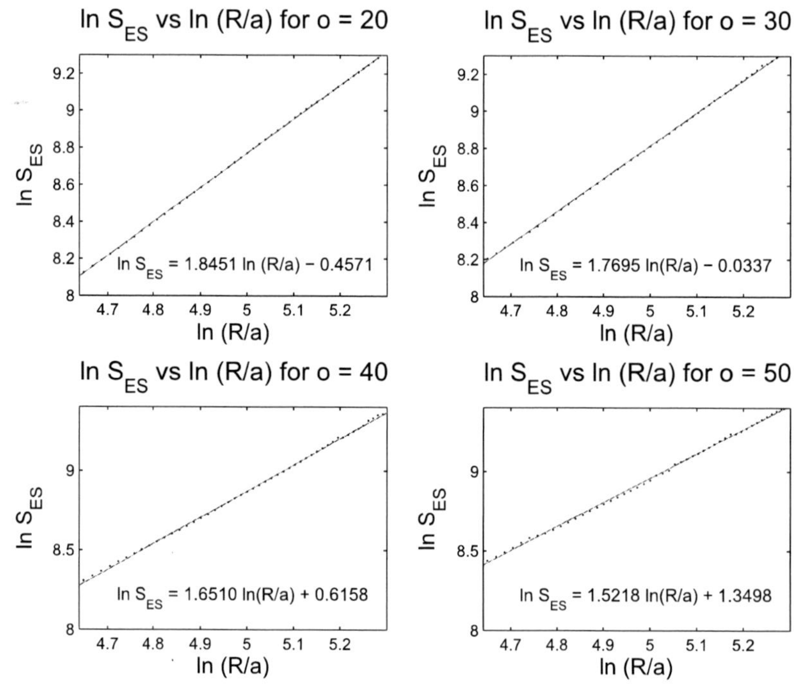

Figure 4. Plot of logarithm of the ES entropy versus $\ln(R/a)$ for $N = 300,\ n = 100 - 200$ for different values of $o = 20, 30, 40, 50$. The data approximately fit to straight lines with slope < 2.

4. The slopes of all these straight lines are less than 2 and the higher the value of o the smaller is the slope. This shows that the ES entropy approximately scales as a power of the area:

$$S_{\text{ES}} \simeq \lambda_0 \left(\frac{R}{a}\right)^{2\mu} = \frac{\lambda_0}{4\pi}\left(\frac{\mathcal{A}}{a^2}\right)^{\mu}. \tag{50}$$

The power μ, however, is always less than unity and decreases with the increase in the number of excitations o. The coefficient λ_0 on the other hand increases with o. Thus contrary to the cases of GS, GCS and SS, the AL is always violated in the case of the 1-particle ES [11].

Now, it is quite interesting to see that even for a small amount of excitation ($o \sim 20$) the entropy-area relationship is so drastically changed that the AL could not be recovered in any limit. Further studies with computations of higher precision ($Pr = 0.01\%$) [14] however reveal that there are slight variations in the linear fits of $\ln S_{\text{ES}}$ vs. $\ln(R/a)$ data for different values of o. In fact, it has been shown that the ES entropy actually approaches the GS entropy (and hence obeys the AL) for very large area [14]. Thus the excitations seem to give rise to some corrections to the AL which are significant for smaller areas, but become negligible for very large areas. These corrections are not manifested by the linear fits of the data (shown in Fig. 4). Therefore a more accurate non-linear fitting is required. We will discuss about these corrections to the AL in the next section where we consider the scalar field to be in a more general state — the MS — which is a linear superposition of GS and 1-particle ES.

6.2. Superposition of Ground and Excited State

The superposition of ground and excited state wave-function $\psi_{\rm MS}$ is given by

$$\psi_{\rm MS}(\hat{x};t) \;=\; \left[c_0\, \psi_{\rm GS}(\hat{x};t) \;+\; c_1\, \psi_{\rm ES}(\hat{x};t)\right] , \tag{51}$$

where $\psi_{\rm GS}$ is the GS wave-function, given by Eq. (25), $\psi_{\rm ES}$ is the ES wave-function, given by Eq. (41), $\hat{x} \equiv \{x_1, \cdots, x_n\}$, and as before $t_j \equiv x_{n+j}$ $(j = 1, \cdots, N-n)$; $t \equiv \{t_1, \cdots, t_{N-n}\} = \{x_{n+1}, \cdots, x_N\}$. We assume that c_0 and c_1 are real constants, and $\psi_{\rm MS}$ is normalized so that $c_0^2 + c_1^2 = 1$. Using Eq. (41), we can write,

$$\psi_{\rm MS}(\hat{x};t) \;=\; [c_0 \;+\; c_1\, f(\hat{x};t)]\, \psi_{\rm GS}(\hat{x};t) \quad , \quad f(\hat{x};t) \;=\; \sqrt{2}\alpha^T K_D^{1/4} U x \;=\; y^T x \,, \tag{52}$$

where the column vector α includes the expansion coefficients defined in the previous section $[\alpha^T = (\alpha_1, \ldots, \alpha_N) = (1/\sqrt{o})(0, \ldots, 0; 1, \ldots, 1)]$, and y is an N-dimensional column vector y defined as

$$y \;=\; \sqrt{2} U^T K_D^{1/4} \alpha \;=\; \begin{pmatrix} y_A \\ y_B \end{pmatrix} \tag{53}$$

y_A and y_B are n- and $(N-n)$-dimensional column vectors, respectively.

The MS density matrix can be expressed as a sum of three terms:

$$\begin{aligned} \rho_{\rm MS}(t;t') \;&=\; \int \prod_{i=1}^{n} dx_i\, \psi_{\rm MS}(\hat{x};t) \psi_{\rm MS}^{\star}(\hat{x};t') \\ &=\; c_0^2\, \rho_{\rm GS}(t;t') \;+\; c_1^2\, \rho_{\rm ES}(t;t') \;+\; c_0 c_1\, \rho_{\rm X}(t;t') \,, \end{aligned} \tag{54}$$

where $\rho_{\rm GS}(t;t')$ is the GS density matrix, Eq. (26), $\rho_{\rm ES}(t;t')$ is the ES density matrix, Eq. (42), and $\rho_{\rm X}(t;t')$ is a cross term due to the superposition of GS and ES. Identifying the matrix Λ, its components, and the constant κ (defined for ES in the previous section), with the column vector y and its components:

$$\begin{aligned} \Lambda \;&=\; \frac{1}{2} y y^T \;=\; \begin{pmatrix} \Lambda_A & \Lambda_B \\ \Lambda_B^T & \Lambda_C \end{pmatrix} , \\ \Lambda_A &= \frac{1}{2} y_A y_A^T \;;\; \Lambda_B = \frac{1}{2} y_A y_B^T \;;\; \Lambda_C = \frac{1}{2} y_B y_B^T \,, \\ \kappa \;&=\; {\rm Tr}(\Lambda_A A^{-1}) \;=\; \frac{1}{2} y_A^T A^{-1} y_A \,, \end{aligned} \tag{55}$$

one can evaluate the cross term $\rho_{\rm X}$ as

$$\begin{aligned} \rho_{\rm X}(t;t') \;&=\; \int \prod_{i=1}^{n} dx_i \left[f(\hat{x};t) + f(\hat{x};t')\right] \psi_{\rm GS}(\hat{x};t)\, \psi_{\rm GS}^{\star}(\hat{x};t') \\ &=\; (y_B - p)^T \left(t + t'\right) \rho_{\rm GS}(t;t') \,, \end{aligned} \tag{56}$$

where p is an $(N-n)$-dimensional column vector defined by

$$p \;=\; B^T A^{-1} y_A\,. \tag{57}$$

The full density matrix for the MS, Eq. (54), thus reduces to

$$\rho_{\rm MS}(t;t') = \left[c_0^2 + c_1^2\kappa\left\{1 + u(t;t')\right\} + c_0 c_1 v(t;t')\right]\rho_{\rm GS}(t;t')\,, \tag{58}$$

where

$$u(t;t') \;=\; -\;\frac{t^T\Lambda_\gamma t + t'^T\Lambda_\gamma t'}{2} \;+\; t^T\Lambda_\beta t' \quad,\quad v(t;t') \;=\; (y_B - p)^T\,(t+t')\,. \tag{59}$$

Defining

$$F(t;t') \;=\; 1 \;+\; \kappa_1 w(t;t') \;+\; \kappa_2 v(t;t') \;+\; \frac{\kappa_2^2}{2}v^2(t;t')\,, \tag{60}$$

where

$$\begin{aligned} w(t;t') \;&=\; -\;\frac{t^T\Lambda_{\gamma'} t + t'^T\Lambda_{\gamma'} t'}{2} \;+\; t^T\Lambda_{\beta'} t'\,,\\ \Lambda_{\beta'} \;&=\; \Lambda_\beta \;-\; 2\kappa_0\left(\Lambda_\beta \;-\; \frac{\Lambda_C}{\kappa}\right) \quad,\quad \Lambda_{\gamma'} \;=\; \Lambda_\gamma \;+\; 2\kappa_0\left(\Lambda_\beta \;-\; \frac{\Lambda_C}{\kappa}\right),\\ \kappa_0 \;&=\; \frac{c_0^2}{\tilde\kappa}\;;\quad \kappa_1 \;=\; \frac{c_1^2}{\tilde\kappa}\;;\quad \kappa_2 \;=\; \frac{c_0 c_1}{\tilde\kappa}\quad,\quad \tilde\kappa \;=\; c_0^2 + c_1^2\kappa\,, \end{aligned} \tag{61}$$

the density matrix (58) can be written as

$$\rho_{\rm MS}(t;t') \;=\; \tilde\kappa\, F(t;t')\,\rho_{\rm GS}(t;t')\,. \tag{62}$$

In the above, $\Lambda_{\beta'}$ and $\Lambda_{\gamma'}$ are $(N-n)\times(N-n)$ matrices, and constants $(\kappa_0, \kappa_1, \kappa_2)$ describe the amount of mixing between the GS and ES.

Now, similar to the case of ES, here also the pre-factor $F(t;t')$ of the Gaussian $\rho_{\rm GS}(t;t')$ in Eq. (62) contains non-exponential terms. Therefore $\rho_{\rm MS}$ cannot be factorized into $(N-n)$ two-HO density matrices. However, as mentioned in sec. 6., that when the vector t^T is outside the maximum t^T_{max}, Eq. (47), corresponding to the 3σ limits, the Gaussian inside $\rho_{\rm GS}(t;t')$ is negligible. Therefore, if the conditions (46) (given in sec. 6.) as well as the conditions

$$\tilde\epsilon_1 \equiv t^T_{max}\Lambda_{\beta'}t_{max} \ll 1\,,\; \tilde\epsilon_2 \equiv t^T_{max}\Lambda_{\gamma'}t_{max} \;\ll 1 \tag{63}$$

are satisfied, then keeping terms up to quadratic order in t, t', the pre-factor $F(t;t')$ can be approximated as

$$F(t;t') \;\approx\; \exp\left[\kappa_1\, w(t;t') \;+\; \kappa_2\, v(t;t')\right]\,. \tag{64}$$

Then by using Eq.(26) for $\rho_{\rm GS}(t;t')$ the (approximated) MS density matrix can be expressed in the form:

$$\rho_{\rm MS}(t;t') \;=\; \tilde\kappa\sqrt{\frac{|\Omega|}{\pi^{N-n}|A|}}\;\exp\left[z(t;t') + \kappa_2 v(t;t')\right]\,, \tag{65}$$

where

$$z(t;t') = -\frac{t^T\gamma' t + t'^T\gamma' t'}{2} + t^T\beta' t' , \tag{66}$$

and the $(N-n)\times(N-n)$ matrices β' and γ' are defined by

$$\begin{aligned} \beta' &= \beta + \kappa_1\Lambda_{\beta'} = \beta + \kappa_1\Lambda_\beta - 2\kappa_0\kappa_1\left(\Lambda_\beta - \frac{\Lambda_C}{\kappa}\right) \\ \gamma' &= \gamma + \kappa_1\Lambda_{\gamma'} = \gamma + \kappa_1\Lambda_\gamma + 2\kappa_0\kappa_1\left(\Lambda_\beta - \frac{\Lambda_C}{\kappa}\right) . \end{aligned} \tag{67}$$

The matrix β' is symmetric while the matrix γ' is not necessarily symmetric.

Now shifting $(N-n)$ variables $t \equiv \{x_{n+1},\cdots,x_N\}$ and $t' \equiv \{x'_{n+1},\cdots,x'_N\}$ by constant values $s \equiv \{s_1,\cdots,s_{N-n}\}$:

$$t \to t + s ; \quad t' \to t' + s , \tag{68}$$

the density matrix (65) becomes

$$\rho_{\text{MS}}(t;t') = \mathcal{N}\ \exp\left[-\frac{t^T\gamma' t + t'^T\gamma' t'}{2} + t^T\beta' t'\right] , \tag{69}$$

where

$$\mathcal{N} = \tilde{\kappa}\sqrt{\frac{|\Omega|}{\pi^{N-n}|A|}}\ \exp\left[-s^T\left(\beta' - \gamma'\right)^T s\right] . \tag{70}$$

The $(N-n)$-dimensional constant column vector s is determined from the equation

$$s^T\left(\beta' - \frac{\gamma' + \gamma'^T}{2}\right) = -\kappa_2\left(y_B - B^T A^{-1} y_A\right) . \tag{71}$$

For either $c_0 = 0$ or $c_1 = 0$, the constant $\kappa_2 = 0$, whence from the above equation (71), we have $s = 0$. Thus it can be verified that the MS density matrix (69) reduces to the GS density matrix (26) when $c_0 = 1, c_1 = 0$, which imply $\beta' = \beta, \gamma' = \gamma$. On the other hand when $c_0 = 0, c_1 = 1$, which imply $\beta' = \beta + \Lambda_\beta, \gamma' = \gamma + \Lambda_\gamma$, the MS density matrix (69) is the same as ES density matrix (42). In general, when both c_0 and c_1 are non-vanishing, then under the shifts $\beta \to \beta', \gamma \to \gamma'$ (where β' and γ' are given by Eqs. (67)) the MS density matrix (69) is of the same form as the GS density matrix (26), up to a normalization factor given above. Such a normalization constant does not affect the entropy computation. Therefore the total MS entropy S_{MS} can be evaluated following the same steps [Eqs. (29) – (32)] as in the case of GS, albeit with the replacements $\beta \to \beta', \gamma \to \gamma'$.

Computation of the entanglement entropy has been done numerically (using MATLAB) in [14], for $N = 300$, $n = 100 - 200$, $o = 30, 40, 50$, and with a precision setting $Pr = 0.01\%$ in each of the following cases:

(i) GS ($c_0 = 1, c_1 = 0$),

(ii) ES ($c_0 = 0, c_1 = 1$),

(iii) an equal mixing (MS_{Eq}) of ES with GS ($c_0 = c_1 = 1/\sqrt{2}$), and

(iv) a high mixing (MS_{Hi}) of ES with GS ($c_0 = 1/2, c_1 = \sqrt{3}/2$).

The conditions (46) as well as (63) are found to be satisfied for the above values of the parameters.

Before proceeding to the results we would like to mention the following: the expectation value of energy, $\mathcal{E}$, for MS turns out to be

$$\mathcal{E} = \langle \psi_{\rm MS} | H | \psi_{\rm MS} \rangle = \mathcal{E}_0 + \frac{c_1^2}{o} \sum_{i=N-o+1}^{N} k_{Di}^{1/2} , \tag{72}$$

where $\mathcal{E}_0 = \frac{1}{2} \sum_{i=1}^{N} k_{Di}^{1/2}$ is the (zero-point) GS energy. The fractional excess of energy over the zero-point energy is therefore given by

$$\frac{\Delta \mathcal{E}}{\mathcal{E}_0} = \frac{\mathcal{E} - \mathcal{E}_0}{\mathcal{E}_0} = \frac{2c_1^2}{o} \left[1 + \frac{\sum_{i=1}^{N-o} k_{Di}^{1/2}}{\sum_{i=N-o+1}^{N} k_{Di}^{1/2}} \right]^{-1} . \tag{73}$$

Now, the value of c_1 is between 0 and 1 and as mentioned earlier $k_{Di} > k_{Dj}$ for $i > j$. Therefore even in the extreme situation $c_1 = 1$, i.e., ES, with a fairly high amount of excitation $o \sim 50$, the fractional change in energy is at most about $\sim 4\%$. Moreover, since there are o number of terms in the sum in the second term of Eq.(72), the excitation energy $(\mathcal{E} - \mathcal{E}_0) \sim 1$ (in units of $1/a$, where a is the lattice spacing). Hence, if we choose $a \sim$ Planck length, then this excitation energy is of the order of Planck energy. As the mass of a semi-classical black hole is much larger than the Planck mass, one may therefore safely neglect the back-reaction of the scalar field on the background.

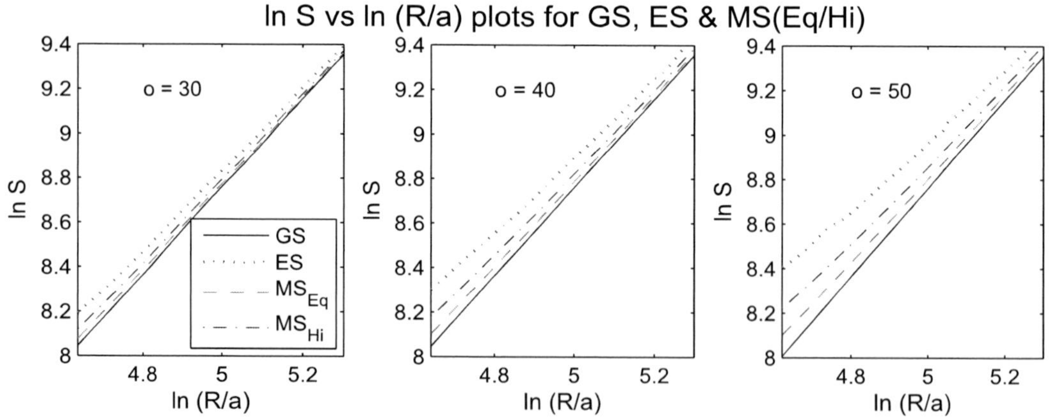

Figure 5. Plots of logarithm of GS, ES and MS (Eq/Hi) entropies versus $\ln(R/a)$ for $N = 300$, $n = 100 - 200$ and $o = 30, 40, 50$. The numerical error in the computation is less than 0.01%.

Fig. 5 shows the plots of the logarithm of the total entropy S vs. $\ln(R/a) = \ln(n + 1/2)$, for the cases of GS, ES and MS (Eq/Hi) with different values of the excitation ($o = 30, 40, 50$). The plot for GS is the linear fit with slope $\simeq 2$ shown earlier. The plots for the MS (Eq/Hi) cases, as well as for ES, are nearly linear for different values of the

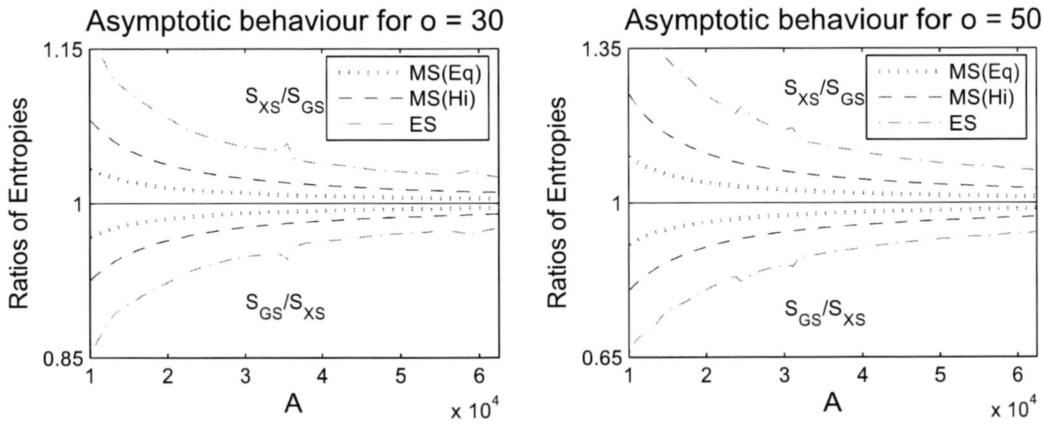

Figure 6. Ratios of GS and MS (Eq/Hi) or ES entropies and their reciprocals plotted against the area $\mathcal{A}$ (in units of a^2) for $o = 30, 50$, to show the asymptotic nature of the MS and ES entropies with respect to the GS entropy. The curves on the upper half (above 1) show the variation of $S_{\rm XS}/S_{\rm GS}$ with $\mathcal{A}$, where XS stands for MS(Eq/Hi) or ES, while the lower curves show the variation of $S_{\rm GS}/S_{\rm XS}$ with $\mathcal{A}$.

excitations $o = 30, 40, 50$ and appear to coincide with the plot for GS for large areas ($\mathcal{A} = 4\pi R^2 \gg a^2$). For a closer examination of this, the ratios $S_{\rm MS}(\rm EqorHi)/S_{\rm GS}$, $S_{\rm ES}/S_{\rm GS}$ and their inverse are plotted against the area $\mathcal{A}$ in Fig. 6. All these ratios approach to unity with increasing area for different excitations ($o = 30, 50$), i.e., the MS (Eq/Hi) and the ES entropies coincide asymptotically with the GS entropy, following the criterion of 'asymptotic equivalence' [18]:

$$\lim_{\mathcal{A}\to\infty} \frac{S_{\rm XS}(\mathcal{A})}{S_{\rm GS}(\mathcal{A})} = 1 \quad ; \quad \lim_{\mathcal{A}\to\infty} \frac{S_{\rm GS}(\mathcal{A})}{S_{\rm XS}(\mathcal{A})} = 1 \quad , \qquad {\rm XS} \equiv {\rm MS\ (Eq\ or\ Hi)\ or\ ES}\,. \tag{74}$$

From Fig. 6 one can also observe that the MS(Eq) entropy is closer to the GS entropy for large $\mathcal{A}$, than the MS(Hi) entropy and the ES entropy, the latter being the farthest. Thus for smaller values of the relative weight c_1 of the mixing of ES with GS the asymptote is sharper.

Fig. 7 shows the best fit ratios of the MS entropies (for equal and high mixings, with $o = 30, 40, 50$) to the GS entropy, which follow a simple formula:

$$\frac{S_{\rm MS}}{S_{GS}} = \tilde{\sigma}_0 + \tilde{\sigma}_1 \left(\frac{\mathcal{A}}{a^2}\right)^{-\tilde{\nu}} , \tag{75}$$

where the values of the fitting parameters $\tilde{\sigma}_0, \tilde{\sigma}_1$ and $\tilde{\nu}$ are shown in Table 1 for different values of $o = 30, 40, 50$. For all these values of o, the parameter $\tilde{\sigma}_0 \approx 1$ in both MS(Eq) and MS(Hi) cases. The parameter $\tilde{\sigma}_1$ is of the order of 10^3 and increases with increasing excitations. The parameter $\tilde{\nu}$ lies between 1 and 1.25 for the above values of o, and also increases with increasing o. Using the expression for the GS entropy, viz., $S_{\rm GS} = n_0(\mathcal{A}/a^2)$, where n_0 is a constant, we can rewrite the above Eq. (75) as

$$S_{\rm MS} = \sigma_0 \left(\frac{\mathcal{A}}{a^2}\right) + \sigma_1 \left(\frac{\mathcal{A}}{a^2}\right)^{-\nu} , \tag{76}$$

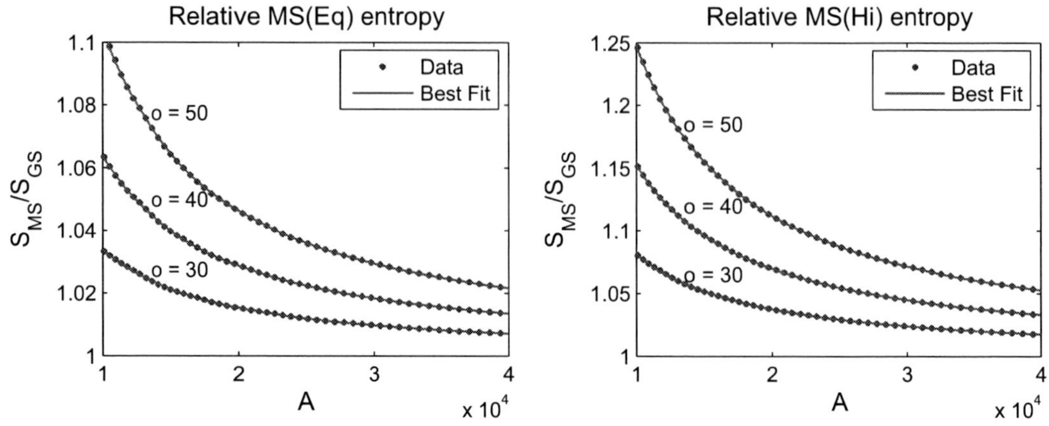

Figure 7. Best fit plots (solid lines) of the relative mixed state entropies ($S_{\rm MS}/S_{\rm GS}$) for equal and high mixings versus the area $\mathcal{A}$ (in units of a^2), for $o = 30, 40, 50$. The corresponding data are shown by asterisks.

Table 1. Values of the parameters of the fit $S_{\rm MS}/S_{\rm GS} = \tilde{\sigma}_0 + \tilde{\sigma}_1 \left(\mathcal{A}/a^2\right)^{-\tilde{\nu}}$ for both MS(Eq) and MS(Hi) cases with excitation $o = 30, 40, 50$.

Fitting Parameters	For MS_{Eq}			For MS_{Hi}		
	$o = 30$	$o = 40$	$o = 50$	$o = 30$	$o = 40$	$o = 50$
$\tilde{\sigma}_0$	1.001	1.002	1.003	1.001	1.004	1.006
$\tilde{\sigma}_1$	1738	4288	8039	2956	7652	14120
$\tilde{\nu}$	1.180	1.210	1.225	1.141	1.178	1.192

where $\sigma_0 = n_0\tilde{\sigma}_0, \sigma_1 = n_0\tilde{\sigma}_1 \propto c_1$ and $\nu = \tilde{\nu} - 1$. The exponent $-\nu$ lies between 0 and -0.25 for both equal and high mixings with the above values of o. It is instructive to stress the implications of the above result:

(i) For the pure vacuum wave-functional, $c_1 = 0$ and $S_{\rm ent}$ is identical to Bekenstein-Hawking entropy. This clearly shows that the entanglement entropy of ground state leads to the area law and the excited states contribute to the power-law corrections.

(ii) For large black-holes, power-law correction falls off rapidly and we recover $S_{\rm BH}$. However, for the small black-holes, the second term dominates and black-hole entropy is no more proportional to area. Physical interpretation of this result is immediately apparent. In the large black-hole (or low-energy) limit, it is difficult to excite the modes and hence, the ground state modes contribute significantly to $S_{\rm ent}$. However, in the small black-hole (or high-energy) limit, larger number of field modes can be excited and hence they contribute significantly to $S_{\rm ent}$.

(iii) The power-law corrections to the Bekenstein-Hawking area law derived here in the context of entanglement of scalar fields have features similar to those derived in the case of brick-wall model [20] and higher-derivative gravity [12]. For instance, it was shown that the entropy of five-dimensional Boulware-Deser black-hole [19] is given by

$$S = \frac{\mathcal{A}}{4} + c\,\mathcal{A}^{1/3} \qquad ; \qquad c = \text{ constant}\,. \tag{77}$$

As in Eq. (76) the above entropy is proportional to area for large horizon radius, however it strongly deviates in the small horizon limit. It is important to note that the corrections to the black-hole entropy are generic and valid even for black-holes in General relativity without any higher curvature terms[4].

7. Location of the Degrees of Freedom

In the previous section, we showed that the entanglement entropy provides generic power-law corrections to the area. We also showed that the quantum DOF that contribute to the area law and the subleading corrections are different. As we had mentioned in the introduction, this leads to another question: *To what extend does the quantum DOF close to or far from the horizon contribute to the black-hole entropy?* In this section, we address this question and show that large contribution ($\sim 97\%$) to the area-law comes from close to the horizon while the subleading contributions has a larger contribution from the regions far from the horizon.

Let us recall the expression for the interaction matrix K, with elements K_{ij} given by Eq. (20), for the system of N HOs. The last two terms which signify the nearest-neighbour (NN) interaction between the oscillators, are solely responsible for the entanglement entropy, i.e., if these two terms are set to zero the entropy vanishes. In order to find which DOF give rise to the entropy or what are their contributions, let us perform the following operations on the matrix K [15]:

Operation 1:

Let us set the off-diagonal elements of K, which signify the NN interactions, to zero (by hand) everywhere except in a 'window', whose center is at a point q. The indices i, j of the matrix K run from $q-s$ to $q+s$, where $s \leq q$, so that the interaction region is restricted to a width of $d = 2s+1$ radial lattice points. For instance, with $s = 1$ the window is of size 3×3, and the matrix K is schematically depicted as:

$$K = \left(\begin{array}{ccc|ccc|c} \times &&&&&& \\ & \times &&&&& \\ && \times &&&& \\ \hline &&& \times & \times & & \\ &&& \times & \times & \times & \\ &&& & \times & \times & \\ \hline &&&&&& \times \end{array}\right). \tag{78}$$

[4] In this context, it should be mentioned that it is not possible to check for logarithmic corrections to the entropy in our analysis, as the numerical error we obtain is much larger than $\ln(n+1/2)$.

Now we let the center q of window to vary from 0 to a value greater than n, and thus allow the window to move rigidly across from the origin to a point outside the horizon at n as shown pictorially in Fig. 8. For every window location (values of q) and fixed window width d, we compute the entanglement entropy $S(q, \text{fixed } d)$ in each of the cases GS, ES and MS (Eq/Hi) and find the percentage contribution of the entropy as a function of q:

$$pc(q) = \frac{S(q, \text{fixed } d)}{S_{\text{total}}} \times 100\,, \tag{79}$$

where S_{total} is the total entropy with all NN interactions, i.e., with the indices i, j running from 0 to N. The variations of $pc(q)$ with q for a window width of $d = 5$ lattice points is shown in Fig. 8, for fixed values $N = 300, n = 100$ in each of the cases GS and ES, MS (Eq/Hi) with $o = 30, 50$.

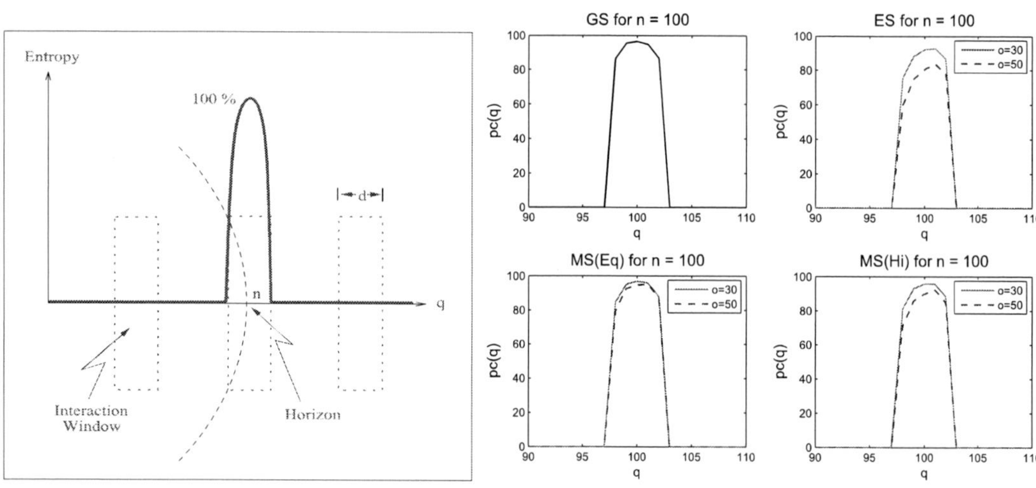

Figure 8. The figure on the left is a pictorial representation of the entropy due to changing positions of the interaction window (green dotted boxes). Contributions to the entropy are observed only when the window includes the horizon at n. The figures on the right are the actual plots of the percentage contribution $pc(q)$ to the total entropy as a function of window position q, for a fixed window size $d = 5$ and fixed $N = 300, n = 100$, in each of cases of GS, ES and MS (Eq/Hi). For ES and MS (Eq/Hi) the solid curve is for $o = 30$ whereas the broken curve is for $o = 50$.

In all the cases of GS, ES and MS (Eq/Hi) there is no contribution to the entropy, i.e., $pc(q) = 0$, when the interaction window does not include the horizon at n. The contributions to the entropy rise significantly when the window includes the horizon, i.e., when the window center q is very close to n. In the case of GS, $pc(q)$ peaks when the window is symmetrically placed between inside and outside, i.e., when $q = n$, and decreases when q moves away from n. In the ES and MS(Eq/Hi) cases, however, the peak tends to shift towards a value $q > n$ and the amplitude of the peak diminish with increasing o and/or the mixing weight c_1. As a result, the entire profile of $pc(q)$ is more and more asymmetric with increasing o and/or c_1, when the window is symmetrically placed between inside and outside of the horizon. The peak is shortest and the profile is most asymmetric for ES with $o = 50$, as shown in Fig. 8. On the whole the above results thus confirm that the entanglement between the scalar field DOF inside and outside the horizon gives rise to the entropy, and the DOF in the vicinity (inside or outside) of the horizon contribute most to the total entropy while the DOF that are far from the horizon contribute a small portion that remains. Such contributions from the far-away DOF increase with increasing excitations and/or amount of mixing of ES with GS. This is indicated by the diminishing maxima of $pc(q)$ in the MS(Eq/Hi) and ES cases, compared to the case of GS. To estimate the contribution of the far-away DOF on the total entropy we perform an alternate operation on the interaction matrix K as described below.

Operation 2:

let us again set the off-diagonal elements of the matrix K to zero (by hand) everywhere except in a 'window' whose outer boundary is the horizon at n. The index i of the elements of the matrix K therefore runs from a point p to n, where $0 \le p \le n$. Now we let the width $d = n - p$ of the window to vary from 0 to n, as shown pictorially in Fig. 9. Computing the entanglement entropy $S(d)$ for every window width d, one then finds the percentage contribution of the entropy as a function of d:

$$pc(d) = \frac{S(d)}{S_{\text{total}}} \times 100\,, \tag{80}$$

where S_{total} is the total entropy which is recovered for the full width $d = n$, i.e., $p = 0$. For convenience we consider here the two extreme cases: GS ($o = 0$) and ES (with $o = 30, 50$). The effects of the DOF on the entropy in the MS (Eq/Hi) cases are intermediate between those for the GS and ES cases, and are therefore of not much interest.

The variation of the percentage contribution $pc(d)$ with the window width d is shown in Fig. 9 for parametric choices of $n = 100, 150, 200$. In the case of GS, almost the entire entropy is recovered within a width of just $d = 3$. This again shows that most of the total GS entropy, which obeys the AL, is contributed by the interaction region that encompassed the DOF very close to the horizon. In the case of ES, for which the AL is violated, it takes d to be as much as $15 - 20$, depending on the excitations $o = 30 - 50$, so that the total ES entropy is recovered. Therefore the DOF that are far-away from the horizon have a greater contribution to the entropy in the case of ES, than for GS, and such contributions increase with increasing excitations o. The location of the horizon (values of n) are also found to affect the entropy contributions. In order to estimate of how much the DOF that are far from the horizon affect the total entropy in the case of ES, as compared to the case of GS, let us consider the percentage increase in entropy for an increment in the interaction region by exactly one lattice point, i.e.,

$$\Delta pc(d) = pc(d) - pc(d-1)\,. \tag{81}$$

This is given by the slope of the $pc(d)$ vs. d plots shown in Fig. 9. In Fig. 10 the variations of $\Delta pc(d)$ with $(n - d)$ are shown for $n = 100, 150, 200$ in the cases of GS and ES (with $o = 30, 50$).

For GS, the entropy increases from 0 to about 85% of the total entropy when the first lattice point just inside the horizon is included in the interaction region. Inclusion of a second lattice point adds another 9%, a third lattice point adds 3%, a fourth lattice point

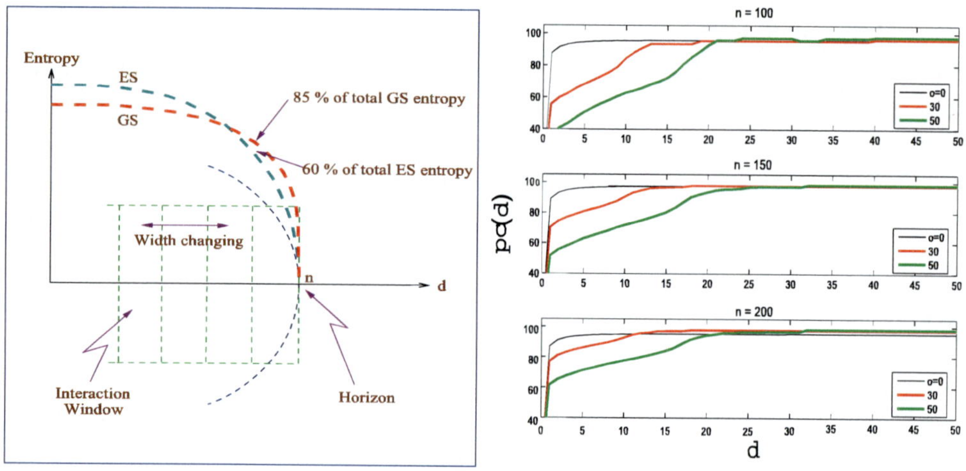

Figure 9. The figure on the left pictorially shows how the contribution to the total entropy increases with increasing width d of the interaction window (green dashed box) in the cases of GS and ES. The figures on the right are the actual plots of the percentage contribution $pc(d)$ to the total entropy as a function of d, for fixed $N = 300, n = 100, 150, 200$, in each of cases of GS ($o = 0$) and ES (with $o = 30, 50$). The solid thin curve is for GS ($o = 0$), whereas the bold light and thick curves are respectively for ES with $o = 30$ and ES with $o = 50$.

adds 1%, and so on. The contributions to the entropy by these additional lattice points farther and farther from the horizon decrease rapidly, and by the time when the $(n/3)^{\rm th}$ is included the increment in entropy is less than 0.01%. This happens for all values of n, i.e., the horizon location does not affect the way by which the DOF contribute to the entropy. For ES, however, the inclusion of the first lattice point inside the horizon raises the entropy from 0 to about $55 - 75\%$ (depending on $n = 100 - 200$) for $o = 30$, and to about $40 - 60\%$ (depending on $n = 100 - 200$) for $o = 50$. The next successive points add about $9\%, 4 - 5\%, 3 - 4\%, \cdots$, depending on $o = 30 - 50$ but fairly independent of n, and the corresponding slope is smaller.

From the above results we thus observe that although most of the entropy is contributed by the DOF close to the horizon, the DOF that are farther away must also be taken into account for the AL to emerge for GS, and the AL plus corrections for ES. With increasing excitations o, the contributions from the far-away DOF become more and more significant, as are the corrections to the AL. Thus the AL may be looked upon as a consequence of entanglement of the DOF near the horizon, whereas the corrections to the AL may be attributed to the contributions to the entropy by the DOF that are far from the horizon [14, 15].

8. Entanglement Entropy of Massive Scalar Field

In all the analysis presented in this review, we have considered an ideal situation where the scalar field are non-interacting and massless. One natural question that arises is what happens if we include interactions or if the field is massive? It is, in general, not possible to

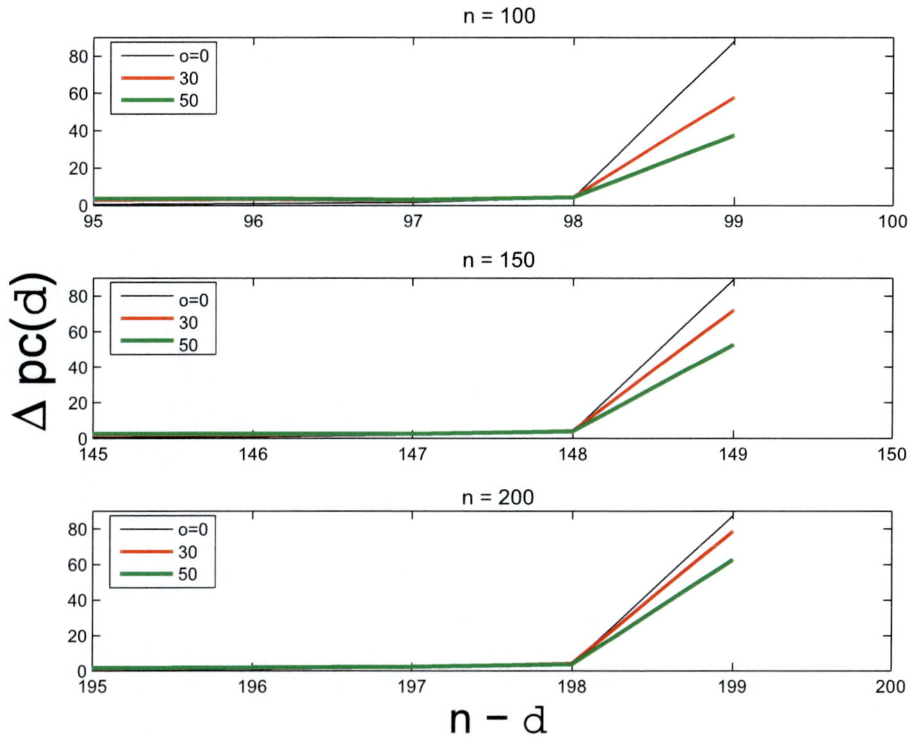

Figure 10. Variations of $\Delta pc(d)$ with $n - t$ in the cases of GS ($o = 0$) and ES (with $o = 30, 50$), and once again for $N = 300, n = 100, 150, 200$. The solid thin curve is for GS ($o = 0$), whereas the bold light and thick curves are respectively for ES with $o = 30$ and ES with $o = 50$.

obtain S_{ent} non-perturbatively for interacting fields and is beyond the scope of this review. In this section, we consider a massive field and show that all the results of the previous sections continue to hold.

The action for a massive scalar field propagating in the background space-time with metric $g_{\mu\nu}$ is given by

$$S = -\frac{1}{2}\int d^4x\,\sqrt{-g}\,\left[g^{\mu\nu}\,\partial_\mu\varphi\,\partial_\nu\varphi + m^2\varphi^2\right] \tag{82}$$

Now, proceeding as before in the case of massless scalar field (see sec. 4.), one can obtain the discretized Hamiltonian for the case of massive scalar field and show that this Hamiltonian resembles that of a $N-$coupled HO, Eq. (19), with an interaction matrix $K^{(m)}$ whose elements $K_{ij}^{(m)}$ are related to the interaction matrix elements K_{ij} [Eq.(20)] for a massless scalar field as

$$K_{ij}^{(m)} = K_{ij} + m^2a^2\,. \tag{83}$$

With this new interaction matrix $K^{(m)}$, one then finds the reduced density matrix following the steps discussed earlier and finally computes the entanglement entropy S_m for the

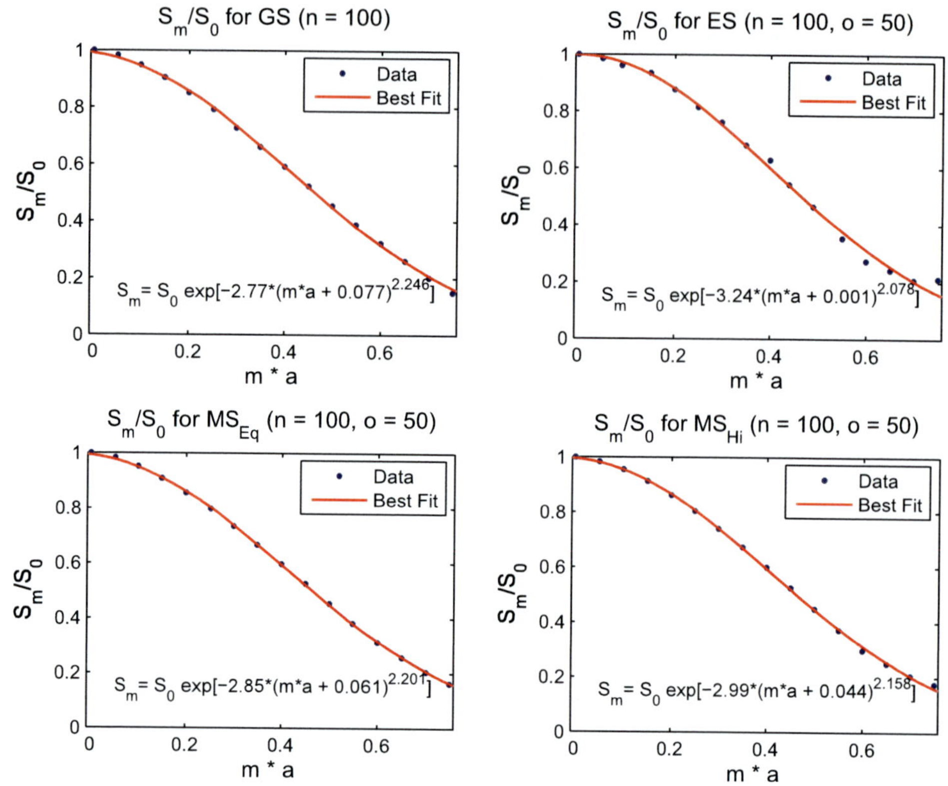

Figure 11. Best fit plots of the relative variation of the total entropy S_m for a massive scalar field (in units of the total entropy S_0 corresponding to a massless scalar field) with the mass m times the lattice spacing a, for fixed $n = 100, o = 30$, in each of cases of GS, ES and MS (Eq/Hi). The corresponding data are shown by asterisks. The fits show an exponential damping of the ratio S_m/S_0 with mass.

massive scalar field.

The variation of S_m/S_0 [where, S_0 is the entropy due to the massless scalar] with $(m \times a)$ is shown in Fig. 11 for the cases of GS, ES and MS(Eq/Hi) for fixed parametric values $N = 300, n = 100, o = 50$. The data fit very well with a Gaussian which shows that S_m falls off exponentially with respect to S_0 as the mass increases:

$$S_m = S_0 \exp\left[-\alpha_1 (ma + \alpha_2)^\lambda\right] \tag{84}$$

where α_1, α_2 and λ are the fitting parameters. Depending on the state (GS, ES or MS), the parameter α_1 varies between 2.77 and 3.24, α_2 is between 0.077 and 0.001 and the power λ is between 2.246 and 2.078. Therefore, although S_m/S_0 approximately scales as $e^{-m^2a^2}$, from the small variation in the power λ one finds that the exponential damping is strongest for GS, and gradually slows down as more and more ES oscillators are mixed with GS, the damping is slowest for the ES case.

Even with a fairly high amount of excitation ($o = 50$) the fitting parameters α_1, α_2 and λ change very little for the different cases GS, MS(Eq/Hi) and ES. Therefore for a particular value of the mass m, the relationship between S_m and (R/a) practically remains the same as that between S_0 and (R/a) in all the cases of GS, MS(Eq/Hi) and ES. The analysis and inferences of the previous sections go through for the massive scalar field as well, resulting in correction terms as obtained before.

9. Conclusions

In the absence of a workable quantum theory of gravity, the best strategy is to slowly build a coherent picture and hope to understand — and, in due course, solve — some of the problems of black-hole thermodynamics. Thus, it is important to explore all possible avenues. Quantum entanglement, as a source of black-hole entropy stands out for its simplicity and generality. The results discussed in this review highlight the nontrivial, and somewhat counterintuitive, facets of quantum entanglement and its role as the source of black-hole entropy. More precisely, assuming the modes evolve adiabatically, we have shown that:

- entanglement leads to generic power-law corrections to the area law
- the quantum degrees of freedom that lead to $S_{\rm BH}$ and subleading corrections are different.
- it is possible to identify the quantum degrees of freedom that contribute to the area law and the subleading corrections.
- the interactions do not change the form of the entanglement entropy for different quantum states.

It is important to note that although the analysis presented here is semiclassical, since, the entanglement is a quantum effect and should be present in any theory of quantum gravity. Hence, the results presented here do have implications beyond the semiclassical regime. There are some new insights which arise in this approach which are worth exploring further:

- Is there a connection between the entanglement entropy and the Noether charge? On the face of it there is no apparent connection: For diffeomorfism invariant theories, like Einstenian gravity, the Noether charge is interpreted as the black-hole entropy [12], implying that the higher order space-time derivatives contribute to the subleading power-law corrections. In the case of entanglement, the subleading contributions arise due to the excited quantum states of the scalar field which exists also for black-holes in Einstein gravity.

  However, it has been shown [21] that the classical conserved charge for a nonminimally coupled matter fields propagating on the fixed curved background is identical to the Noether charge defined in Ref. [12]. This raises the following question: Can the excited states of the quantum scalar field be mapped to the non-minimal coupling of the field? If yes, then conserved charge defined in Ref. [21] can be related to the Noether charge. This is currently under investigation [22].

- Any approach that aspires to explain black hole entropy from fundamental principles must provide a natural explaination for the factor $1/4$ in the Bekenstein-Hawking entropy. However, in the case of entanglement entropy, the proportionality constant in the relation $S = 0.3(R/a)^2$ for GS obtained in ref. [10] differs from the $1/4$ in the Bekenstein-Hawking relation [Eq.(2)]. This discrepancy persists for MS and ES. A probable reason behind this mismatch is the dependence of the pre-factor on the type of the discretization scheme. For example, another discretization scheme, resulting in the NN interactions between four or more immediate neighbors, would result in a different pre-factor. Is it at all possible to obtain the Bekenstein-Hawking value?

- In this review, we have assumed that the quantum modes evolve adiabatically thus neglecting the contribution of the late-time modes leading to the Hawking particles. What happens if we relax this assumption? Will it change the relation (76)?

  The analysis in the previous section suggests that it may not, if we treat the late-time modes perturbatively. In Sec. (4.), we showed explicitly that the time-dependent Hamiltonian (11) becomes a free field Hamiltonian (15) for any Lemâitre time τ_0. At any time $\tau = \tau_0 + \delta\tau$, the Hamiltonian (11) can be written as

$$H_{BH} \simeq H_F + \delta H \tag{85}$$

  where δH is the small perturbation and includes the interaction. In this case, if $(\delta H)/H_F \ll 1$, the relation (76) will not change. However, it will be interesting to investigate the effects at late times where the above perturbation expansion fails.

- Recently, Pleino et al [23] provided analytical proofs of numerical results of Bombelli et al [9] and Srednicki [10] and showed that the entropy-area relation do not depend on the shape of the traced out volume [23]. It will be interesting to do such an analysis for excited states.

- Can a temperature emerge in the entanglement entropy scenario, and if so, then along with the current entropy, will it be consistent with the first law of black hole thermodynamics? Are the second and third laws of thermodynamics valid for this entropy? Can the entanglement of scalar fields help us to understand the evolution or dynamics of black-holes and the information loss problem? We hope to report on these in future.

Acknowledgments

The works of SD and SSu are supported by the Natural Sciences and Engineering Research Council of Canada. SSh is being supported by the Marie Curie Incoming International fellowship IIF-2006-039205.

A. Appendix: Why Consider Scalar Fields?

In this appendix, we discuss the motivation for considering massless/massive scalar fields for the entanglement entropy computations, from the perspective of gravitational metric perturbations in asymptotically flat spherically symmetric space-times.

Let us consider the Einstein-Hilbert action with a positive cosmological constant $|\Lambda|$:

$$S_{EH}(\bar{g}) = M_{\rm Pl}^2 \int d^4x\sqrt{-\bar{g}}\left[\bar{R} - 2|\Lambda|\right] . \tag{86}$$

Let us decompose the metric $\bar{g}_{\mu\nu}$ in terms of a background metric $g_{\mu\nu}$ and fluctuations $h_{\mu\nu}$:

$$\bar{g}_{\mu\nu} = g_{\mu\nu} + h_{\mu\nu} . \tag{87}$$

Assuming $h_{\mu\nu}$ to be small, and expanding the action keeping only the parts quadratic in $h_{\mu\nu}$, gives [24]

$$S_{EH}(g,h) = -M_{\rm Pl}^2 \int d^4x\sqrt{|g|}\left[2\,\gamma^{\alpha}_{\mu\nu}\,\gamma^{\mu\nu}_{\ \ \alpha} + \frac{1}{4}\nabla_\mu\tilde{h}\nabla^\mu\tilde{h} + \frac{|\Lambda|}{2}h_{\mu\nu}\tilde{h}^{\mu\nu}\right] . \tag{88}$$

where

$$\tilde{h}_{\mu\nu} \equiv h_{\mu\nu} - \frac{1}{2}g_{\mu\nu}h^{\alpha}_{\alpha}\,, \qquad \tilde{h} \equiv \tilde{h}^{\mu}_{\mu}\,, \tag{89}$$

$$\gamma^{\alpha}_{\mu\nu} \equiv \frac{1}{2}(\nabla_\mu\tilde{h}^{\alpha}_{\nu} + \nabla_\nu\tilde{h}^{\alpha}_{\mu} - \nabla^\alpha\tilde{h}_{\mu\nu}) . \tag{90}$$

One can easily verify that the above action (88) is invariant under the infinitesimal gauge transformation $h_{\mu\nu} \to h_{\mu\nu} + \nabla_{(\mu}\xi_{\nu)}$ when the background metric $g_{\mu\nu}$ satisfies the vacuum Einstein's equation in presence of the cosmological constant $|\Lambda|$. The gauge arbitrariness can be removed by imposing the harmonic gauge condition $\partial_\mu\tilde{h}^{\mu\nu} = 0$ [24].

Now by keeping only the first derivatives of $h_{\mu\nu}$, the action (88) further reduces to [25]:

$$S_{EH}(g,h) = -\frac{M_{\rm Pl}^2}{2}\int d^4x\sqrt{|g|}\;[\nabla_\alpha h_{\mu\nu}\nabla^\alpha h^{\mu\nu} + |\Lambda|h_{\mu\nu}h^{\mu\nu}]. \tag{91}$$

This corresponds to the action for a massive spin-2 field $h_{\mu\nu}$ propagating in the background $g_{\mu\nu}$, the mass being given in terms of the cosmological constant $|\Lambda|$.

In the weak field limit $h^{\mu\nu}$ can be approximated as a plane-wave perturbation with a particular frequency, i.e.,

$$h_{\mu\nu} = M_{\rm Pl}^{-1}\epsilon_{\mu\nu}\varphi(x^\mu)\,, \tag{92}$$

where $\epsilon_{\mu\nu}$ is the constant polarization tensor. Consequently, the above action (91) reduces to a form which is the same as the action for a massive scalar field φ propagating in the background $g_{\mu\nu}$:

$$S_{EH}(g,h) = -\frac{1}{2}\int d^4x\sqrt{|g|}\;\left[\partial_\alpha\varphi\partial^\alpha\varphi + |\Lambda|\varphi^2\right] . \tag{93}$$

Now, one may further note that in four-dimensional spherically symmetric space-times, the metric perturbations are of two kinds — axial and polar [26–28]. The equations of motion of both these perturbations are scalar in nature and are related to each other by a unitary transformation [26]. The equations of motion of the axial perturbations are identical with those of a test, massless scalar field propagating in the black-hole background:

$$\Box\varphi \equiv \frac{1}{\sqrt{-g}}\partial_\mu\left(\sqrt{-g}g^{\mu\nu}\partial_\nu\varphi\right) = 0 . \tag{94}$$

Hence, by computing the entanglement entropy of the scalar fields one can obtain the entropy of a class of metric perturbations of the background space-time. Of course, such a computation would not account for the entropy of all perturbations, because a generic perturbation is a superposition of the plane wave modes and the entanglement entropy is a non-linear function of the wave-function[5]. Nevertheless, scalar fields are expected to shed important light on the role of entanglement in the area law.

B. Appendix: Hamiltonian of Scalar Fields in Black-Hole Space-Times

In this appendix, we find the expression for the Hamiltonian of a scalar field propagating in a static spherically symmetric space-time and show that for a particular time slicing this Hamiltonian reduces to that of a scalar field in flat space-time.

Let us consider the line-element for a general four-dimensional spherically symmetric space-time:

$$ds^2 \quad = \quad -A(\tau,\xi)\,d\tau^2 + \frac{d\xi^2}{B(\tau,\xi)} + \rho^2(\tau,\xi)\left(d\theta^2 + \sin^2\theta d\phi^2\right), \tag{95}$$

where A, B, ρ are continuous, differentiable functions of (τ, ξ). The action for the scalar field φ propagating in this space-time is given by

$$\begin{aligned} S &= -\frac{1}{2}\int d^4x\,\sqrt{-g}\,g^{\mu\nu}\,\partial_\mu\varphi\,\partial_\nu\varphi \qquad (96)\\ &= -\frac{1}{2}\sum_{lm}\int d\tau d\xi\Big[-\frac{\rho^2}{\sqrt{AB}}(\partial_\tau\varphi_{lm})^2 + \sqrt{AB}\rho^2\,(\partial_\xi\varphi_{lm})^2 + \,l(l+1)\sqrt{\frac{A}{B}}\,\varphi_{lm}^2\Big]\,. \end{aligned}$$

where we have decomposed φ in terms of the real spherical harmonics $(Z_{lm}(\theta,\phi))$:

$$\varphi(x^\mu) = \sum_{lm}\varphi_{lm}(\tau,\xi)Z_{lm}(\theta,\phi)\,. \tag{97}$$

Following the standard rules, the canonical momenta and Hamiltonian of the field are given by

$$\begin{aligned} \Pi_{lm} &= \frac{\partial\mathcal{L}}{\partial(\partial_\tau\varphi_{lm})} = \frac{\rho^2}{\sqrt{AB}}\,\partial_\tau\varphi_{lm}\,, \qquad H = \sum_{l,m} H_{lm}\,, \qquad (98)\\ H_{lm}(\tau) &= \frac{1}{2}\int_\tau^\infty d\xi\left[\frac{\sqrt{AB}}{\rho^2}\Pi_{lm}^2 + \sqrt{AB}\,\rho^2(\partial_\xi\varphi_{lm})^2 + l(l+1)\sqrt{\frac{A}{B}}\,\varphi_{lm}^2\right]. \end{aligned}$$

The canonical variables (φ_{lm}, Π_{lm}) satisfy the Poisson brackets

$$\begin{aligned} \{\varphi_{lm}(\tau,\xi), \Pi_{lm}(\tau,\xi')\} &= \delta(\xi-\xi')\,,\\ \{\varphi_{lm}(\tau,\xi), \varphi_{lm}(\tau,\xi')\} &= 0 = \{\Pi_{lm}(\tau,\xi), \Pi_{lm}(\tau,\xi')\}\,. \end{aligned} \tag{99}$$

[5]It has been recently shown that the value of the entanglement entropy is independent of the number of field species [29].

In the time-dependent Lemaître coordinates [25, 31] the metric components of the line-element (95) are given by

$$A(\tau,\xi)=1 \quad ; \qquad B(\tau,\xi)=\frac{1}{1-f(r)} \quad ; \qquad \rho(\tau,\xi)=r\,, \tag{100}$$

where $r = r(\tau, \xi)$. The line-element in the Lemaître coordinates is related to that in the time-independent Schwarzschild coordinates, viz.,

$$ds^2=-f(r)dt^2+\frac{dr}{f(r)}+r^2\left(d\theta^2+\sin^2\theta d\phi^2\right) \quad ; \qquad f(r=r_h)=0 \tag{101}$$

by the following transformations [31]:

$$\tau=t\pm\int dr\frac{\sqrt{1-f(r)}}{f(r)} \quad ; \qquad \xi=t+\int dr\frac{[1-f(r)]^{-1/2}}{f(r)}\,. \tag{102}$$

Unlike the line-element in Schwarzschild coordinates, the line-element in Lemaître coordinates is not singular at the horizon r_h. Moreover, the coordinate ξ (or, τ) is space(or, time)-like everywhere, whereas r(or, t) is space(or, time)-like only for $r > r_h$.

In the Lemaître coordinates the general Hamiltonian (99) takes the form

$$H_{lm}(\tau)=\frac{1}{2}\int_\tau^\infty d\xi\left[\frac{\Pi_{lm}^2}{r^2\sqrt{1-f(r)}}+\frac{r^2\left(\partial_\xi\varphi_{lm}\right)^2}{\sqrt{1-f(r)}}+l(l+1)\sqrt{1-f(r)}\,\varphi_{lm}^2\right], \tag{103}$$

which depends explicitly on the Lemaître time.

Choosing now a fixed Lemaître time ($\tau = \tau_0 = 0$, say), the relations (102) lead to:

$$\frac{d\xi}{dr}=\frac{1}{\sqrt{1-f(r)}}\,. \tag{104}$$

If we set $d\theta = d\phi = 0$, then for the fixed Lemaître time τ_0 it follows that $ds^2 = d\xi^2/B(\tau_0, \xi) = dr^2$, i.e., the covariant cut-off is $|ds| = dr$. Substituting the above relation (104) in the Hamiltonian (103) we get

$$H_{lm}(0)=\frac{1}{2}\int_0^\infty dr\left[\frac{\Pi_{lm}^2\,r^{-2}}{1-f(r)}+r^2\left(\partial_r\varphi_{lm}\right)^2+l(l+1)\,\varphi_{lm}^2\right], \tag{105}$$

where the variables (φ_{lm}, Π_{lm}) satisfy the relation:

$$\{\varphi_{lm}(r),\Pi_{lm}(r')\}=\sqrt{1-f(r)}\delta(r-r'). \tag{106}$$

Performing the following canonical transformations

$$\Pi_{lm}\to r\sqrt{1-f(r)}\,\Pi_{lm}\;;\;\varphi_{lm}\to\frac{\varphi_{lm}}{r} \tag{107}$$

the full Hamiltonian reduces to that of a free scalar field propagating in flat space-time [32]

$$H=\sum_{lm}\frac{1}{2}\int_0^\infty dr\left\{\pi_{lm}^2(r)+r^2\left[\frac{\partial}{\partial r}\left(\frac{\varphi_{lm}(r)}{r}\right)\right]^2+\frac{l(l+1)}{r^2}\,\varphi_{lm}^2(r)\right\}. \tag{108}$$

This happens for *any* fixed value of the Lemaître time τ, provided the scalar field is traced over either the region $r \in (0, r_h]$ or the region $r \in [r_h, \infty)$. Note that the black-hole singularity can be entirely avoided for the latter choice, and for evaluating time-independent quantities such as entropy, it suffices to use the above Hamiltonian.

The approach here differs from that of Ref. [30] where the authors divide the exterior region $r \geq r_s$ into two by introducing an hypothetical spherical surface and obtain the entanglement entropy of that surface. In contrast, we consider the complete $r \geq r_s$ region and obtain the entropy for the black hole horizon.

References

[1] J. D. Bekenstein, *Lett. Nuovo Cimento* **4**, 737 (1972); *Phys. Rev.* **D7**, 2333 (1973); *Phys. Rev.* **D9**, 3292 (1974); *Phys. Rev.* **D12**, 3077 (1975).

[2] J. M. Bardeen, B. Carter and S. W. Hawking, *Comm. Math. Phys.* **31**, 161 (1973).

[3] S. W. Hawking, *Nature* **248**, 30 (1974); S. W. Hawking, *Commun. Math. Phys.* **43**, 199 (1975);

[4] E. W. Kolb and M. S. Turner, *The Early Universe*, Addison-Wesley (1990).

[5] R. M. Wald, *Living Rev. Rel.* **4**, 6 (2001) [arxiv: gr-qc/9912119]. W. Israel, *Lect. Notes Phys.* **617**, 15 (2003); D. N. Page, *New J. Phys.* **7**, 203 (2005) [arxiv: hep-th/0409024].

[6] A. Strominger and C. Vafa, *Phys. Lett.* **B379**, 99 (1996); A. Ashtekar, J. Baez, A. Corichi and K. Krasnov, *Phys. Rev. Lett.* **80**, 904 (1998); S. Carlip, ibid 88, 241301 (2002); A. Dasgupta, *Class. Quant. Grav.* **23**, 635 (2006) [arxiv: gr-qc/0505017].

[7] R. Brustein and A. Yarom, *Nucl. Phys.* **B709**, 391 (2005); R. Brustein et al, *JHEP* **0601**, 098 (2006); *JHEP* **0704**, 086 (2007); D. Marolf and A. Yarom, *JHEP* **0601**, 141 (2006); S. Ryu and T. Takayanagi, *JHEP* **0608**, 045 (2006).

[8] G. 'tHooft, *Nucl. Phys.* **B256**, 727 (1985); V. P. Frolov and D. V. Fursaev, *Phys. Rev.* **D56**, 2212 (1997).

[9] L. Bombelli, R. K. Koul, J. Lee and R. Sorkin, *Phys. Rev.* **D34**, 373 (1986).

[10] M. Srednicki, *Phys. Rev. Lett.* **71**, 666 (1993) [arxiv: hep-th/9303048].

[11] S. Das and S. Shankaranarayanan, *Phys. Rev.* **D73**, 121701 (2006) [arxiv:gr-qc/0511066].

[12] R. M. Wald, *Phys. Rev.* **D48**, R3427 (1993) [arxiv: gr-qc/9307038].

[13] S. Carlip, *Class. Quant. Grav.* **17**, 4175 (2000), gr-qc/0005017; S. Das et al, *Class. Quant. Grav.* **19**, 2355 (2002), hep-th/0111001; R. C. Myers and J. Z. Simon, *Phys. Rev. D* **38**, 2434 (1988); A. Sen, arxiv:0708.1270.

[14] S. Das, S. Shankaranarayanan and S. Sur, *Phys. Rev.* **D77**, 064013 (2008) [arxiv: 0705.2070]; arxiv: 0711.3164.

[15] S. Das and S. Shankaranarayanan, *Class. Quant. Grav.* **24**, 5299 (2007) [arxiv: gr-qc/0703082]; S. Das, S. Shankaranarayanan and S. Sur, arxiv: 0708.2098 [gr-qc].

[16] M. Ahmadi, S. Das and S. Shankaranarayanan, *Can. J. Phys.* **84**, 493 (2006) [arxiv: hep-th/0507228]; S. Das and S. Shankaranarayanan, *J. Phys. Conf. Ser.* **68**, 012015 (2007) [arxiv: gr-qc/0610022].

[17] P. Tommasini, E. Timmermans and A. F. R. de Toledo Piza, arxiv: quant-ph/9709052.

[18] See for example A. Erdelyi, *Asymptoic Analysis*, Dover Publications (1956); G. H. Hardy and E. M. Wright, *An Introduction to the Theory of Numbers*, Clarendon Press, Oxford (1975); N. G. de Bruijn, *Asymptotic Methods in Analysis*, Dover Publications (1981); for a general definition of asymptotic equivalence see also at http://thesaurus.maths.org/.

[19] T. Jacobson and R. C. Myers, *Phys. Rev. Lett.* **70**, 3684, (1993); A. Paranjape, S. Sarkar and T. Padmanabhan, *Phys. Rev.* **D74**, 104015 (2006) [arxiv: hep-th/0607240].

[20] S. Sarkar, S. Shankaranarayanan and L. Sriramkumar, arxiv: 0710.2013 [gr-qc].

[21] V. P. Frolov and D. V. Fursaev, *Class. Quant. Grav.* **15**, 2041 (1998) [arXiv:hep-th/9802010].

[22] S. Das, S. Shankaranarayanan and S. Sur, *Work in progress.*

[23] M.B. Plenio et al, *Phys. Rev. Lett.* **94** (2005) 060503; M. Cramer et al, arXiv:quant-ph/0505092.

[24] G. 't Hooft and M. J. G. Veltman, *Ann. Poincaré Phys. Theor.* **A20**, 69 (1974); N. H. Barth and S. M. Christensen, *Phys. Rev.* **D28**, 1876 (1983).

[25] L. D. Landau, E. M. Lifshitz, *Classical Theory of Fields*, Course of Theoretical Physics, Volume 2, Pergamon Press, New York (1975).

[26] See for example S. Chandrasekhar, *The Mathematical Theory of black-holes*, Clarendon Press, Oxford (1992); K. D. Kokkotas, B. G. Schmidt, *Living Rev. Rel.* **2**, 2 (1999) [arxiv: gr-qc/9909058].

[27] H. Kodama, A. Ishibashi, *Prog. Theor. Phys.* **111**, 29 (2004) [arxiv: hep-th/0308128]; S. Das, S. Shankaranarayanan, *Class. Quant. Grav.* **22**, L7 (2005) [arxiv: hep-th/0410209].

[28] F. J. Zerelli, *Phys. Rev.* **D2**, 2141 (1970); V. Moncrief, *Ann. Phys. (N. Y.)* **88**, 323 (1974).

[29] R. Brout, arXiv:0802.1588 [gr-qc].

[30] S. Mukohyama et al, *Phys. Rev.* **D58**, 064001 (1998).

[31] S. Shankaranarayanan, *Phys. Rev.* **D67**, 084026 (2003) [arxiv: gr-qc/0301090].

[32] K. Melnikov and M. Weinstein, *Int. J. Mod. Phys.* **D13** 1595 (2004) [arxiv: hep-th/0205223].

In: Horizons in World Physics, Volume 268
Editors: M. Everett and L. Pedroza, pp. 247-277
ISBN 978-1-60692-861-5

Chapter 7

INTERPOLATING GAUGES, PARAMETER DIFFERENTIABILITY, WT-IDENTITIES AND THE ϵ-TERM

Satish D. Joglekar*
Department of Physics, I. I. T. Kanpur , Kanpur 208016, India

Abstract

Evaluation of variation of a Green's function in a gauge field theory with a gauge parameter θ involves field transformations that are (close to) singular. Recently, we had demonstrated [hep-th/0106264] some unusual results that follow from this fact for an interpolating gauge interpolating between the Feynman and the Coulomb gauge (formulated by Doust). We carry out further studies of this model. We study properties of simple loop integrals involved in an interpolating gauge. We find several unusual features not normally noticed in covariant Quantum field theories. We find that the proof of continuation of a Green's function from the Feynman gauge to the Coulomb gauge *via* such a gauge in a gauge-invariant manner seems obstructed by the lack of differentiability of the path-integral with respect to θ (at least at discrete values for a specific Green's function considered) and/or by additional contributions to the WT-identities. We show this by the consideration of simple loop diagrams for a simple scattering process. The lack of differentiability, alternately, produces a large change in the path-integral for a small enough change in θ near some values. We find several applications of these observations in a gauge field theory. We show that the usual procedure followed in the derivation of the WT-identity that leads to the evaluation of a gauge variation of a Green's function involves steps that are not always valid in the context of such interpolating gauges. We further find new results related to the need for keeping the ϵ-term in the in the derivation of the WT-identity and and a nontrivial contribution to gauge variation from it. We also demonstrate how arguments using Wick rotation cannot rid us of these problems. This work brings out the pitfalls in the use of interpolating gauges in a clearer focus.

*E-mail address: sdj@iitk.ac.in

1. Introduction

The standard model [1] is a non-abelian gauge theory possessing a nonabelian gauge-invariance. The consequences of the gauge-invariance have been formulated as the WT-identities [2, 3] and are very important to the discussion of renormalization and unitarity of the gauge theories[4, 1]. At the practical as well as formal level, they are important in the discussion of gauge-independence of observables. While the WT-identities in their usual form that is relevant to the discussion of renormalizability (i.e. structure of counterterms etc.) are formulated via the usual (constant) BRS transformation, the form of the WT-identities relevant to the discussion of gauge-independence is formulated by considering a *field-dependent* gauge transformation[3] or its equivalent in the BRS formulation [These have been called IFBRS (Infinitesimal Field-dependent BRS) transformations[5]]. Recently, we found [5] reason to be cautious about the use of these field dependent transformations as they are (close to) singular and have to be treated very carefully. We had found, (and also given its justification), that while such a procedure that uses these IFBRS transformations does not seem to lead to any obvious trouble *within the class* of the Lorentz-type gauges, it does indeed spell an unexpected trouble for a class of interpolating gauges interpolating between the Lorentz and the Coulomb gauge . We had found that in view of the "singular" nature of the transformation involved, a careful treatment of the path-integral *including a correct ϵ-term* throughout was imperative and we had further found that quite unexpected results follow from this treatment. These results are further summarized in Sec. 2. (We expect this phenomenon to be of a scope more general than the context in which it was analyzed). One of the purposes of this work is to analyse these results in detail to shed new light.

We shall now elaborate on the importance and scope of the subject matter discussed in this work. Calculations in the standard model, a non-abelian gauge theory, have been done in a variety of gauges depending on the ease and convenience of calculations [6, 7, 19]. Many different gauges have also been used in formal treatments in different contexts. For example axial gauges have been used in the treatment of Chern-Simon theory, planar gauges in the perturbative QCD, radial gauges in QCD sum-rules and Coulomb gauges in the confinement problem in QCD [9]. Superstring theories also use to advantage both the covariant and the light-cone treatments [10]. One of the important questions, far from obvious, has been whether the results for physical observables are, in fact, independent of the choice of the gauge used in calculations. While it has naively been assumed that this must do so, it is quite another matter to actually prove the gauge independence of observables in general[1]. A good deal of literature has been devoted to this question [9] directly or indirectly.

The approach of interpolating gauges has been used to give a definition of gauges other than the Lorentz gauges [14, 9]. It has also been employed in formal arguments that attempt to show the gauge-independence of observables in gauges so connected to, say, the Lorentz gauges[14, 11]. The basic idea behind interpolating gauges is to formulate the gauge theory in a gauge for which the gauge function $F[A(x);\alpha]$ depends on one or more parameters α

[1]As the path-integral in Lorentz-type gauges are well-defined, a little thought will show that in the path-integral framework this is really a question of whether and how path-integrals in other sets of gauges can be defined in a manner consistent with the Lorentz gauges.

in such a manner that for different values of the parameters we recover gauge theories in different gauges. For example, the gauge function $F[A(x);\theta]$ used by Doust [12] to connect the Coulomb and the Feynman gauge is given by

$$F[A,\theta] = [\theta\partial^0 A_0 - \frac{1}{\theta}\partial_i A_i] \tag{1}$$

where for θ=1, we recover the Feynman gauge and for $\theta\rightarrow$ 0, we recover the Coulomb gauge. Similarly, one could interpolate between the axial and the Lorentz type gauges by a gauge function such as

$$F[A,\kappa,\lambda] = \frac{1}{\sqrt{\lambda}}[(1-\kappa)\partial^\mu A_\mu + \kappa\eta.A] \tag{2}$$

where for κ=0, we recover a Lorentz-type gauge and for κ=1, we recover the axial gauge in the $\lambda\rightarrow$ 0 limit. Such interpolating gauges have been employed in attempts to prove independence of observables on the choice of the gauge. The arguments in such proofs proceed along the same lines as those that prove the gauge independence of physical observables under, say, a variation of the gauge parameter in the Lorentz type gauges. It is here that we wish to first introduce a note of caution that while such procedures may work within the class of Lorentz gauges, the results of [5] indicate that this may not be true for interpolating gauges from one class to another as elaborated below. [The problem of definition of gauges other than the Lorentz has in fact another solution that proceeds via a careful way to link various pairs of gauges. This has already been introduced [15] and results evaluated from these [16] for various noncovariant gauges].

In ref. [5], we had established several new observations regarding the interpolating gauges. These observations apply to the type of the interpolating gauges such as those given by (1) [and possibly also to a large class of similar interpolating gauges] but they do not affect the results while dealing with the usual class of Lorentz-type gauges (in an unbroken theory at least). These observations pertain to the role of the ϵ-term in the path-integrals for such interpolating gauges. They are:

(i) While discussing the gauge variation, (e.g. varying θ in (1)), the variation of the ϵ-term must be taken into account, if the gauge-independence of the expectation value of a gauge-invariant operator is to be at all preserved.

(ii) When this variation of the ϵ-term under $\theta\rightarrow\theta+\delta\theta$ is taken into account, the net effect on the propagator is NOT an infinitesimal one; but a drastic one if $\frac{\delta\theta}{\varepsilon}$ is sufficiently large[2]. These observations were further employed to imply that an interpolating gauge as in (1) *with any simple* ϵ*-term* cannot interpolate gauge-invariantly between two gauges.

While the above observations in [5] could be interpreted only in a negative light regarding the viability of using the interpolating gauges, the study of interpolating gauges in [5] also raises several unusual questions of a *general* nature about the derivation and the usage of the WT-identities in the gauge theories especially when they concern the variation of Green's functions with a gauge parameter. It is believed that these questions have enough significance by themselves; independent of the context they arose from and could have applications in gauge field theory in general. This article brings out these questions

[2] We note that $\frac{\delta\theta}{\varepsilon}$ is a dimensionful quantity; so the exact meaning of this qualitative characterrization involves other kinematical quantities[5].

into a clearer focus by a further study of the example in [5]. When we write down the WT-identities, we do not normally take account of the variation of the ϵ-term in the path-integral. The results of Ref. [5] lead one to strongly suspect that this procedure to be a simplification valid (probably) only for the class of the covariant Lorentz gauges. While dealing with the θ-variation of the Green's functions in gauges such as those in Eq. (1), we find that we cannot drop the effect of the ϵ-term. Further, even if we were to take into account the effect of the θ-variation $[\theta \longrightarrow \theta + \delta\theta]$ of the ϵ-term , observation (ii) above suggests that we cannot possibly regard the effect as "infinitesimal" if $\frac{\delta\theta}{\varepsilon}$ is sufficiently[3] large [5].

These results point out to the fact that an infinitesimal variation in θ [viz. $\theta \longrightarrow \theta + \delta\theta$], may sometimes produce an "out of proportion" effect in Green's functions in such formulations. (In other words, the Green's functions may not be differentiable at some points). In such cases the WT-identities derived by treating all variations as infinitesimal and, in particular, ignoring the variation of the ϵ-term may not always be the correct procedure.

In view of these observations in [5], we undertook to analyse the example in [5] in more details so that any unusual features of the gauge theories in such gauges stand out. A close look at the analysis makes one believe that the scope of the observations made here and in [5] may be much more wide than the specific context in which it was analysed.

In this work, therefore, we analyse in a greater detail some relevant simple examples where there is reason to suspect new features not addressed to so far. Using these, we aim to address the question of the usage of WT-identities in the discussion of the gauge-dependence of Green's functions and observables in the interpolating gauges.

In view of the unusualness of conclusions arrived at, we find it desirable to analyze the issues in a fine detail so that there are no obvious loopholes. We shall therefore first analyze a simple example by stages.

One of the essential points in [5] was that as $\theta \longrightarrow \theta+\delta\theta$, the variation in the ϵ-term, even though infinitesimal of $O[\varepsilon\delta\theta]$ formally, can in fact lead to a change in the propagator of much more significant kind. This arose essentially from the fact that the effect of this term in some kinematical region blows up. In this work, we want to analyze, in some greater detail, effects of these kinds and further correlate these to the results of Ref. [5] and the expectations raised by it.

We now state the plan of the paper. In section 2, we shall introduce the notations and summarize the conclusions of Ref. [5]. In section 3, we shall pinpoint the potential sources of trouble possibly requiring caution. We analyze it from the point of view of the validity of the Taylor expansion of a propagator such as

$$D_{00} = \frac{1}{\theta^2 k_0^2 - |\mathbf{k}|^2 + i\varepsilon} \tag{3}$$

around $\theta = \theta_0$ and the problems that it can lead to for a k for which $\theta_0^2 k_0^2 - |\mathbf{k}|^2 \approx 0$. We suspect two kinds of troubles: (i) One with differentiability in θ as $\epsilon \longrightarrow 0$. (ii) Second with order of the differentiation $\frac{\partial}{\partial\theta}$ and the limit $\epsilon \longrightarrow 0$.

In section 4, we consider for illustration purposes, a simple model (noncovariant ϕ^4-theory) where such a phenomenon can be analyzed in detail. We, in fact, confirm the

[3] See the earlier footnote regarding this qualitative characterization.

suspicions raised in the Sec 3. In section 5, we consider the various orders of limits/ differentiations occurring in the definition of gauge-variation of S-matrix and Green's functions. We analyze the procedure normally used in this connection keeping in mind for contrast the observations made in the section 4. Here, we point out that there are several delicate situations in the derivation of WT identities and the discussion of gauge dependence. These are the situations where (i) Order of limit $\epsilon \to 0$ and differentiation with θ, (ii) Expansion of an exponential with an "infinitesimal" appearing exponent (iii) Keeping track of the ϵ-term and its effects etc may have to be treated with care.

In section 6, we address to the question of the θ-dependence of S-matrix elements and Green's functions in the interpolating gauge of Eq. (1). We focus our attention on a particular contribution to a simple 1-loop diagram. We show that there exists a value of $\theta \in (0,1)$ where this contribution is not differentiable . What is worse, is that we find that the limits $\epsilon \to 0$ and the differentiation do not commute at this point. In fact the right derivative $\frac{\partial}{\partial\theta}\|_{\theta_0^+}$ in fact goes to infinity. Thus, it is generally not correct to assume that an infinitesimal variation in θ will produce an infinitesimal variation in the path-integral as is done in the derivation of a WT-identity. The simple example dealt with here will also enable us to see that such situations will also exist in more complicated loop diagrams. In section 7, we extend the above argument to a class of off-shell Green's functions.

In Section 8, we shall analyse the results of section 6 in a direct manner. One of the suspicions one may have had is that one should not have the problems enumerated above because one could have formulated the field theory in Euclidean space to begin with ; thus avoiding any necessity of an ϵ-term. The analysis of this section will help us understand why this procedure cannot rid us of the problems. Here, we find how the non-differentiability in θ is connected to the presence of a branch-point in the complex s'-plane (here, $s' = \theta^2 p_0^2 - |\mathbf{p}|^2$, see section 4 for more on its definition). We also find out as to why these problem cannot be avoided in the Euclidean formulation as they will pop up while carrying out the analytic continuation from the Euclidean formulation.

In section 9, we derive a result which makes a contact between the result in [5] and the result in section 6 of this work. In section 10, we shall examine the procedure we normally follow in the derivation of the WT-identities. We point out, with reasons, the places that require a careful treatment. In section 11, we add several comments. We comment on the extension of the result about non-differentiability of the path-integral. We also comment on the use of wavepckets for external lines. In appendix A, we shall present a simple example of an off-shell Green's function where we explicitly confirm a nondifferentiable behavior in θ. In section 12, we summarise our conclusions.

This work brings out some of the pitfalls one has to face while formulating a gauge-invariant field theory using interpolating gauges.

As a final comment, we note that despite the unusual nature of conclusions, this work does not require more than a usual knowledge of Quantum Field Theory and algebra.

2. Summary of some results and Notations

In this section, we shall summarize some results from the past works of references [12] and [5] that are needed for our purpose. In the process, we shall also introduce our notations.

In this work, we propose to discuss, in the path integral framework, some difficulties encountered in the use of the interpolating gauges that interpolate between pairs of gauges. Interpolating gauges are introduced by considering gauge functionals that depend on one or more parameters. We shall, therefore, consider the Faddeev-Popov effective action [FPEA] with a local gauge function $F[A(x);\alpha]$ which may depend on several parameters, collectively denoted by α . We denote this FPEA by $S_{eff}[A,c,\overline{c};\alpha]$ which is given by

$$S_{eff}[A,c,\overline{c};\alpha] = S_0[A] + S_{gf}[A;\alpha] + S_{gh}[A,c,\overline{c};\alpha] \tag{4}$$

with

$$S_{gf}[A;\alpha] = -\frac{1}{2}\int d^4xF[A,\alpha]^2 \tag{5}$$

and[4]

$$S_{gh} = -\int d^4xd^4y\overline{c}^{\alpha}(x)M^{\alpha\beta}[x,y;A;\alpha]c^{\beta}(y) \tag{6}$$

with

$$\int d^4yM^{\alpha\beta}[x,y;A;\alpha]c^{\beta}(y) = \int d^4y \; \frac{\delta F^{\alpha}[A(x);\alpha]}{\delta A^{\gamma}_{\mu}(y)}D^{\gamma\beta}_{\mu}[A(y)]c^{\beta}(y) \tag{7}$$

and

$$D^{\alpha\beta}_{\mu}[A] = \delta^{\alpha\beta}\partial_{\mu} + g\, f^{\alpha\beta\gamma}A^{\gamma}_{\mu} \tag{8}$$

Here, $f^{\alpha\beta\gamma}$ are the antisymmetric structure constants of a semi-simple gauge group.

$S_{eff}[A,c,\overline{c};\alpha]$ is invariant under the BRS transformations :

$$\begin{aligned} \delta A^{\alpha}_{\mu}(x) &= D^{\alpha\beta}_{\mu}[A(x)]c^{\beta}(x)\delta\Lambda \\ \delta c^{\alpha}(x) &= -\tfrac{1}{2}gf^{\alpha\beta\gamma}c^{\beta}(x)c^{\gamma}(x) \quad \delta\Lambda \end{aligned}$$

$$\delta\overline{c}^{\alpha}(x) = F[A(x);\alpha]\delta\Lambda \tag{9}$$

In this work, we shall, in particular, utilize the example of the interpolating gauge used by Doust to interpolate between the Feynman and the Coulomb gauge. This example proves to be simple enough and yet bring out several problems associated with these techniques, some of which were demonstrated in [5]. It has (as a special case of his)

$$F[A,\theta] = [\theta\partial^0 A_0 - \frac{1}{\theta}\partial_i A_i] \tag{10}$$

We shall now summarize the results of [5]. In this work we considered the interpolating gauges, as for example those with an F given by (10), together with *a conventional ϵ-term* viz.

$$-i\varepsilon\int d^4x[\frac{1}{2}A_{\mu}A^{\mu} - \overline{c}c] \tag{11}$$

[or a suitable modification that does not change *signs* of terms in (11)] and discussed *whether such a formulation can really interpolate in a gauge-invariant way*. For this purpose, we considered the vacuum expectation value of a gauge invariant operator $O[A]$:

$$< O[A] > = \frac{<< O[A] >>}{<< 1 >>} \tag{12}$$

[4] The ghost action is always arbitrary upto a constant and, in particular, an overall sign. The following is a convention we make.

with

$$
\begin{aligned}
<< O[A] >>|_{\alpha_0} &= \int D\phi\, O[A] \exp\{iS_0[A] - \frac{i}{2}\int d^4x F[A(x), \alpha_0]^2 + \\
& iS_{gh}[A, c, \overline{c}; \alpha_0] + \int d^4x \varepsilon[\frac{1}{2}A^2 - \overline{c}c]\}
\end{aligned} \tag{13}
$$

We then considered the question as to whether the above expression will necessarily imply the independence of the expectation value with the parameter α. To see if this is so, we considered a field transformation of a infinitesimal field-dependent BRS [IFBRS]-type (see e.g.[5]; these are also spelt out in section 10) that leads to

$$
\begin{aligned}
<< O[A] >>|_{\alpha_0} &= \int D\phi\, O[A] \exp\{iS_0[A] - \frac{i}{2}\int d^4x F[A(x), \alpha_0 - \delta\alpha]^2 + \\
& iS_{gh}[A, c, \overline{c}; \alpha_0 - \delta\alpha] + \int d^4x \varepsilon[\frac{1}{2}A^2 - \overline{c}c] + \varepsilon\delta R\}
\end{aligned} \tag{14}
$$

We now note that the right hand side has an effective action evaluated at the parameter $\alpha_0 - \delta\alpha$; but at the same time the ϵ-term has now changed it to

$$
\int d^4x \varepsilon[\frac{1}{2}A^2 - \overline{c}c] \Longrightarrow \int d^4x \varepsilon[\frac{1}{2}A^2 - \overline{c}c] + \varepsilon\delta R \tag{15}
$$

with

$$
\delta R = -\varepsilon i \int d^4z \overline{c} \frac{\partial F}{\partial \alpha} |_{\alpha_0} \int d^4x [\partial.A - F[A, \alpha]c] \delta\alpha \tag{16}
$$

The effect of this term on the free propagator was then considered in the context of the gauges of (10) and next it was found that this term, rather than having an infinitesimal effect on the propagator, in fact alters completely the pole structure of the propagator for $\delta\alpha >> \varepsilon^5$. We then concluded that the small variation in ϵ-term with θ has a catastrophic effect: it the gauge boson propagator structure from the causal one to a mixed one even for a small change $\delta\theta$. We further demonstrated that there was no modification of the ϵ-term that would allow us an escape in this gauge. We however found no such effect from the ϵ-term for the set of Lorentz gauges as the gauge parameter λ is varied: the pole structure of the propagator does not alter by such a term. This was verified for the unbroken gauge theory.

3. Need for Caution in Field Theory Path-Integrals

3.1. The Origin of Need for Caution

The path-integrals as used in perturbative quantum field theory are perturbations over infinite dimensional Gaussian integrals. To begin with let us consider a simple one-variable Gaussian integral:

$$
\int dx \exp\{iax^2 - \varepsilon x^2\} \tag{17}
$$

[5] See the earlier footnote regarding this qualitative characterization.

with 'a' real which, for $\epsilon > 0$, as we know is

$$\frac{1}{\sqrt{-i\pi(a - i\varepsilon)}} \tag{18}$$

For ϵ sufficiently small, we can expand this as

$$\frac{1}{\sqrt{-i\pi a}}(1 + \frac{i\varepsilon}{2a}) \tag{19}$$

which holds provided $\epsilon < a$ and may even ignore the ϵ-term in (19). We normally take for granted such approximations. In field theory, we may come upon situations where a may be very small compared to ϵ (or may even vanish) and there an approximation of this kind may break down[6]. Moreover, there are situations where there may be no escape.

3.2. Field Theory Example

Just to illustrate the above point in the context of a simple field theory first, in a somewhat exaggerated fashion, we begin with a trivial observation in the context of the $\lambda\phi^4$ -theory in the path-integral formulation. We consider the generating functional $W_c[J]$ for the Green's functions in the Minkowski space:

$$W_c[J] = \int D\phi \exp\{iS_0[\phi] - \varepsilon \int d^4x\phi^2/2\} \tag{20}$$

where the suffix 'c' stands for the fact that $W_c[J]$ generates the causal Green's functions as ensured by the correct ϵ-term. We may re-express:

$$W_c[J] = \int D\phi \exp\{iS_0[\phi] + \varepsilon \int d^4x\phi^2/2 - 2\varepsilon \int d^4x\phi^2/2\} \tag{21}$$

We may, then, believe that the exponential $\exp\{-\varepsilon \int d^4x\phi^2\}$ can be expanded in powers of the "small" exponent $\{-\varepsilon \int d^4x\phi^2\}$[7].

$$W_c[J] = \int D\phi \exp\{iS_0[\phi] + \varepsilon \int d^4x\phi^2/2\}[1 - \varepsilon \int d^4x\phi^2 + ...] \tag{22}$$

$$\equiv W_{a.c.}[J] - \int D\phi \exp\{iS_0[\phi] + \varepsilon \int d^4x\phi^2/2\}\varepsilon \int d^4x\phi^2 + ...] \tag{23}$$

where the suffix a.c. refers to "anti-causal"[8] Green's functions.

To see if such a procedure is *always* a valid one, we consider the above relation in the tree approximation where everything is expected to be well-defined. We may then naively expect the second and higher terms in (23) to vanish as $\epsilon \to 0$. We would then obtain an absurd conclusion

[6]At this point, it may be argued that a field theory can be formulated in the Euclidean space where no epsilon term is requied. We shall discuss this question in section 8 in further details.

[7]Here, we ignore the possible subtleties that could be introduced by ultraviolet divergences. We could for example stick to tree level.

[8]We shall refer to the Green's functions with $\epsilon \to -\varepsilon$ as the "anti-causal" Green's functions.

$$\lim_{\varepsilon \to 0} W_c[J] = \lim_{\varepsilon \to 0} W_{a.c.}[J] \tag{24}$$

The above relation, in particular, implies for the propagator

$$D_c(x-y) = D_{ac}(x-y) \tag{25}$$

which is incorrect since, for example, for $x_0 > y_0$, the former propagates +ve frequency modes and the latter the negative frequency modes.

The above example brings out the fact that such an exponential, with an exponent "small" in appearance, may not always be amenable to an expansion. This happens, essentially because, the quadratic form in the path integral goes over field configurations for which the "main term" may in fact be small compared to the "small term"; a situation very similar to Eq. (19) being used under the wrong conditions $a < \varepsilon$.

To locate the mathematical error in the above argument in an alternate manner, consider the 2-point function in the tree approximation generated by $W_c[J]$ of Eq. (23). In momentum space, it reads,

$$\begin{aligned}\frac{1}{k^2 - m^2 + i\varepsilon} &= \frac{1}{k^2 - m^2 - i\varepsilon + 2i\varepsilon} \\ &= \frac{1}{k^2 - m^2 - i\varepsilon} + \frac{-2i\varepsilon}{k^2 - m^2 - i\varepsilon}\frac{1}{k^2 - m^2 - i\varepsilon} + \ldots\ldots\end{aligned} \tag{26}$$

The equation (23) in the present context corresponds to this expansion. We immediately recognize that the above Taylor expansion holds only if $|2i\varepsilon| < |k^2 - m^2 - i\varepsilon|$. Thus, the above procedure is not valid for a 4-dimensional volume in the $k-$space $\sim \epsilon \times$ R^3.

Now the pertinent question is whether there is any place in field theoretic calculations where such approximations are actually made and whether there are cases where such points of mathematical rigor *have* to be paid attention to.

We shall show, in section 6, that while considering the gauge-parameter variation of Green's functions in the interpolating gauges, we may have to pay special attention to this point. There, we shall point out examples where an infinitesimal change in a parameter leads to a disproportionate change in a Green's function. In section 10, we shall also discuss the role of ϵ-terms (εR) in the WT-identities. We had emphasized in Ref. [5] that such terms may have to be paid special attention to and we may not ignore effects arising from them when a gauge parameter is varied [$\epsilon R \to \varepsilon(R + \delta R)$]. The effect was, in fact, such as to alter the boundary condition on the propagator. We shall find that we may not be able to treat the exponents of a term of (ϵR) as amenable to expansion.

We know several cases where the propagator *denominators* depend on a parameter. For example, the gauge propagator in the R$_\xi-$gauges [13] has the form:

$$D_{\mu\nu} = \frac{g_{\mu\nu} - \frac{k_\mu k_\nu}{k^2 - \frac{M^2}{\xi} + i\varepsilon}(1 - \frac{1}{\xi})}{k^2 - M^2 + i\varepsilon} \tag{27}$$

and the associated ghost and the unphysical scalar fields have a similar ξ-dependence in the denominator. Similarly [a form of] an interpolating gauge that interpolates between the

Coulomb and the Feynman gauge used by Doust [12] has the propagator for (0,0) component

$$D_{00} = \frac{1}{\theta^2 k_0^2 - |\mathbf{k}|^2 + i\varepsilon} \tag{28}$$

with a similar dependence in the spatial components. It is expected that the various other interpolating schemes [14, 17] will also have propagator denominators depending on interpolating parameters in such sensitive a manner.

A typical Feynman diagram, in the context of the above example (28), depends on θ through such propagators. While considering the parameter variation of a Feynman diagram with the gauge parameter θ, we are in effect expanding each propagator in a Taylor series around θ_0 in a series such as

$$D_{00} = \frac{1}{\theta^2 k_0^2 - |\mathbf{k}|^2 + i\varepsilon} \tag{29}$$

$$= \frac{1}{\theta_0^2 k_0^2 - |\mathbf{k}|^2 + i\varepsilon} + \frac{1}{\theta_0^2 k_0^2 - |\mathbf{k}|^2 + i\varepsilon} \bullet \frac{-2\theta_0 \delta\theta k_0^2}{\theta_0^2 k_0^2 - |\mathbf{k}|^2 + i\varepsilon} + \ldots\ldots \tag{30}$$

and picking the second term on the right side to compute the θ- variation. We, however, note that this Taylor series is valid only if

$$\left|\frac{-2\theta_0 \delta\theta k_0^2}{\theta_0^2 k_0^2 - |\mathbf{k}|^2 + i\varepsilon}\right| < 1 \tag{31}$$

We, thus, note several points:

[a] We have to be careful while using the above Taylor series (or a procedure equivalent to it) in formal manipulations since it may not always hold for arbitrary $\delta\theta$;

[b] For $\delta\theta$ sufficiently large [with some small fixed ϵ] , the above condition (31) is not valid for a 4-dimensional volume $\sim \delta\theta$ X R^3 in the k-space.

[c] For a fixed $\epsilon > 0$, and a finite k_0 , there is a nonzero range of $\delta\theta$ for which the above condition is necessarily fulfilled. *Nonetheless, this range* $\rightarrow 0$ *, as* $\epsilon \rightarrow 0$*. Recall that to evaluate the S-matrix elements, we always take the limit* $\epsilon \rightarrow 0$.

[d] We note in passing that in the 4-dimensional subspace: $\theta_0^2 k_0^2$-$|\mathbf{k}|^2 \simeq 0$, the above ratio in Eq. (31) is $\sim \frac{\delta\theta}{\varepsilon}$. And this is precisely the parameter on which an unexpected dependence was found [5] in the discussion of the gauge variation of a path-integral in this sort of an interpolating gauge. We also note that for such a k, the radius of convergence of the series (30) is $\sim \frac{\delta\theta}{\varepsilon}$. This is a simple and clear illustration of how the two independent small parameters $\delta\theta$ and ϵ become entangled!

[e] The naive expectation that ϵ is just a spectator in the entire discussions of WT-identities and of gauge-independence in gauge field theories and it need not be paid special attention to does not always seem valid. According to the results of [5], this is the case, probably only for the class of the Lorentz gauges.

[f] The point [c] above suggests the possibility that there could arise a difficulty in the definition of a derivative with respect to θ in loop integrals as $\epsilon \rightarrow 0$ and that the behavior of loop integrals could possibly depend on the orders of $\frac{d}{d\theta}$ and the limits $\epsilon \rightarrow 0$. We shall confirm these suspicions in Sections 4 and 6.

At this point, one may wonder whether working with a Wick-rotated Euclidean field theory will not rid us of all such problems as then ϵ would be redundant. We shall clarify this point in the Section 8. It turns out that this is not always possible.

4. Direct Analysis of Nondifferentiability of a Generating Functional

The purpose of this section is to make clear, in a direct fashion, the difficulty in defining the partial derivative of a generating functional [see e.g. eq. (33) below] with respect a parameter α, on which it depends, when the dependence on α is such that a propagator denominator depends sensitively on it in some domain in the momentum space. As expected from the previous section, this happens when the propagator singularities depend on α.

We shall find it useful to illustrate the point by considering an artificial but a simple model that captures, in a direct manner, the essential point made in the previous section. Rather than get into complications of a generating functional of a gauge theory in the beginning itself, we shall consider a scalar field theory with a specific (but noncovariant) kinetic energy term. The generating functional of Green's functions is given by

$$W[J,\alpha] = \lim_{\varepsilon \to 0} W[J,\alpha,\varepsilon] \tag{32}$$

$$= \lim_{\varepsilon \to 0} \int D\phi \exp\{i \int d^4x[\frac{1}{2}\phi(\alpha\partial_0^2 - \nabla^2 - m^2 + i\varepsilon)\phi + \frac{\lambda}{4}\phi^4] + i\int d^4xJ\phi\};\ \alpha > 0 \tag{33}$$

We thus note that to compute $\frac{\partial W[J,\alpha]}{\partial\alpha}$, we must first evaluate

$$\lim_{\varepsilon \to 0} \{W[J,\alpha+\delta\alpha,\varepsilon] - W[J,\alpha,\varepsilon]\} \tag{34}$$

which requires the evaluation of

$$\lim_{\varepsilon \to 0} \int D\phi \exp\{i \int d^4x[\frac{1}{2}\phi(\alpha\partial_0^2 - \nabla^2 - m^2 + i\varepsilon)\phi + \frac{\lambda}{4}\phi^4] + i\int d^4xJ\phi\} \bullet \left(\exp[i\delta\alpha \int d^4x\phi\partial_0^2\phi] - 1\right) \tag{35}$$

We wish to pose the following question: can one, for *any* small enough $\delta\alpha$, *and irrespective of* ϵ, expand the exponential, *within the path integral*, as

$$\exp[i\delta\alpha \int d^4x\phi\partial_0^2\phi] \approx 1 + O[\delta\alpha] \tag{36}$$

To see this, we first consider the bare two point function of the expression (35), in the momentum space:

$$\frac{1}{(\alpha+\delta\alpha)k_0^2 - |\mathbf{k}|^2 - m^2 + i\varepsilon} - \frac{1}{\alpha k_0^2 - |\mathbf{k}|^2 - m^2 + i\varepsilon} \tag{37}$$

We note that we can generally Taylor expand the first term as

$$\frac{1}{\alpha k_0^2 - |\mathbf{k}|^2 - m^2 + i\varepsilon} + \frac{1}{\alpha k_0^2 - |\mathbf{k}|^2 - m^2 + i\varepsilon} \bullet \frac{-\delta\alpha k_0^2}{\alpha k_0^2 - |\mathbf{k}|^2 - m^2 + i\varepsilon} + ... \quad (38)$$

provided $|\alpha k_0^2 - |\mathbf{k}|^2 - m^2 + i\varepsilon| > |\delta\alpha k_0^2|$. We now recall the discussion in the previous section and note that, evidently, the magnitude of ϵ comes into play in being able to carry out the expansion when k is such that $\alpha k_0^2 - |\mathbf{k}|^2 - m^2 \simeq 0$!

Now, the fact that *the Taylor expansion of (38) fails as* $\epsilon \rightarrow 0$ *in a subspace does seem to have an effect in a one-loop diagram.* To illustrate this effect, we wish to consider an $s-$channel[9] diagram for the process $\phi\phi \rightarrow \phi\phi$ in the one loop approximation. This process involves the integral[10]

$$I(p, m, \alpha, \varepsilon) = i \int \frac{d^4k}{(\alpha k_0^2 - |\mathbf{k}|^2 - m^2 + i\varepsilon)[\alpha(k+p)_0^2 - |\mathbf{k}+\mathbf{p}|^2 - m^2 + i\varepsilon]} \quad (39)$$

We shall find it convenient to relate this to the "fish diagram" in the usual covariant ϕ^4-theory [18]. We define,

$$p'_\mu = (\sqrt{\alpha}p_0, \mathbf{p}); \qquad k'_\mu = (\sqrt{\alpha}k_0, \mathbf{k}) \quad (40)$$

so that, we have

$$I(p, m, \alpha, \varepsilon) = \frac{1}{\sqrt{\alpha}} F(p', m, \varepsilon) \quad (41)$$

where F stands for the "fish diagram" amplitude in the ϕ^4-theory:

$$F(p, m, \varepsilon) = i \int \frac{d^4k}{(k_0^2 - |\mathbf{k}|^2 - m^2 + i\varepsilon)[(k+p)_0^2 - |\mathbf{k}+\mathbf{p}|^2 - m^2 + i\varepsilon]} \quad (42)$$

Now, we know how this is evaluated and know the analytic properties of the amplitude as seen from any text-book [18] .

We define $s = p^2$ and $s' = p'^2$. F can be looked upon as a function of a complex variable $s = p^2$. The analytic properties of $F(p, m, \varepsilon)$ on the real-s axis depend on whether $s > 4m^2$ or $s < 4m^2$. Thus, the analytic properties of $I(p, m, \alpha, \varepsilon)$ will depend on whether $s' = \alpha p_0^2 - |\mathbf{p}|^2 > 4m^2$ or $s' < 4m^2$. For $s' < 4m^2$, we can perform a Wick rotation and evaluate the integral, by say, a cutoff method[11]. The result is [here, A is a real constant]:

$$\begin{aligned} I(p, m, \alpha, \varepsilon) = & \quad \tfrac{1}{\sqrt{\alpha}} A \textstyle\int_0^1 dx \ln \frac{[m^2 - x(1-x)s' - i\varepsilon]}{\Lambda^2} & (43) \\ = & \quad \tfrac{1}{\sqrt{\alpha}} A \textstyle\int_0^1 dx \ln \frac{[m^2 - x(1-x)(\alpha p_0^2 - |\mathbf{p}|^2) - i\varepsilon]}{\Lambda^2} & (44) \end{aligned}$$

[9]We shall be interested later in the *imaginary* part of the amplitude. Only the s-channel diagram can have this imaginary part for the physical amplitude.

[10]This work involves noncovariant formulations. Thus several results will necessarily be explicitly Lorentz frame dependent.

[11]Later, we shall be focussing on the imaginary part, which is finite.Hence the details of regularization does not matter.

Now for $\Re e\, s' < 4m^2$, this quantity is analytic, and it has a branch-cut from $s' = 4m^2$ along the real s' axis [actually, slightly below the real axis]. We, now, will drop the real term $\sim \ln \Lambda^2$ and focus attention on the finite part $\frac{A}{\sqrt{\alpha}} I'(p, m, \alpha, \varepsilon)$. We note that the logarithm has a phase varying from -π to π . We then have,

$$\begin{aligned} I'(p, m, \alpha, \varepsilon) &= \int_0^1 dx \frac{1}{2} \ln\{[m^2 - x(1-x)(\alpha p_0^2 - |\mathbf{p}|^2)]^2 + \varepsilon^2\} \\ &+ \int_0^1 dx[-\pi i + i \arctan\{\frac{\varepsilon}{x(1-x)(\alpha p_0^2 - |\mathbf{p}|^2) - m^2}\}] \end{aligned} \tag{45}$$

Here, we have chosen the range of arctan as $[0,\pi)$ for our convenience. [Note that this definition differs from the principal branch arctan]. We note, in particular that

$$\lim_{\varepsilon \to 0} Im I'(p, m, \alpha, \varepsilon) = \int_0^1 dx\{-\pi i \Theta[B]\} = -\pi i \Delta x \tag{46}$$

with Θ[B] as the step function and

$$B = x(1-x)(\alpha p_0^2 - |\mathbf{p}|^2) - m^2 = x(1-x)s' - m^2 \tag{47}$$

and Δx is the interval over which $B > 0$. Now, for the s-channel process $\phi\phi \rightarrow \phi\phi$, we evaluate the expression of (35) in the one loop approximation. We, then, have for the difference,

$$\begin{aligned} & Im\left\{ \lim_{\varepsilon \to 0} I'(\alpha + \delta\alpha, p, m, \varepsilon) - \lim_{\varepsilon \to 0} I'(\alpha, p, m, \varepsilon) \right\} \\ = \; & Im\left\{ \Pi_i (p_i^2 - m_i^2) FT \frac{\delta^4}{\delta J^4} \lim_{\varepsilon \to 0} << \exp[i\delta\alpha \int d^4x \phi \partial_0^2 \phi] - 1 >>|_{J=0} \right\} \end{aligned} \tag{48}$$

where, $\frac{\delta^4}{\delta J^4}$ is a brief way to express the act of extracting the 4-point function, FT stands for the Fourier transform. We now choose α=α_0 such that $s' = 4m^2$and choose a $(\alpha_0 + \delta\alpha)$ such that $s' > 4m^2$. We now take the imaginary part of both sides. We note:

(i) I'(α_0+$\delta\alpha$) has a nonzero imaginary part as $\epsilon \rightarrow 0$.

(ii)I'(α_0^-) has *no* imaginary part as $\epsilon \rightarrow 0$. Here, α_0^- stands for the limit from left $\alpha \rightarrow \alpha_0$. This is so since B of (47) is necessarily negative for $\alpha < \alpha_0$.

(iii)The difference of the imaginary parts is *not* proportional to $\delta\alpha$; a simple calculation shows that for small $\delta\alpha$, it is proportional to $\sqrt{\delta\alpha}$. To see this, we note that with $\alpha = \alpha_0 + \delta\alpha$, $s' = 4m^2 + \delta\alpha p_0^2$; and for $x \sim \frac{1}{2}$

$B = -4m^2(x - \frac{1}{2})^2 + x(1-x)\delta\alpha p_0^2 \simeq -4m^2(\Delta x)^2 + \frac{1}{4}\delta\alpha p_0^2$ and thus, $B > 0$ over a range if x, viz. $2\Delta x \sim \sqrt{\delta\alpha}$.

(iv) On account of (iii), the difference of the imaginary parts cannot be understood as

$$Im\Pi_i(p_i^2 - m_i^2)FT\frac{\delta^4}{\delta J^4}\begin{array}{c}lim\\ \varepsilon \to 0\end{array} << [i\delta\alpha\int d^4x\phi\partial_0^2\phi] >>|_{J=0} \tag{49}$$

which is proportional to $\delta\alpha$.

(v)In fact, a simple analysis will reveal that

$$\Pi_i(p_i^2 - m_i^2)FT\frac{\delta^4}{\delta J^4}\begin{array}{c}lim\\ \varepsilon \to 0\end{array} << [i\delta\alpha\int d^4x\phi\partial_0^2\phi] >>|_{J=0;\alpha\to\alpha_0^-} \tag{50}$$

does not have an imaginary part. This is seen as follows: The expression (50) is understood as being proportional to $\delta\alpha\frac{\partial}{\partial\alpha}\begin{array}{c}lim\\ \varepsilon \to 0\end{array} I(p, m, \alpha, \varepsilon)\ |_{\alpha_0^-}$ which is real.

(vi) We further note that as a result of (iii),

$$\frac{\partial}{\partial\alpha}\begin{array}{c}lim\\ \varepsilon \to 0\end{array} I'(\alpha, p, m, \varepsilon)$$

is ill-defined at this value of α. We shall discuss this in much more detail in section 6 where we will employ discussion carried out in this section in the context of a gauge field theory 1-loop calculation.

Finally, we add a comment on whether we could avoid such problems if we do the field theory keeping ϵ small but nonzero. We then note that derivative of I with α will exist, and so will the Taylor series of I around $\alpha = \alpha_0$. But now the derivative will be a very large number $\to \infty$ as $\epsilon \to 0$; and the radius of convergence of the Taylor expansion will be very small.So we may not calculate, by a Taylor expansion, the value of $I(\alpha_0 + \delta\alpha)$ from that for $I(\alpha_0)$ if "$\delta\alpha >> \varepsilon$" . Moreover, we should note that the limit $\epsilon \to 0$ is required for a physical (unitary) theory because even in a $\lambda\phi^4$-theory with $L' = L + i\varepsilon\phi^2$, a Hermitian Hamiltonian (and a unitary S-matrix) is obtained only as $\epsilon \to 0$.

We also note the reason we consider this question at all. In a gauge field theory, we derive the WT-identity for gauge variation, we need to confirm whether such variations will be infinitesimal of first order always.We shall find that this is not always so in the interpolating gauges and we find some unexpected results from it in sections 9-10.

5. Delicate Limits in the Discussion of Gauge-Independence

In the previous sections, we noted places in Quantum field theory calculations where care may me needed. We wish now to note down the procedure that we normally follow for establishing the WT-identity used in evaluating gauge variation of Green's functions. We do this with a view to see if any of the steps could fail in gauges such those under consideration. We shall then discuss several points regarding the derivation in Sections 6-11.

We consider the generating functional of the unrenormalized connected Green's functions $Z[J, \theta, \varepsilon]$. We obtain the (renormalized) S-matrix elements from it by the operation Θ: defined as a succession of the following:

$$S(p_i, \theta) = \Theta Z[J, \theta, \varepsilon] \equiv \begin{array}{c}lim\\ \varepsilon \to 0\end{array} S(p_i, \theta, \varepsilon) \tag{51}$$

$$\Theta \equiv \lim_{\varepsilon \to 0} \lim_{p_i^2 \to m_i^2 - i\varepsilon} \Pi_i (p_i^2 - m_i^2 + i\varepsilon) \frac{\phi_i(p_i)}{\sqrt{Z_i}} F.T.[\frac{\delta^n}{\delta J^n(x)}] \tag{52}$$

where the nth order functional derivative acting on $Z[J, \theta, \varepsilon]$ at $J = 0$ is written as $[\frac{\delta^n}{\delta J^n(x)}]$ for brevity; $F.T.$ stands for the Fourier transform; $Z_i(\theta)$'s stand for the mass-shell renormalization constants and $\phi_i(p_i)$ is the physical wavefunction[12]. The gauge independence of $S(p_i, \theta, \varepsilon)$ is expressed by the requirement

$$\frac{\partial}{\partial \theta} S(p_i, \theta) = \frac{\partial}{\partial \theta} \lim_{\varepsilon \to 0} S(p_i, \theta, \varepsilon) = 0 \tag{53}$$

It is to be particularly noted that the order of the limit ($\epsilon \to 0$) and differentiation $\frac{\partial}{\partial \theta}$ is as follows: $\frac{\partial}{\partial \theta} \lim_{\varepsilon \to 0}$

We shall, now, recapitulate how one actually obtains the gauge-dependence of the on-shell physical Green's functions in practice:

(a)We start from $Z[J, \theta, \varepsilon]$ and evaluate $\frac{\partial}{\partial \theta} Z[J, \theta, \varepsilon]$ for a fixed ϵ.

(b)We then use the WT identities to simplify the quantity under question [3].

(c)We find the part of this that contributes to the *on-shell physical* Green's functions. In other words, we evaluate

$$\Theta \frac{\partial}{\partial \theta} Z[J, \theta, \varepsilon] = \lim_{\varepsilon \to 0} \lim_{p_i^2 \to m_i^2 - i\varepsilon} \Pi_i (p_i^2 - m_i^2 + i\varepsilon) \frac{\partial}{\partial \theta} \frac{\phi_i(p_i)}{\sqrt{Z_i}} F.T.[\frac{\delta^n}{\delta J^n(x)}] Z[J, \theta, \varepsilon] \tag{54}$$

We wish to note several points in the above derivation:
[1] We first assume that the limit $\epsilon \to 0$ and the differentiation $\frac{\partial}{\partial \theta}$ can be interchanged.
[2] We evaluate then $\frac{\partial W[J, \theta, \varepsilon]}{\partial \theta}$ by

$$\frac{\partial W[J, \theta, \varepsilon]}{\partial \theta} = \frac{\partial}{\partial \theta} \int D\phi \exp\left(iS_{eff}[\phi, \theta] + i \int d^4x J\phi \right) \tag{55}$$

. In doing so, it is being tacitly assumed that this derivative always exists even as $\epsilon \to 0$. We have already seen a counter-example[13] in Sec.4 and we shall examine this question in Section 6 further.

[12] We note the somewhat unconventional appearance of truncation factor $(p_i^2 - m_i^2 + i\varepsilon)$ and the on-shell limit $p_i^2 \to m_i^2 - i\varepsilon$. Before we let $\varepsilon \to 0$, we have to do this to correctly truncate the external line propagators. Also, in the discussion of gauge-independence, we need to remove any pole-less contributions to the $\frac{\partial}{\partial \theta} G$; and these are then correctly removed by the above on-shell limit $p_i^2 \to m_i^2 - i\varepsilon$. The parameter ϵ is let go to zero only in the end.

[13] At this point, it may be thought that it is not important whether the derivative $\frac{dI}{d\theta}$ exists at a point θ_0; it is only required that $\int \frac{dI}{d\theta} d\theta$ exists over a small interval covering θ_0. After all, we may make $\frac{dI}{d\theta}$ finite by keeping ϵ finite till end (or possibly by using suitable wavepackets for external lines: see section 11). But this does not evade the problem in section 10 about the contribution of the ϵ-term. We find the language of "non-differentiability" as the best way to exhibit the problem.

[3] Equivalently, the derivative is evaluated by the assumption that

$$\frac{\partial}{\partial\theta}\int D\phi \exp\left(iS_{eff}[\phi,\theta]+i\int d^4xJ\phi\right) \tag{56}$$

can be evaluated by expressing it as

$$i\int D\phi\frac{\partial}{\partial\theta}S_{eff}[\phi,\theta]\exp\left(iS_{eff}[\phi,\theta]+i\int d^4xJ\phi\right) \tag{57}$$

In this we make an assumption that

$$\int D\phi \exp\left(iS_{eff}[\phi,\theta]+i\int d^4xJ\phi\right)\{\exp(iS_{eff}[\phi,\theta+\delta\theta])-\exp(iS_{eff}[\phi,\theta])\} \tag{58}$$

can always be written as

$$i\int D\phi\frac{\partial}{\partial\theta}S_{eff}[\phi,\theta]\delta\theta\exp\left(iS_{eff}[\phi,\theta]+i\int d^4xJ\phi\right) \tag{59}$$

even as $\epsilon \longrightarrow 0$. In this connection, we draw attention to the example in the section 4.

[4] In writing down the WT-identities, we do not keep track of the contribution to it from the ϵ-term. We recall that recently, we have noted that [5] a careful attention to the ϵ-term has to be given in the case of gauges other than the Lorentz gauges.

[5] We then simplify the resultant expression by the use of the WT-identities. We then pick out the part of this that contributes to the truncated physical Green's functions.

We shall analyze these points in detail in the next section.

6. An Explicit Example

In this section, we shall make a number of obervations that have a potential bearing on the derivation of WT identities for evaluating gauge-dependence of Green's functions.

We start with the example of the gauge theory in an interpolating gauge of Eq. (10) so defined as to interpolate between the Feynman and the Coulomb gauge [12]. For simplicity, we shall first focus our attention on a particular integral[14] that appears in a contribution to the 4-point function (of gluons) in the one loop approximation. It appears in the one loop diagram with two 4-point vertices and 2 A_0 internal lines[15]. The integral reads

$$I(p,\theta,\varepsilon)=i\int\frac{d^4k}{(\theta^2k_0^2-|\mathbf{k}|^2+i\varepsilon)[\theta^2(k+p)_0^2-|\mathbf{k}+\mathbf{p}|^2+i\varepsilon]} \tag{60}$$

To correlate the above integral with with that of Section 4 [See eq. (39)], we shall find it convenient to rather consider an integral with a *mass* included[16]. We consider,

$$I(p,m,\theta,\varepsilon)=i\int\frac{d^4k}{(\theta^2k_0^2-|\mathbf{k}|^2-m^2+i\varepsilon)[\theta^2(k+p)_0^2-|\mathbf{k}+\mathbf{p}|^2-m^2+i\varepsilon]} \tag{61}$$

[14]It is of course true that very often the *total* amplitude in a gauge theory has properties different from a specific piece of it; this being due to cancellations of terms.

[15]We are focussing attention on a part of the contribution due to time-like gluons.

[16]We could regard this as an infrared regularization of the amplitude in question. We shall, however, eventually focus on the imaginary part of (60) and this is infrared finite here.

We shall then explicitly demonstrate that, in the physical region, i.e. for $p^2 > 4m^2$, there exists a value of θ viz. $\theta_0 \in [0,1]$ such that at this point

(a) The derivative $\frac{\partial}{\partial\theta} \lim_{\varepsilon \to 0} I(p,m,\theta,\varepsilon)$ does not exist: i.e.

$$\frac{\partial}{\partial\theta} \lim_{\varepsilon \to 0} I(p,m,\theta,\varepsilon) \mid_{\theta_0^-} \neq \frac{\partial}{\partial\theta} \lim_{\varepsilon \to 0} I(p,m,\theta,\varepsilon) \mid_{\theta_0^+} \tag{62}$$

(b) What is worse is that the imaginary part of the left hand side vanishes at this point, whereas the imaginary part of the right hand side $\to$ infinity. Thus, this quantity has infinite discontinuity.

(c)Moreover, we find that

$$\lim_{\varepsilon \to 0} \frac{\partial}{\partial\theta} I(p,m,\theta,\varepsilon) - \frac{\partial}{\partial\theta} \lim_{\varepsilon \to 0} I(p,m,\theta,\varepsilon) \mid_{\theta_0^\pm} \neq 0$$

where the first term on the left hand side is evaluated for the both limits $\theta \to \theta_0^\mp$.

To see these results, we shall find it convenient to make use of the earlier discussion in Section 3. We find from (44)

$$I(p,m,\theta,\varepsilon) = \frac{A}{\theta} \int_0^1 dx \ln \frac{[m^2 - x(1-x)(\theta^2 p_0^2 - |\mathbf{p}|^2) - i\varepsilon]}{\Lambda^2} \tag{63}$$

We are presently interested in discussing the evaluation of the θ-derivative and also as to what happens in the two different orders of the limit and the derivative. So we will drop the term $\sim \ln \Lambda^2$ and focus attention on the finite part $\frac{A}{\theta} I'(p,m,\theta,\varepsilon)$. We note that the logarithm has a phase varying from $-\pi$ to π for $\Re es' > 4m^2$. We then have, using (45),

$$\begin{aligned} I'(p,m,\theta,\varepsilon) &= \int_0^1 dx \frac{1}{2} \ln\{[m^2 - x(1-x)(\theta^2 p_0^2 - |\mathbf{p}|^2)]^2 + \varepsilon^2\} \\ &+ \int_0^1 dx[-\pi i + i \arctan\{\frac{\varepsilon}{x(1-x)(\theta^2 p_0^2 - |\mathbf{p}|^2) - m^2}\}] \end{aligned} \tag{64}$$

We shall now focus attention on the imaginary part of I'. We have,

$$ImI' = \int_0^1 dx\{-\pi + \arctan \frac{\varepsilon}{B}\} \tag{65}$$

where, as defined in Section 4,

$$B(\alpha,\theta,p) \equiv [-m^2 + x(1-x)(\theta^2 p_0^2 - |\mathbf{p}|^2)] = [-m^2 + x(1-x)s'] \tag{66}$$

Now,

$$\lim_{\varepsilon \to 0} ImI' = -\int_0^1 dx \pi \Theta(B) \tag{67}$$

We, thus, find

$$\frac{\partial}{\partial\theta} \lim_{\varepsilon \to 0} ImI' = -\pi \int_0^1 dx \delta(B) \frac{\partial B}{\partial\theta} \tag{68}$$

We note

$$\delta(B) = \delta(-y^2 s' + (\frac{1}{4}s' - m^2)) = \frac{1}{|s'|}\delta(y^2 - a^2) \tag{69}$$

with $y = x - \frac{1}{2}$; and

$$a^2 = \frac{s' - 4m^2}{4s'}; \quad a \equiv +\frac{\sqrt{s' - 4m^2}}{2\sqrt{s'}} \tag{70}$$

For $s' < 4m^2$, $a^2 < 0$ and the δ-function does not contribute to the integrand. For $s' > 4m^2$, we use

$$\delta(y^2 - a^2) = \frac{1}{2a}\{\delta(y-a) + \delta(y+a)\} \tag{71}$$

and

$$B = (\frac{1}{4} - y^2)s' - m^2 \tag{72}$$

$$\frac{\partial B}{\partial \theta} = 2\theta p_0^2(\frac{1}{4} - y^2) \tag{73}$$

to find

$$\frac{\partial}{\partial \theta} ImI' = -\pi \int_0^1 dx \delta(B)\frac{\partial B}{\partial \theta} = \frac{-2\pi\theta p_0^2}{2as'} \int dy(\frac{1}{4} - y^2)\{\delta(y-a) + \delta(y+a)\} \tag{74}$$

i.e.

$$\frac{\partial}{\partial \theta} \lim_{\varepsilon \to 0} ImI' == \frac{-2\pi\theta p_0^2}{\sqrt{s' - 4m^2}\sqrt{s'}}(\frac{1}{2} - 2a^2) \tag{75}$$

We thus find that as $\theta \to \theta_0$, where $s'[\theta_0] = 4m^2$, $a \to 0$ and

$$\lim_{\theta \to \theta_0^+} \frac{\partial}{\partial \theta} \lim_{\varepsilon \to 0} ImI' \to \infty \tag{76}$$

On the other hand,

$$\lim_{\theta \to \theta_0^-} \frac{\partial}{\partial \theta} \lim_{\varepsilon \to 0} ImI' = 0 \tag{77}$$

Thus, the imaginary part of I' is not differentiable at $\theta = \theta_0$.Furthermore, we note that for a given $\epsilon > 0$,

$$\left\{ \lim_{\theta \to \theta_0^+} \frac{\partial}{\partial \theta} ImI' - \lim_{\theta \to \theta_0^-} \frac{\partial}{\partial \theta} ImI' \right\} = 0 \tag{78}$$

with each given by

$$\lim_{\theta \to \theta_0^+} \frac{\partial}{\partial \theta} ImI' = \lim_{\theta \to \theta_0^-} \frac{\partial}{\partial \theta} ImI' = \int_0^1 dx\{-\frac{\varepsilon}{B^2 + \varepsilon^2}\frac{\partial B}{\partial \theta}\}\|_{\theta_0} \tag{79}$$

Thus,

$$\lim_{\varepsilon \to 0} \left\{ \lim_{\theta \to \theta_0^+} \frac{\partial}{\partial \theta} ImI' - \lim_{\theta \to \theta_0^-} \frac{\partial}{\partial \theta} ImI' \right\} = 0 \tag{80}$$

Finally, we shall give the treatment for $m^2 = 0$ exactly, which differs somewhat from the above. In this case, B of (66) becomes,

$$B(\alpha, \theta, p) \equiv x(1-x)(\theta^2 p_0^2 - |\mathbf{p}|^2) = x(1-x)s' \tag{81}$$

and thus (67) is modified to,

$$\begin{aligned} \lim_{\varepsilon \to 0} ImI' &= -\int_0^1 dx \pi \Theta(s') \\ &= -\pi\Theta(\theta^2 p_0^2 - |\mathbf{p}|^2) = -\pi\Theta(\theta - \frac{|\mathbf{p}|}{p_0}) \end{aligned} \tag{82}$$

and thus, (75) is modified to (note: we assume $0 \leq \theta \leq 1$),

$$\frac{\partial}{\partial\theta} \lim_{\varepsilon \to 0} ImI' = -\pi\delta(\theta - \frac{|\mathbf{p}|}{p_0}) \tag{83}$$

7. Extension to a Green's Function

In view of the fact that we have considered only a *part* of a contribution to *one* of the diagrams for the 4-point function, to illustrate the point, one might immediately suspect that the problem could disappear in the entire contribution to the S-matrix element. We note first that WT-identities for gauge variations are equations written down for *Green's functions* and not just the S-matrix elements. Moreover, what we wish to drive at at the moment is that the *path-integral itself is not a differentiable function of* θ as a result of the problems mentioned in the past sections [for a further elaboration, please see section 11.1]. For this purpose, it proves sufficient if the problem with differentiability persists at least at the level of Green's functions, even if it were to vanish from S-matrix elements from possible mutual cancellations between diagrams. In this section, we wish to make several remarks in this connection.

We first show, in an obvious way, that this problem with differentiabilty persists for a contribution to an *off-shell* Green's function of the type considered earlier in section 6. An inspection of the diagrams contributing to the Green's functions will then make it seem extremely unlikely that *at the level of off-shell Green's functions* there could persist a cancellation; even if there were one for S-matrix elements. [We have verified this point for the simpler example outlined in the Appendix A].

To see this, we recall that this diagram depends only on $s' = \theta^2 p_0^2 - |\mathbf{p}|^2$ and not on individual 4-momenta p_1,$p_2$and p_3. We could thus choose any *off-shell* $p_1$and p_2 consistent with the condition that s is real with $s > 4m^2$ and θ_0 is such that $s' = \theta_0^2 p_0^2 - |\mathbf{p}|^2 = 4m^2$. Then for all such *off-shell* choices of $p_1$and p_2, the above problem with the differentiability will persist for this contribution.

In appendix A, we shall consider the example of a scalar QED and consider the off-shell 1-loop Green's function for the process $\phi\phi \to \phi\phi$. We have verified that the imaginary part of such a process [with some further restrictions on momenta, spelt out in the appendix] does have a discontinuous behaviour on account of the threshold due to unphysical photons and moreover the threshold for these does depends on θ. These are the essential ingredients that lead to non-differentiabilty.]

8. A Close Analysis of the Example in Section 6

In this section, we shall qualitatively try to understand why the derivative with respect to θ does not exist at some point and why the order of differentiation $\frac{\partial}{\partial\theta}$ and the limit $\lim_{\varepsilon \to 0}$ makes a significant difference there. We can understand this with the help of a close look at the analyticity properties of the integral $I'(p, m, \theta, \varepsilon)$ in the Section 6. We recall

$$I'(p, m, \theta, \varepsilon) = \int dx \ln\{m^2 - x(1-x)(\theta^2 p_0^2 - |\mathbf{p}|^2) - i\varepsilon\} \tag{84}$$

The analyticity properties of the integrand above, for a given p depend upon x and θ. For a real $s = p^2 < 4m^2$ and a $\theta \in$[0,1], $s' = p'^2 < 4m^2$ also; and hence for all $x \in$[0,1] and all $\theta \in$[0,1], the integrand is analytic even as $\varepsilon \to 0$. A small [enough] variation of θ will not alter these facts and hence there is no problem occurring in the order of limits $\delta\theta \to 0$ and $\varepsilon \to 0$ in this unphysical region. (Besides, in the unphysical region, the entire calculation can be done in the Euclidean field theory which would not need ϵ).

We shall, of course, be interested in this issue in the *physical* region $s = p^2 > 4m^2$. For a given $s > 4m^2$, there is a range of $\theta \in [0, 1]$, such that there $s' > 4m^2$ and a range of $x \in [0, 1]$ exists for which the integrand is close to the branch-point. Also, for any $\theta \in [0, 1]$, $s' > 4m^2$ *for all* s *greater than a certain lower bound* (See section 11 for more details). Thus, this issue of analytic properties cannot be avoided for any physical region if one is to interpolate between the Feynman and the Coulomb gauge.

Next, we note that for a fixed p, the variables $s'$and θ are related: a small variation in θ induces a small variation in s' which is the variable relevant to the analytic properties of the integrand. Thus, for example, we choose a value of $\theta = \theta_0$ such that s' is slightly smaller than $4m^2$, then for a range of x $\varepsilon[0, 1]$ this point is very close to the branch-point of the integrand in the complex s'-plane. Then the Taylor expansion of the integrand in $\theta - \theta_0$ is then valid only with a small radius of convergence *that is directly dependent [proportional] to* ϵ*!* In fact, (84) will show that the entire integral cannot be expanded in such a Taylor series except for small[17] enough $(\theta - \theta_0) \sim \varepsilon$.

We thus see that the fact that the loop-integrand could not be Taylor expanded for a large enough $(\theta - \theta_0)$ in a *part* of the 4-dimensional space has, in fact, a reflection on the entire integral. This was anticipated in Section 3.

The above demonstration now shows why the order of the two limits discussed earlier matters near $\theta = \theta_0$. If we were to keep ϵ fixed then there is a small enough radius $(\theta - \theta_0)$ for which the Taylor expansion holds and the limit $(\theta - \theta_0) \to 0$ can in fact be taken. This procedure can give ultimately

$$\lim_{\varepsilon \to 0} \frac{\partial}{\partial\theta} I(p_i, \theta, \varepsilon)$$

On the other hand, if we are to let $\epsilon \to 0$ first, then the radius of convergence of the Taylor expansion in θ itself shrinks to zero and we cannot know how to obtain $\frac{\partial}{\partial\theta} \lim_{\varepsilon \to 0} I(p_i, \theta, \varepsilon)$

[17] We note that if we took any $\theta > \theta_0$, the integrand is close to the branch-cut for a range of x. It may appear that the integral should then be problematic for all such θ and not just for $\theta = \theta_0$. This is not the case, at least in this example. We thus end up with just one troublesome point for *this* diagram that however varies with Lorentz frame (for this, see section 11).

by differentiation of the integrand and subsequent use of the WT identities. Evaluation of the gauge dependence requires us to evaluate the latter object.

We shall now comment on the treatment of the field theory using the Wick rotation. It may be thought that the troubles in section 4 and 6 are artificial because after all we could carry out the Wick rotation in (39) and (60) and thus go to the Euclidean field theory that does not require ϵ. Thus, it may be argued that the problems that depend on the limits involving ϵ and those that depend on the size of ϵ should be artificial. Now, the field theory of Section 4 and that used in Section 6 are in fact well behaved with respect to the operation $\frac{\partial}{\partial\theta}$ in the unphysical region. To see, how the problem could arise in the physical region, we recall the fact that I' is a function of $s' = \theta^2 p_0^2 - |\mathbf{p}|^2$ and it has a branch point at $s' = 4m^2$. Of course, for $\Re es' < 4m^2$, the function is analytic function of s' and therefore a differentiable function of θ (with p_μ fixed). But, as $\epsilon \to 0$, the derivative with respect to s' and hence with respect to θ [for a fixed p_μ] cannot exist at the branch point. Thus, the fact that the generating functional appears (formally) differentiable in θ in the Euclidean region is no mystery ; but by no means guarantees good behavior everywhere in the physical region and that is where we are interested in the gauge-independence issue.

9. Relation between the Altered Propagator Structure in [5] and Present Results

In reference [5], we had discussed the gauges interpolating between the Coulomb and the Feynman gauge with a simple ϵ-term. We had noted some unusual features of what happens when the interpolating parameter θ, in the gauge interpolating between the Coulomb and the Feynman gauge, is varied. We had shown that it was important to pay particular attention to the ϵ-term in the discussion of the gauge-independence. We had considered how the ϵ term should be modified with θ if we are to keep the vacuum expectation value of a gauge-invariant operator unchanged. We had further shown that as a result of this modification in the ϵ-term, the free propagator undergoes a radical change in form as the parameter θ is varied through[18] a $\delta\theta >> \epsilon$. We had argued, in fact, from this observation that such interpolating gauges that assume any standard (fixed) ϵ-term cannot preserve gauge-independence as θ is varied and thus do not interpolate correctly between the Feynman and the Coulomb gauge; while on the other hand trying to modify the ϵ-term as required for gauge-independence leads to pathological behavior in the path-integral.

In this section, we shall make a contact between the concrete results obtained in this work in Sec. 6 and this result obtained earlier in [5].

We shall establish a simple result to begin with. We shall then relate it to the question of the variation of the 4-point function with respect to θ discussed in section 6. It reads

$$\lim_{\varepsilon \to 0} I"(\theta, CC) = \lim_{\varepsilon \to 0} I"(\theta, \frac{1}{2}[C+A]\frac{1}{2}[C+A]) - \pi i \int_0^1 dx \Theta[B] \quad (85)$$

Here we have defined, in all cases employed below,

$$I(\theta) = \frac{A}{\theta} I' \text{ and } ImI(\theta) = \frac{A}{\theta} I" \quad (86)$$

[18] The relation between $\delta\theta$ and ϵ given here is a abbreviated way of expressing the condition between these two quantities of different dimensions. See [5] for the exact condition.

and further $I(\theta, \frac{1}{2}[C+A]\frac{1}{2}[C+A])$ refers to the (truncated) fish-diagram amplitude obtained by taking the propagator as $\frac{1}{2}$(C+A) in place of C. Here, C refers to the causal propagator and A to the "anti-causal" propagator (anti-causal means the one with $i\varepsilon \to -i\varepsilon$ in C). The above relation, as we shall later see from Eq. (93) , represents in a different manner, the rapid effect of variation with respect to θ (i.e. $\frac{\partial I}{\partial \theta}\mid_{\theta_0^+} \to \infty$) which is represented by a replacement of the propagator "C" by "$\frac{1}{2}(C+A)$" .

To prove this result, we consider the fish amplitude I with propagators $\frac{1}{2}(C+A)$. We define respectively $I(\theta, CC)$, $I(\theta, CA)$ as the amplitude in question with both the propagators taken as causal, and with one causal and one anti-causal . Then

$$I'(\theta, CA) = I'(\theta, AC) \tag{87}$$

and

$$I"[\theta, \frac{1}{2}(C+A)\frac{1}{2}(C+A)] = \frac{1}{4}\{I"[\theta, CC] + I"[\theta, AA] + 2I"[\theta, CA]\} \tag{88}$$

We can then show that as $\epsilon \to 0$,

$$\lim_{\varepsilon \to 0} (I"[\theta, AA] - I"[\theta, CC]) = 0 \tag{89}$$

$$\lim_{\varepsilon \to 0} (I"[\theta, CA] - I"[\theta, CC]) = 2\pi i \int_0^1 dx\Theta[B] \tag{90}$$

Thus, we then have as $\epsilon \to 0$,

$$\lim_{\varepsilon \to 0} I"[\theta, \frac{1}{2}(C+A)\frac{1}{2}(C+A)] = \lim_{\varepsilon \to 0} I"[\theta, CC] + \pi i \int_0^1 dx\Theta[B] \tag{91}$$

We shall now apply this to the amplitude of (63). Let θ_0 be such that

$$s' = \theta_0^2 p_0^2 - |\mathbf{p}|^2 = 4m^2$$

We shall let $\theta = \theta_0 + \delta\theta$. We shall assume that $\delta\theta >> \varepsilon$. To understand the significance of the above result of Eq. (91), in the present context, we compare it with the result that expresses the discontinuous behavior of I at $\theta = \theta_0$. It reads,

$$\lim_{\varepsilon \to 0} I"(\theta, CC) = \lim_{\varepsilon \to 0} I"(\theta_0, CC) - \pi i \int_0^1 dx\Theta[B] + O[\delta\theta] \tag{92}$$

A comparison of Equations (91) and (92) shows that as $\epsilon \to 0$,

$$\lim_{\varepsilon \to 0} I"(\theta_0, CC) = \lim_{\varepsilon \to 0} I"[\theta, \frac{1}{2}(C+A)\frac{1}{2}(C+A)] + O[\delta\theta] \tag{93}$$

The above equation is an alternate way of representing the nonsmooth behavior of I: As θ is varied from θ_0 to $\theta_0 + \delta\theta$, I varies drastically so that the disproportionate change in $I"$ viz. $\{\lim_{\varepsilon \to 0} I"(\theta, CC) - \lim_{\varepsilon \to 0} I"(\theta_0, CC)\}$ is alternately expressed as that represented by a drastic change in the propagator structure from $C \to \frac{1}{2}[C+A]$. Then the two quantities $I(\theta_0, CC)$ and $I[\theta, \frac{1}{2}(C+A)\frac{1}{2}(C+A)]$, evaluated with *different propagators* for neighboring θ's, differ only by a term of $O[\delta\theta]$.

This conclusion can be understood well in the light of the work of reference [5]. There it was found that when $\theta \to \theta + \delta\theta$ and *the corresponding the gauge variation of the* ϵ-*term is also taken into account*, the propagator changed from $C \to \frac{1}{2}[C + A]$ provided $\delta\theta >> \varepsilon$. The above conclusion (93) is a reflection of this. Normally, the WT-identities are taken to imply that under $\theta \to \theta + \delta\theta$, the corresponding change in a Green's function is an infinitesimal of $O(\delta\theta)$. The above example explicitly shows that that need not be so, and the Eq. (93) further shows that a non-trivial change in the Green's function can be correlated to the change in the ϵ-term arrived at in [5]. Only after this "large" change has been removed, then the *residue* is of $O(\delta\theta)$. We shall have more to say about this in the next section where we will compare the result (93) with that of the carefully evaluated WT-identity. [We note that the eq. (93) deals only with the *imaginary* parts of the amplitude I'. It is only the imaginary part of I' that shows a nonsmooth behavior with θ. For the calculation within the actual context of a gauge theory, the real part has to be dealt with for set of all diagrams together].

10. Expansion of an Exponential with an "Infinitesimal" Exponent

In this section, we shall establish a further contact of the results in the earlier sections with the work of reference [5]. In reference [5], we had established several new observations in the context of the interpolating gauges such as those considered in earlier sections. The ones relevant here are:

(a) In the treatment of gauge variation, it was necessary to take into account the ϵ-term carefully.

(b)The gauge variation of the ϵ-term had to be fully taken into account *and could not be expanded out as an infinitesimal exponent.*

To establish a contact between the analysis of earlier sections and these results, we shall now follow a procedure parallel to that in [5].

We now consider the generating functional

$$W[J; \theta - \delta\theta] = \int D\phi \exp\{iS_{eff}[A, c, \overline{c}; \theta - \delta\theta] + \varepsilon R + i \int d^4x J^\mu A_\mu\} \tag{94}$$

where S_{eff} refers to the effective action in the interpolating gauge that interpolates between the Coulomb and the Feynman gauge. We imagine performing the transformation as in [5],

$$A'^{\alpha}_{\mu}(x) - A^{\alpha}_{\mu}(x) \equiv \delta A^{\alpha}_{\mu}(x) = iD^{\alpha\beta}_{\mu} c^{\beta}(x) \int d^4z \overline{c}(z) \frac{\partial F^{\gamma}[A(z); \theta]}{\partial\theta} |_{\theta} \delta\theta \tag{95}$$

$$\delta c^{\alpha}(x) = -i\frac{1}{2} g f^{\alpha\beta\gamma} c^{\beta}(x) c^{\gamma}(x) \int d^4z \overline{c}(z) \frac{\partial F^{\gamma}[A(z); \theta]}{\partial\theta} |_{\theta} \delta\theta \tag{96}$$

$$\delta\overline{c}^{\alpha}(x) = iF^{\alpha}[A(x); \theta] \int d^4z \overline{c}(z) \frac{\partial F^{\gamma}[A(z); \theta]}{\partial\theta} |_{\theta} \delta\theta \tag{97}$$

As shown in [5], to preserve the vacuum-expectation-value of a gauge-invariant operator under this transformation, it is required that the ϵ-term is suitably changed from $\varepsilon R \rightarrow \epsilon(R+\delta R)$ [For a definition of δR, see Eq.(16)]. It is then easy to show that [5]

$$W[J;\theta-\delta\theta] = \int D\phi \exp\{iS_{eff}[A,c,\overline{c};\theta-\delta\theta]+\varepsilon R+i\int J^{\mu}A_{\mu}d^4x\} \quad (98)$$

$$= \int D\phi' \exp\{iS_{eff}[A',c',\overline{c}';\theta]+\varepsilon[R+\delta R] +i\int J^{\mu}[A'_{\mu}-\delta A_{\mu}]d^4x\} \quad (99)$$

We now suppress the primes in (99), and rewrite the two equations as

$$\int D\phi \exp\{iS_{eff}[A,c,\overline{c};\theta-\delta\theta]+\varepsilon R+i\int J^{\mu}A_{\mu}\} = \int D\phi \exp\{iS_{eff}[A,c,\overline{c};\theta]+\varepsilon[R+\delta R]+i\int d^4x J^{\mu}[A_{\mu}-\delta A_{\mu}] \quad (100)$$

Or written differently,

$$<< \exp\{\frac{1}{2}i\int d^4x\Delta F^2 - i\Delta S_{gh}\} >>|_{\theta} = << \exp\{\varepsilon\delta R - i\int d^4x J^{\mu}\delta A_{\mu}\} >>|_{\theta} \quad (101)$$

where, we have introduced the short-hand notation

$$<< O[\phi] >>|_{\theta} \equiv \int D\phi O[\phi]\exp\{iS_{eff}[A,c,\overline{c};\theta]+\varepsilon R+i\int d^4x J^{\mu}A_{\mu}\} \quad (102)$$

and

$$\Delta F^2 \equiv F^2(\theta)-F^2(\theta-\delta\theta);\ \Delta S_{gh} = S_{gh}(\theta)-S_{gh}(\theta-\delta\theta)$$

The relation (101) as it stands is undoubtedly correct. In the light of results of Sections 4 and 6 and of reference [5], we however ask whether the simplified version of this result, obtained by expanding the exponential with an "infinitesimal" exponents, viz.

$$<< \{\frac{1}{2}i\int d^4x\Delta F^2 - i\Delta S_{gh}\} >>|_{\theta} = << \{\varepsilon\delta R - i\int d^4x J^{\mu}\delta A_{\mu}\} >>|_{\theta} \quad (103)$$

i.e.

$$<< \{\frac{1}{2}i\int d^4x\Delta F^2 - i\Delta S_{gh} - \varepsilon\delta R + i\int d^4x J^{\mu}\delta A_{\mu}\} >>|_{\theta} = 0 \quad (104)$$

and also the one obtained by dropping the $\varepsilon\delta R$ term altogether in (104), viz.

$$<< \{\frac{1}{2}i\int d^4x\Delta F^2 - i\Delta S_{gh} + i\int d^4x J^{\mu}\delta A_{\mu}\} >>|_{\theta} = 0 \quad (105)$$

(which we normally understand as the WT-identity relevant to the evaluation of the θ-dependence of Green's functions), actually holds in this form. We consider this issue in light of the results in section 4, where we found that

$$<< \exp[i\delta\alpha\int d^4x\phi\partial_0^2\phi] - 1 >> \quad (106)$$

could not always be interpreted as

$$<< [i\delta\alpha \int d^4x\phi\partial_0^2\phi] >> \tag{107}$$

and it was later explicitly understood during the discussion regarding the example considered in section 6. We had also seen in reference [5] that the effect of the modification of the ϵ-term was by no means always infinitesimal: It, in fact, lead to a unexpected modification of the propagator structure. In the normal usage of the WT-identity [2,3], we not only carry out this expansion, but in fact *ignore* the $\epsilon\delta$R term altogether[19].

We shall now try to apply the WT-identity (101) to the special case of the 4-point function considered in Section 6. We shall look at the left hand side of (101)in this context. To this 4-point function in the 1-loop approximation, there are a number of diagrams contributing. The effect of diagrams contributing to this term is to give the change in the 4-point function in the one loop approximation when the parameter θ is changed from θ-$\delta\theta$ to θ. Among this difference there is the contribution which is exactly of the form of the difference

$$\lim_{\varepsilon \to 0} \{I(\theta) - I(\theta - \delta\theta)\}$$

As seen, this difference is not understood as $\delta\theta\frac{dI}{d\theta}$ for a process at such[20] a $\theta - \delta\theta = \theta_0$ that $s'(\theta_0) = 4m^2$. At this point, it may appear that as we are taking only one of the contributions to the S-matrix elements, such discontinuities will cancel out when all contributions are taken into account. Here, we would like to draw attention to the remarks made in section 7 that such problems are by no means specific to the contribution to the S-matrix element, they are also present in the off-shell Green's functions where we *do not* expect fortuitous cancellations of this kind to happen. [This fact has explicitly been verified in the example considered in Appendix A].

Now, this contribution $[I(\theta) - I(\theta - \delta\theta)]$ cannot be expanded as $\frac{dI}{d\theta}\delta\theta$; so that it is not possible to expand out (at least) this contribution to$<< \exp\{\frac{1}{2}i\int d^4x\Delta F^2\} - 1 >>|_\theta$ as arising from $<< \frac{1}{2}i\int d^4x\Delta F^2\} >>|_\theta$as then it would have to be proportional to $\delta\theta$. It is the latter term that is normally taken as one of the terms in the WT-identity after assuming that such an expansion does indeed hold!

We next move onto the term$<< \exp\{\varepsilon\delta R\} >>|_\theta$on the right hand side of (101). We would normally assume that we can expand $\exp\{\varepsilon\delta R\}$ as

$$\exp\{\varepsilon\delta R\} \simeq 1 + \varepsilon\delta R$$

and we normally drop the term $\varepsilon\delta R$ on the right hand side of the WT-identity (104). We had however seen an unusual effect of this term in [5] in the context of the present example of the interpolating gauge, which was neither infinitesimal nor ignorable (See section 2 for more details). In fact, according to [5], we *cannot* always treat this term as of a lesser

[19]For unbroken gauge theories in the Lorentz class of gauges, we believe this is justified either by noting that no modification of the pole structrure happens in this case by the ϵ-term variation $\epsilon\delta$R or by independent arguments.

[20]For the compatibility of notations with that used earlier in this section, we have somewhat altered our conventions for $\delta\theta$.

order compared to the original ϵ-term, εR, in the exponent: its effects could be drastic enough to alter the propagator structure for $\delta\theta >> \varepsilon$. (This was essentially because $\varepsilon\delta R$ can contribute to an inverse propagator in such a way that the contribution blows up in a sensitive kinematical region and can overwhelm the ϵR term itself. The net ϵ-term then determines the new propagator structure). Thus, it is by no means obvious that the expansion of $\exp\{\varepsilon\delta R\}$ can be carried out nor is it obvious that its effects can be ignored as it is normally done.

We shall explicitly show this in working for the example we have worked out in section 9. To see this, we restructure Eq.(93) as follows:

$$\begin{aligned}
&\lim_{\varepsilon \to 0} I"(\theta_0, CC) - \lim_{\varepsilon \to 0} I"(\theta, CC) \\
&= \left\{ \lim_{\varepsilon \to 0} I"[\theta, \frac{1}{2}(C+A)\frac{1}{2}(C+A)] - \lim_{\varepsilon \to 0} I"(\theta, CC) \right\} \\
&+O[\delta\theta]
\end{aligned} \tag{108}$$

We now compare the above equation with (101). We note that the left hand side is a particular contribution to $<< \exp\{\frac{1}{2}i\int d^4x\Delta F^2\} - 1 >>|_\theta$ from the diagram considered. The curly bracket on the right hand side is the corresponding ("large") contribution from $<< \exp(\varepsilon\delta R) - 1\} >>|_\theta$and has arisen from the seemingly "infinitesimal" exponent $\varepsilon\delta R$[21]. .We note that it is this contribution that carries in it the "drastic" change in I. *It is the* residue *that now is* $O(\delta\theta)$ and (when such residues are now collected for all contributing diagrams) it can be identified with the usual infinitesimal change from $<< i\int d^4xJ^\mu\delta A_\mu\} >>|_\theta$that we normally associate with the gauge-variation of a Green's function via WT-identity.

To summarize, we have written out the rigorous WT-identity (101) as would follow from the path-integral that takes into account the ϵ-term. We went on to discuss whether the simplifications one usually makes in it are always valid in the context of such interpolating gauges.We therefore applied the WT-identity (101) to the (off-shell and *truncated*) 4-point function of the gauge theories in 1-loop approximation. We write (101) as

$$<< \exp\{\frac{1}{2}i\int d^4x\Delta F^2 - i\Delta S_{gh}\} - \exp\{\varepsilon\delta R\} >>|_\theta = << -i\int d^4xJ^\mu\delta A_\mu\} >>|_\theta \tag{109}$$

Contribution to the left hand side of (109) comes from a set of diagrams in pairs (the first term giving the gauge variation of an original diagram to the Green's function and the second corresponds to the variation from an appropriate modification of the propagators $C \to \frac{1}{2}(C+A)$ in *that* diagram). We focused attention on a particular pair of contributions to the two terms coming from two time-like gluons intermediate state. We showed that [for specific kinematical relations between p_i's and θ] this contribution to neither $<< \exp\{\frac{1}{2}i\int d^4x\Delta F^2\} >>|_\theta$nor$<< \exp\{\varepsilon\delta R\} >>|_\theta$be expanded out as $1 + O(\delta\theta)$. We further noted that the $\exp\{\varepsilon\delta R\}$term could not be dropped out of the WT-identity. We found further that *when the difference between these contributions is taken, that difference*

[21]The contribution from $<< \exp(\varepsilon\delta R) - 1\} >>|_\theta$ to a truncated diagram having internal gauge boson lines is obtained as the difference of the diagram evaluated with $\frac{1}{2}(C+A)$ as the propagator and the one with C as the propagator [5].

is $O(\delta\theta)$. We expect a similar result to hold for such pairs of terms arising from different diagrams and confirm the conclusion presented here for the entire 4-point function.

11. Further Comments

We shall now add several comments.

11.1. Extension of Difficulties Associated with Definition of $\frac{\partial}{\partial\theta}$

In section 6, we discussed a specific simple contribution to the 4-point function and studied its properties as θ is varied. We found that for any physical amplitude, this contribution was not differentiable with θ at some value of θ. Not only did the derivative not exist, the left and the right derivatives differ by infinity. Moreover, we showed that the order of the limits $\epsilon \to 0$ and differentiation $\frac{\partial}{\partial\theta}$ mattered at this point . We argued that the same effect also holds for a similar contribution to an off-shell Green's function. From its appearance, the scope of this result appears limited. In this section, we shall make a comment that suggests that the scope of this specific result itself is in fact wider than stated so far and show that the difficulties encountered in Section 6 are by no means confined to a particular value of θ given $s > 4m^2$. [In addition, of course, we would encounter similar problems with other Green's functions which we have not dealt with in this work].

We shall give a simple argument to see that the difficulty cannot be confined to a particular value of θ for a given process. We do this by considering different Lorentz frames.

We note that for a given s sufficiently larger than $4m^2$ (we shall soon be more specific about how much larger) there exists a Lorentz frame where $s' = 4m^2$ for *any* $\theta \in (0, 1]$. To see this, we note that the existence of the solution for

$$s' = \theta^2 p_0^2 - |\mathbf{p}|^2 = 4m^2$$

with

$$s = p_0^2 - |\mathbf{p}|^2 > 4m^2$$

simply requires that

$$s - 4m^2 > (1 - \theta^2)p_0^2 = (1 - \theta^2)(s + |\mathbf{p}|^2)$$

This leads to

$$s > \frac{4m^2}{\theta^2} + \frac{(1 - \theta^2)}{\theta^2}|\mathbf{p}|^2 \geq \frac{4m^2}{\theta^2}$$

$$s > \frac{4m^2}{\theta^2}$$

also proves to be a sufficient condition for the existence of a Lorentz frame with $s' = 4m^2$.

Thus, the difficulties we encountered in Section 6 will be encountered in *some frame* for *any* $\theta \in (0, 1]$ provided s is sufficiently large.

We now consider the generating functional for the theory expanded in terms of the gauge field Green's functions in momentum space as:

$$Z[J, \theta] = \sum_n \int \prod_i d^4p_i J(p_i)\delta^4(\sum_i p_i)\widetilde{G_c}^{(n)}(p_1, ...p_n; \theta) \tag{110}$$

Focusing our attention on the $n = 4$ term, for the present, we note that given *any* $\theta \in (0, 1)$, there exists a $\widetilde{G_c}^{(4)}(p_1, ...p_4; \theta)$ which is not differentiable there. In fact, this is true for all $\widetilde{G_c}^{(4)}(p_1, ...p_4; \theta)$ for which p_1, p_2, p_3 lie in a certain volume in an 11-dimensional subspace of the 12-dimensional momentum space. This makes $Z[J, \theta]$ nondifferentiable[22] everywhere in $\theta \in (0, 1)$.

11.2. Wave-Packets

If one does not use sharply defined external line momenta but rather wave-packets, the value of s and also $s'$for the given diagram in Section 6 are "smeared". Then, with appropriate choice of wave-packets, it is possible to allow for the definition of $\frac{\partial}{\partial \theta} I' \mid_{smeared}$ even around the region of the unphysical particle threshold. However, this is no more helpful than making $\frac{\partial}{\partial \theta} I'$ finite by keeping $\epsilon > 0$ till the end (See section 5). In either case, the discussion in section 10 remains valid: the ϵ-term continues to contribute the same way except that both the terms on the left hand side of (109) are now simultaneously smeared. The important conclusions are unaltered.

12. Conclusions

We shall now summarize our conclusions.We pointed out that interpolating gauges necessarily contain parameter-dependent denominators. We considered the special case of the interpolating gauge of Doust that interpolates between the Feynman and the Coulomb gauge. We drew attention to the fact that the path-integrals in such gauges do lead to Green's functions that are not differentiable functions of the variable parameter θ. We dealt with a specific contribution to a 4-point function in 1-loop approximation in detail. In connection with this, we established several results. We found that this was not differentiable at some value of the interpolating parameter $\theta \in (0, 1)$. In fact, we showed that the path-integral that generates such diagrams has this obstruction at *every* $\theta \in (0, 1)$. We further showed that this amplitude at such a $\theta = \theta_0$ is not a differentiable function of θ, and consequently, the gauge variation around this point is not of $O(\delta\theta)$ as the parameter is varied from $\theta_0 \to \theta_0 + \delta\theta$. This contradicts the assumption one makes in the derivation of the WT-identity. We further found that in the neighborhood of such point, *it was necessary to keep the contribution from the variation of the ϵ-term* in the WT-identity. Both of these contributions were " large" in the neighborhood of this point. We further made the connection with the results of Ref.[5].

Appendix A

In this appendix, we shall give an explicit example to substantiate the claims in Section 7 for the off-shell Green's functions. It will prove simpler to deal with an abelian gauge theory: we shall consider scalar electrodynamics given by the Lagrangian density,

[22]Here, we are assuming that we can use sources $J(p_i)$ that can create sharply defined momentum states (plane wave states). We can also relax this assumption and use only wavepackets for external lines. For this, please refer to the following subsection 11.2 .

$$\mathcal{L} = (\partial_\mu \phi^* + ieA_\mu \phi^*)(\partial^\mu \phi - ieA_\mu \phi) - m^2 \phi^* \phi - \frac{1}{2} F[A,\theta]^2 - \frac{1}{4} F_{\mu\nu} F^{\mu\nu}$$

For future use, we note the propagator for the gauge boson[12]:

$$G_{\mu\nu} = G^{tr}_{\mu\nu} + G^{T}_{\mu\nu} + G^{L}_{\mu\nu}$$

with

$$G^{tr}_{\mu\nu} = \frac{i g_{\mu i} g_{\nu j}}{k^2 + i\varepsilon} (\delta_{ij} - \frac{k_i k_j}{|\mathbf{k}|^2})$$

$$G^{T}_{\mu\nu} = -\frac{i g_{\mu 0} g_{\nu 0}}{(\theta^2 k_0^2 - |\mathbf{k}|^2 + i\varepsilon)}$$

$$G^{L}_{\mu\nu} = \theta^2 \frac{i g_{\mu i} g_{\nu j} k^i k^j}{(\theta^2 k_0^2 - |\mathbf{k}|^2 + i\varepsilon)} \frac{1}{|\mathbf{k}|^2}$$

are respectively, the transverse, the time-like, and the longitudinal propagators. We note that both the $G^{T}_{\mu\nu}, G^{L}_{\mu\nu}$ have the same pole structure, while the transverse part has only a usual pole at $k^2 + i\varepsilon = 0$.

We consider the 1-loop contributions to the process:

$$\phi^+ \phi^- \to \phi^+ \phi^-$$

through 2-photon exchange. We let the 4-momenta of the 4 particles be $p_i; i = 1,2,3,4$ respectively.We shall not necessarily require these to be on-shell but later choose them suitably.

We consider the imaginary part of the forward amplitude for this process with $p_i^2 \leq m^2$and $s > 0$ and $u < 0$. In this case, the contribution to the imaginary part comes only from the two photons in the intermediate states and there are three possibilities:(1) both photon propagators are G^{tr};(2) One photon propagator is $G^{tr}$and the other is $G^{(T,L)}$; and (3)both the photon propagators are $G^{(T,L)}$. It turns out that the threshold for the cases (1) and (2) are at $(p_1+p_2)^2 = (k_1+k_2)^2 = 0$ and thus are independent of θ. Only the threshold for the third case is at $\theta^2(p_{10}+p_{20})^2 - |\mathbf{p}_1 + \mathbf{p}_2|^2 = 0$ as also found in Section 6. This θ-dependent threshold gives trouble with differentiation.

We have verified that the imaginary part arising from the case (3) vanishes for the on-shell amplitude in the one-loop approximation; which is guaranteed here from the tree WT-identities (*terms in which do not yet depend on the gauge)*. We have also verified that the imaginary part from the case (3) does not vanish near threshold for the off-shell amplitude.

Acknowledgments

I would like to acknowledge financial support from Department of Science and Technology, Government of India in the form of a grant for the project No. DST-PHY-19990170. I would also like to thank Prof. Tulsi Dass for a critical reading of the manuscript.

References

[1] See. e.g. T.P.Cheng and L-F.Li *Gauge theories of Elementary Particle Physics* (Clarendon, Oxford, 1984)

[2] G. 't Hooft and M.Veltman, *Nucl.Phys.B* **50**, 318 (1972).

[3] B. W. Lee, and J. Zinn Justin, *Phys.Rev.D* **7**, 1049 (1973).

[4] C.Itzykson and J.Zuber, *Quantum Field Theory* (McGraw-Hill, New York, 1985)

[5] S. D. Joglekar, *EPJdirect* **C12**, 1-18, (2001).

[6] See e.g. S.D.Joglekar in "*Finite field-dependent BRS transformations and axial gauges*", invited talk at the conference titled "Theoretical Physics today: Trends and Perspectives" held at IACS Calcutta, April 1998, appeared in *Ind. J.Phys.***73B**, 137(1999)[also see references in 7, 8 below].

[7] G. Leibbrandt, *Noncovariant Gauges* (World Scientific, Singapore, 1994).

[8] A. Bassetto, G. Nardelli, and R. Soldati, *Yang-Mills Theories in Algebraic Noncovariant Gauges* (World Scientific, Singapore, 1991).

[9] See e.g. the references in [7] and [8].

[10] See e.g M.B.Green, J.Schwarz and E.Witten in "*Superstring Theory*" Volumes 1, 2 ; Cambridge University Press, Cambridge (1987)

[11] W.Kummer hep-th/0104123; L. Baulieu and D. Zwanziger *Nucl.Phys. B* **548**, 527(1999) and references therein.

[12] P. Doust, *Ann.Phys.* **177**, 169-288 (1987).

[13] K.Fujikawa, B.W.Lee and I. Sanda *Phys. Rev.* **D6**, 2923, (1972)

[14] A. Burnel Phys. Rev.D 36, 1846 (1988); *ibid.* 36, 1852 (1988).

[15] S.D.Joglekar and B.P.Mandal, *Phys. Rev.* **D51**, 1919 (1995); R.S.Bandhu and S. D. Joglekar, *J.Phys. A-Math and General* **31**, 4217 (1998);S. D. Joglekar and A. Misra, *J. Math. Phys.***41**, 1555 (2000);S. D. Joglekar and A. Misra, *Int. J. Mod. Phys.* **A15** (2000);S.D.Joglekar, *Int. J. Mod. Phys.* **16**,5043, (2001).

[16] S. D. Joglekar and A. Misra, Int. J. Mod. Phys. A15 (2000);S. D. Joglekar and A. Misra, *Mod. Phys. Letters A* **14**, 2083, (1999);S.D.Joglekar and B.P.Mandal: hep-th/0105042 (to appear in *Int. J. Mod. Phys.* **17**(2002))

[17] L. Baulieu and D. Zwanziger Nucl.Phys. B 548, 527(1999) and references therein.

[18] P. Ramond, *Field Theory: A modern Primer* (The Benjamin/Cummings Publishing company, Inc, London, 1981).See e.g. G. Leibbrandt in "*Non-covariant Gauges*", World Scientific, Singapore Chapter 10 and references therein; D.Zwanziger *Prog.Theor.Phys.Suppl.* **131**:233(1998)

[19] See e.g. A. Bassetto, G. Nardelli and R. Soldati in "Yang-Mills Theories in Algebraic Non-covariant Gauges", World Scientific, Singapore, 1991 and references therein and also references in [7] above.

In: Horizons in World Physics, Volume 268
Editors: M. Everett and L. Pedroza, pp. 279-305
ISBN: 978-1-60692-861-5

Chapter 8

EXPLOITATION OF A SIMPLE INTEGRATED HEATER FOR ADVANCED QCM SENSORS

Antonella Macagnano, Simone Pantalei and Emiliano Zampetti
Institute for Microelectronics and Microsystems of Rome, Rome, Italy*

Abstract

Quartz crystal microbalances (QCMs) are commonly well-known as high-resolution mass-sensitive transducers. They are known to be, in fact, a versatile category of physical, biological or chemical sensors sensitive to the mass of molecular analytes. They are composed of piezoelectric crystal plus at least one layer of organic coating to both improve and tune chemical sensitivity. In literature bio-inspired sensor systems, mimicking natural olfaction, have been engineered housing selected arrays of chemical sensors based on quartz crystal microbalances, in suitable measuring chambers. Potentials and limitations of such a system have been deeply studied. In this chapter, interesting advances to quartz performances have been introduced by designing and developing a proper integrated micro-heater on a quartz side. The temperature influence has been investigated in the resulting fundamental frequency of the crystal plate as well as in ad-desorption mechanisms occurring between some different chemical layers above the plate surfaces and selected volatile organic compounds. Three different suitable micro-heater designs have been projected to characterize the functioning, the stability, the desorption time and the sensing performances in general (sensitivity, selectivity, drift, noise, etc.) of selected chemical layers covering the quartzes. The micro-heaters designs have been supported by simulations performed by Finite Element Method (FEM) software in order to analyze the temperature distribution on the whole quartz plate and get a thermal ramp. An electronic circuit interface has been designed to control the developed micro-heater, setting the desired working temperature. Moreover, when environmental temperature and humidity are constant, if the flow is changing, the temperature controller regulates the working temperature either increasing or decreasing the heater supplied power to set the desired temperature. Based on this information, it's possible to estimate the flow velocity value. This is an useful parameter in order to calibrate the sensors, overall if they have to be employed as elements of an "electronic nose" chamber where the flow velocity adopted could vary. The potentialities of such a thermo-ruled artificial olfactory system have been analyzed, too, to support some peculiar sensing strategies.

---

* www.artov.imm.cnr.it

Temperature Effects on Quartz Microbalance

The behavior of a QCM sensor is strongly dependent on the working temperature. The reason lies on the chemical and physical principles of QCM working. Generally a QCM based on chemical sensor is made up of a properly sensing section, the chemical interacting material (CIM), where both adsorption and desorption processes take place, and a mechanical section composed by the transducing layer. This last section is merely the resonating quartz that converts the information of the adsorbed mass on its surface into a variation of the oscillating frequency.

The temperature modulates both of the two sections, tuning the binding properties of the chemical interactive material as well as affecting the physical characteristics of the quartz plate. Even a variation of few centigrade degrees may result in an outstanding variation of the resonating frequency.

The thermal variations due to both the external environment and the entering gas flow (temperature and velocity) may result in an uncompensated change of temperature of the quartz.

In a quartz crystal plate, having the diameter very large when compared with the thickness (ideally infinite), the fundamental and all the nth order harmonic oscillating frequencies are

$$f_n = \frac{nv}{2h}, n = 1, 3, 5, \ldots \tag{1}$$

where h is the thickness of the plate and v is the velocity of the longitudinal or shear wave that propagates in the plate (along the thickness direction). The velocity of propagation of the wave in this direction is equal to

$$v = \sqrt{\frac{c_{ij}}{\rho}} \tag{2}$$

where c_{ij} is the *stiffness coefficient*, that is the stress-strain ratio associated with the propagating wave, and ρ is the quartz density. The correct stiffness coefficient is c_{11} when the quartz is X-cut, and it is c_{66} when the quartz is Y-cut. When the quartz is cut along an axis parallel neither to X nor Y axes (as it is for the AT-cut one), the correct value of c_{ij} is a function of both the stiffness coefficients and the angle coming from the crystallographic axes and the direction of propagation of the wave.

Joining 1 and 2 equations, the resulting equation for the resonant frequencies of an infinite quartz plate oscillating in a thickness shear mode is

$$f_n = \frac{n}{2h}\sqrt{\frac{c_{ij}}{\rho}}, n = 1 ,3, 5, \ldots \tag{3}$$

All the terms in the equation are constant, excepting h, so it is possible to define the frequency constant K:

$$K = \frac{1}{2}\sqrt{\frac{c_{ij}}{\rho}} \quad , \quad f_n = n \cdot \left(\frac{K}{h}\right) \tag{4}$$

The frequency constant, divided by the plate thickness, is the fundamental frequency value of the crystal plate.

Since c_{ij}, ρ, and h, are temperature depending, the resulting oscillating frequency is depending on it, too. Considering the three terms linearly related to the temperature, it is possible to determine the first order temperature coefficient of frequency [Bechmann et al., 1962].

Writing the logarithmic form to solve the equation (4) and differentiating it to the temperature, it results to be:

$$\begin{gathered} \log f_n = \log\left(\frac{n}{2}\right) - \log h + \frac{1}{2}\log c_{ij} - \frac{1}{2}\log\rho \\ T_f = \frac{1}{f}\frac{df}{dT} = \frac{d\log f_n}{dT} = -\frac{1}{h}\frac{dh}{dT} + \frac{1}{2}\cdot\frac{1}{c_{ij}}\frac{dc_{ij}}{dT} - \frac{1}{2}\cdot\frac{1}{\rho}\frac{d\rho}{dT} \end{gathered} \tag{5}$$

The temperature coefficient T_f, usually expressed in parts per million per Celsius degree [ppm · °C^{-1}], indicates the relative variation of the frequency from the starting value caused by a change of the temperature. Equation 5 means that the temperature coefficient is independent on the harmonic order n, but this is true only as a first approximation. The quartz plates, in fact, are not infinite, such as the temperature coefficient is slightly dependent on the harmonic number. This is the reason why plates of a given size are cut at distinct angles in order to operate at different harmonic modes.

Equation 5 shows that the temperature coefficient is depending on temperature effects on thickness, stiffness coefficient c_{ij}, and density [Ballato et al. 1975, Kosinski et al. 1992, Sekimoto et al. 1998].

The temperature coefficients of linear expansion, describing the expansion of the quartz crystal into perpendicular and parallel directions to the Z-axis, are $\alpha_1 = 14.3 \cdot 10^{-6}$ and $\alpha_3 = 7.8 \cdot 10^{-6}$ [°C^{-1}]. The temperature coefficient of linear expansion in the thickness direction is given by:

$$T_h = \alpha' = \alpha_1 \cos^2\theta + \alpha_3 \sin^2\theta \tag{6}$$

where θ is the angle between the thickness direction and the Y-axis.

The value of T_ρ can be calculated considering a small cube of quartz, having mass m and dimensions x, y, z.

The density of the cube is then:

$$\rho = \frac{m}{x \cdot y \cdot z} \tag{7}$$

Calculating the logarithmic derivative, it is possible to work out the temperature coefficient of the density:

$$T_\rho = \frac{1}{\rho} \cdot \frac{d\rho}{dT} = \frac{1}{m} \cdot \frac{dm}{dT} - \frac{1}{x} \cdot \frac{dx}{dT} - \frac{1}{y} \cdot \frac{dy}{dT} - \frac{1}{z} \cdot \frac{dz}{dT}$$

$$\frac{dm}{dT} = 0, \quad \frac{dx}{dT} = \frac{dy}{dT} = \alpha_1, \quad \frac{dz}{dT} = \alpha_3 \tag{8}$$

$$T_\rho = -(2\alpha_1 + \alpha_3) = -36.4 \cdot 10^{-6} \left[{}^\circ C^{-1}\right]$$

In order to find the contribution of the temperature coefficient of the elastic modulus (the stiffness coefficient c_{ij}) it is possible to use the values determined by Mason, reported in Table 1, below.

All the temperature coefficients reported in the table are negative except Tc_{66}, so it is possible to find a zero temperature coefficient for the quartz.

Table 1. Temperature coefficients of the elastic constants of quartz, as determined by Mason.

Temperature coefficients of the stiffness coefficients of quartz					
Tc_{11}	-46.5	$\cdot 10^{-6}$ °C^{-1}	Tc_{33}	-205	$\cdot 10^{-6}$ °C^{-1}
Tc_{12}	-3300	$\cdot 10^{-6}$ °C^{-1}	Tc_{44}	-166	$\cdot 10^{-6}$ °C^{-1}
Tc_{13}	-700	$\cdot 10^{-6}$ °C^{-1}	Tc_{66}	+164	$\cdot 10^{-6}$ °C^{-1}
Tc_{14}	-90	$\cdot 10^{-6}$ °C^{-1}			

In order to determine the temperature coefficient for a quartz plate where the direction of the propagating wave is not parallel to one of the crystallographic axes, it is necessary to calculate the temperature coefficient of the appropriate stiffness coefficient.

Considering a rotated Y-cut quartz crystal plate of an angle θ about the X-axis (see Figure1b) the value of the stiffness coefficient c_{66} (that is c'_{66}) in the new coordinate system can be calculated by a simple matrix algebra.

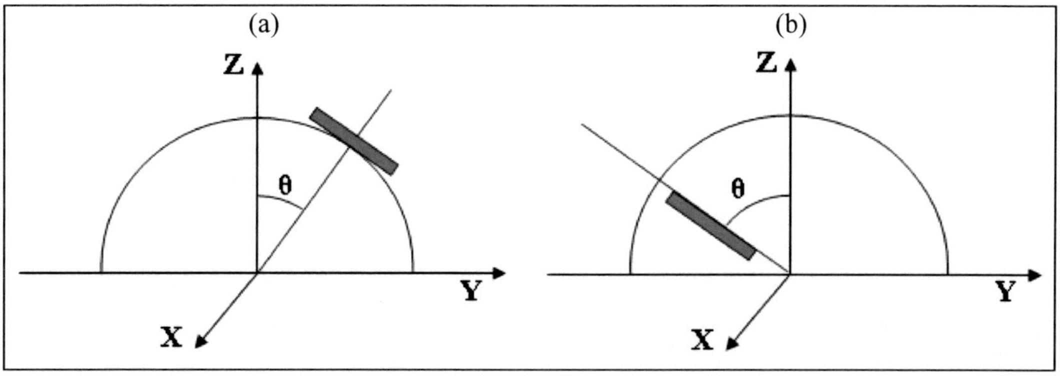

Figure 1. Conventions in use to specify the rotation of the Y-cut family. Notation used by the IEEE (a) and by Bond (b).

If $\mathbf{a}$ is the matrix performing a rotation of an angle θ about the X-axis, the stress matrix $\mathbf{X}$ is transformed in the new coordinate system as:

$$\mathbf{X}' = \mathbf{aX} \tag{9}$$

while the strain matrix is:

$$\mathbf{x}' = \mathbf{a}_\mathbf{c}^{-1}\mathbf{x} \quad , \quad \mathbf{x} = \mathbf{a}_\mathbf{c}\mathbf{x}^{-1} \tag{10}$$

where $\mathbf{a}_\mathbf{c}^{-1}$ is the reciprocal of the conjugate matrix $\mathbf{a}$. If $\mathbf{c}$ is the stiffness matrix, it results:

$$\begin{aligned} &\mathbf{X} = \mathbf{cx} \\ &\mathbf{X}' = \mathbf{acx} \\ &\mathbf{X}' = (\mathbf{aca}_\mathbf{c}) \cdot \mathbf{x}' \quad , \quad \mathbf{X}' = \mathbf{c}'\mathbf{x}' \end{aligned} \tag{11}$$

$$\mathbf{c}' = \mathbf{aca}_\mathbf{c} \tag{12}$$

The element c'_{66} is the result of the multiplication:

$$\mathbf{c}' = \begin{bmatrix} 1 & 0 & 0 & 0 & 0 & 0 \\ 0 & c^2 & s^2 & 2sc & 0 & 0 \\ 0 & s^2 & c^2 & -2sc & 0 & 0 \\ 0 & -sc & sc & -s^2 & 0 & 0 \\ 0 & 0 & 0 & 0 & c & -s \\ 0 & 0 & 0 & 0 & s & c \end{bmatrix} \cdot \begin{bmatrix} c_{11} & c_{12} & c_{13} & c_{14} & 0 & 0 \\ c_{12} & c_{11} & c_{13} & -c_{14} & 0 & 0 \\ c_{13} & c_{13} & c_{33} & 0 & 0 & 0 \\ c_{14} & -c_{14} & 0 & c_{44} & 0 & 0 \\ 0 & 0 & 0 & 0 & c_{44} & c_{14} \\ 0 & 0 & 0 & 0 & c_{14} & c_{66} \end{bmatrix} \cdot \begin{bmatrix} 1 & 0 & 0 & 0 & 0 & 0 \\ 0 & c^2 & s^2 & -sc & 0 & 0 \\ 0 & s^2 & c^2 & sc & 0 & 0 \\ 0 & 2sc & -2sc & -s^2 & 0 & 0 \\ 0 & 0 & 0 & 0 & c & s \\ 0 & 0 & 0 & 0 & -s & c \end{bmatrix}$$

Where s and c stand for sin(θ) and cos(θ) respectively. Taking the last row and the last column element of the product matrix, c'_{66} results to be:

$$c'_{66} = c_{44}s^2 + 2c_{14}sc + c_{66}c^2 \tag{13}$$

The values of c'_{66} for each angle of rotation θ about the X-axis can be found using the equation 13 and the values of the elastic constants, referred to Table 2 as reported by Bechmann and Mason [Beckman 1958, Mason 1950].

Table 2. Stiffness and compliance coefficients of alpha quartz calculated by Bechmann and Mason.

Elastic coefficients of α-quartz							
	Stiffness (Stress-Strain)				*Compliance (Strain-Stress)*		
	Bechmann	Mason			Bechmann	Mason	
C_{11}	86.74	86.05	10^9 N·m^{-2}	S_{11}	12.77	12.79	10^{12} m^2·N^{-1}
C_{12}	6.99	5.05		S_{12}	-1.79	-1.53	
C_{13}	11.91	10.45		S_{13}	-1.22	-1.10	
C_{14}	-17.91	18.25		S_{14}	+4.50	-4.46	
C_{33}	107.2	107.1		S_{33}	9.60	9.56	
C_{44}	57.94	58.65		S_{44}	20.04	19.78	
C_{66}	39.88	40.5		S_{66}	29.12	28.65	

Finally, it results to be the first order temperature coefficient of frequency:

$$T_f = -(\alpha_1 c^2 + \alpha_3 s^2) + \frac{1}{2}\frac{s^2 c_{44} Tc_{44} + c^2 c_{66} Tc_{66} + 2sc c_{14} Tc_{14}}{c_{44} s^2 + 2c_{14} sc + c_{66} c^2} + \\ + \frac{1}{2}\left(36.4 \cdot 10^{-6}\right) \tag{14}$$

These parameters are plotted in the Figure 2:

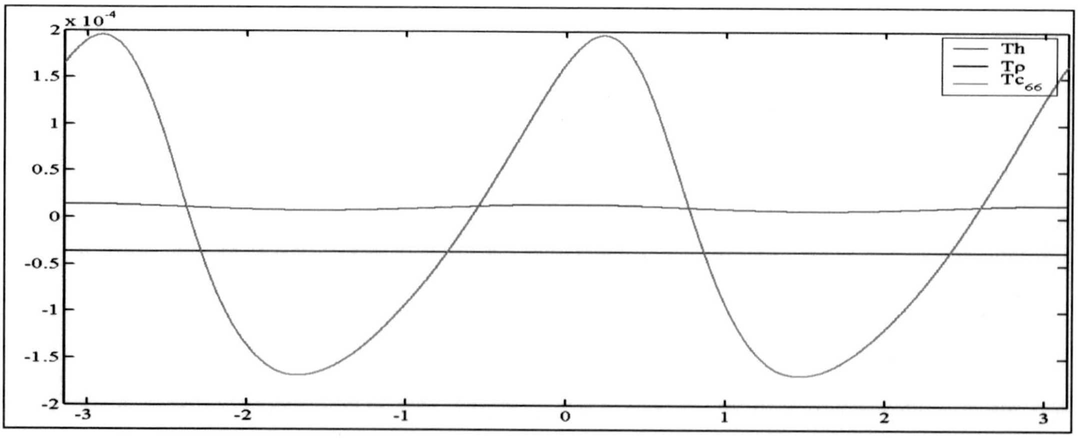

Figure 2. Plot of the temperature coefficients of thickness linear expansion, density and stiffness coefficient versus the rotation angle θ. Measure unit is [°C^{-1}].

In the Figure 2 is depicted the plot of the three different contributions to the overall temperature coefficient, versus the angle of rotation of the quartz plate about the X-axis. Plotting the equation to T_f, the angle values, wherein the temperature coefficient is equal to 0, are θ = -35° 15' (corresponding to the AT-cut) and θ = +49° (corresponding to the BT-cut).

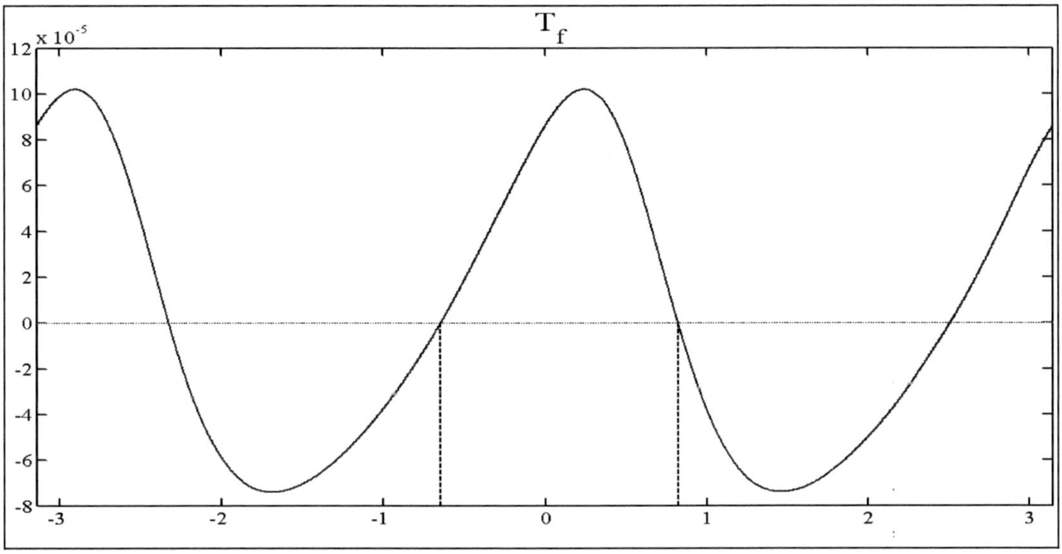

Figure 3. First order temperature coefficient of frequency, expressed in [°C^{-1}], versus the rotation angle θ. Values of θ for a zero temperature coefficient are about -35° 15' (AT-cut) and +49° (BT-cut).

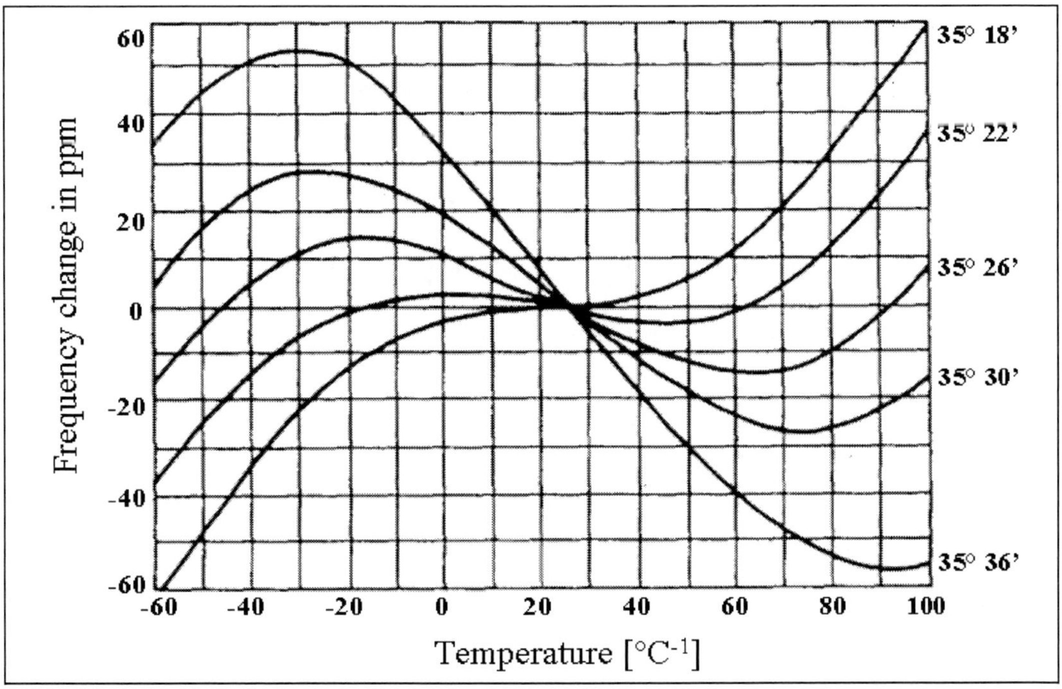

Figure 4. Frequency change (ppm) versus temperature [°C] curve of AT-cut quartz operating at 100 MHz on the fifth harmonic mode, with respect to different orientation angles.

The temperature coefficient of frequency has been derived considering the values of the stiffness coefficient at room temperature, and the linear coefficient of expansion. However, all the quantities used are temperature dependent, so it results to be T_f.

Bechmann [Bechmann 1960, Bechmann et al. 1962] has derived the expressions for the second and third order power series expansion for the stiffness coefficient of quartz. So, it is possible to calculate the temperature coefficient of frequency for rotated Y-cut quartzes in a range of temperature between -196°C and 170 °C.

In general, experimental data are used, showing the frequency-temperature curves for AT and BT-cut quartzes, due to the right orientation angle depending on the nominal frequency, the harmonic order, the diameter-thickness ratio and the operating temperature range. Figure 4 shows an example of an experimental frequency-temperature curve for AT-cut quartzes having thickness of about 84 μm (corresponding to a nominal fundamental frequency of 20 MHz) and a diameter of 0.95 cm, with a diameter-thickness ratio greater than 100, operating at 100 MHz in fifth harmonic mode.

Temperature Effects on the Desorption Processes

Quartz Microbalances can be used as transducers [Ballantine et. al. 1996, Lu et al. 1984, Lucklum et al. 2000] for chemical sensors because of the linear relationship existing between mass added on quartz surfaces and their fundamental resonance frequencies (Sauerbrey equation). A significant feature of these sensors lays on their capability to work at room temperature, avoiding the stressing mechanisms induced by high temperatures. Usually their surfaces are coated by thin films of chemical or bio-chemical molecules [O'Sullivan et al. 1999] being able to interact with environmental analytes and causing decreasing of the quartz oscillation frequency. The complete recovery of the fundamental frequency after a measurement is depending on both the strength of physical-chemical bounds occurring between the samples and the chemical membrane and the concentration of the analytes. In the latter, in fact, the adsorption/desorption times are related to the diffusion mechanisms throughin the bulk. The recovery time is therefore a fundamental parameter of QCM chemical sensor performances.

The Sauerbrey equation relates the mass variation Δm to the series resonant frequency variation Δf of a coated quartz as follows:

$$
\begin{aligned}
&\Delta f = k \cdot \Delta m_{des} \\
&\Delta m_{des} = MW \cdot \Delta n_{des} \\
&\frac{\Delta f}{\Delta t} = \frac{k \cdot MW \cdot \Delta n_{des}}{\Delta t} \qquad \textit{where } k \textit{ is a constant and } MW \textit{ is the molecular weight} \\
&\frac{df}{dt} = \frac{MW \cdot k \cdot dn_{des}}{dt}
\end{aligned}
\tag{15}
$$

The rate of desorption, R_{des}, of an adsorbate from a surface can be expressed in the Arrhenius form:

$$R_{des} = \frac{dn_{des}}{dt} = k_{des} \cdot N^x \ , \quad k_{des} = A \cdot e^{\left(-\frac{E_a^{des}}{RT}\right)} \tag{16}$$

N is the surface concentration of the adsorbed species, x is the kinetic order of desorption, E_a^{des} is the desorption activation energy, R is the gas constant, T is the working temperature and A is a pre-exponential factor. The resulting desorption frequency variation is:

$$\frac{df}{dt} = MW \cdot k \cdot \frac{dn_{des}}{dt} = MW \cdot k \cdot A \cdot N^x \cdot e^{\left(-\frac{E_a^{des}}{RT}\right)} \tag{17}$$

This equation shows that when the temperature T increases, the frequency variation increases too. This effect was related to the recovery time limit. In fact, heating the QCM surfaces with a suitable integrated microheater, the temperature is increased too. The temperature variation on QCM surface is proportional to the electric power supplied to the heater, according to the Joule's law.

FEM Analisys of the Heaters

Typical heating devices consist of electrically resistive layers deposited on the surface of the substrate to be heated. The layer causes Joule heating when a voltage drops across the resistor. The heat produced depends on the properties of the layer and, at the steady state, it is dissipated to the surrounding air on the upside and to the quartz plate on the downside of the layer. The quartz plate itself dissipates to the surrounding air, while the thickness borders are considered thermally insulated. The heat flux to the surroundings is modelled using heat transfer film coefficient, h [W·m^{-2}], where h (5 W·m^{-2}) is the natural convection.

The properties of the materials used for the electric and thermal simulations are listed in Table 3.

Table 3. List of the electric and thermal properties of the materials used for the simulations.

	Electric and thermal properties			
Quartz	density	ρ	2200	[Kg·m^{-3}]
	thermal conductivity	k	1.4	[W·(m·K)$^{-1}$]
	heat capacity	C_p	730	[J·(Kg·K)$^{-1}$]
Gold	electrical conductivity	σ	45.6·10^{6}	[Ω^{-1}·m^{-1}]

The first simulated device consists of a gold thin film layer deposited on the surface of the quartz. The gold thickness is 1500 Å, the electrode radius is 5 mm and the two contacts wide is 2 mm. The quartz plate has a diameter of 7.95 mm and is about 80 μm thick (corresponding to 20 MHz fundamental oscillating frequency). Figure 5 depicts the geometry of the device together with the corresponding generated mesh.

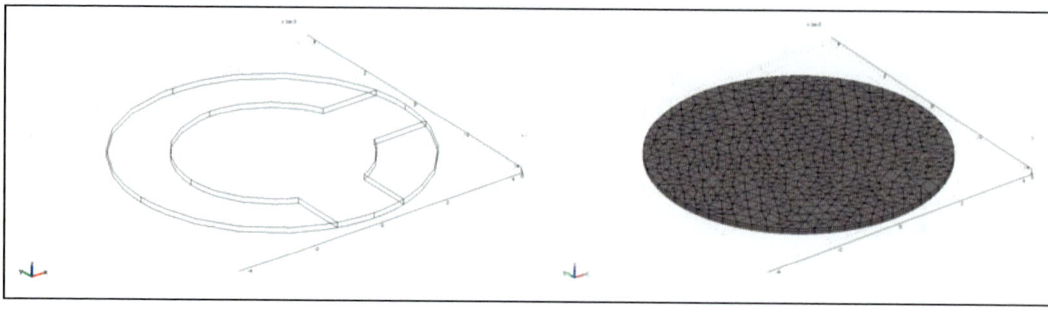

Figure 5. Geometry and mesh used for the simulations of the first proposed device.

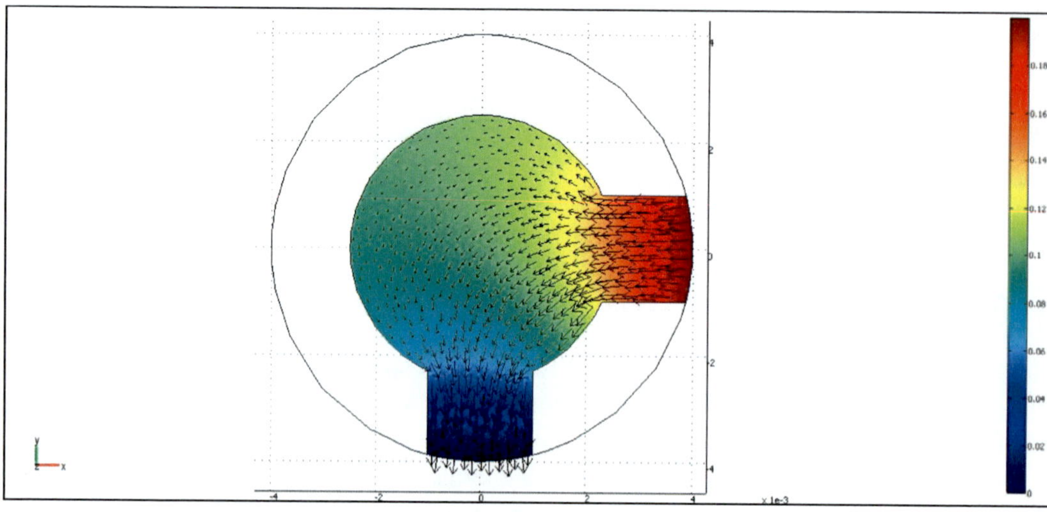

Figure 6. Electric potential distribution [V] and total current density [$A \cdot m^{-2}$] along the gold electrode, corresponding to an input voltage of 0.2V.

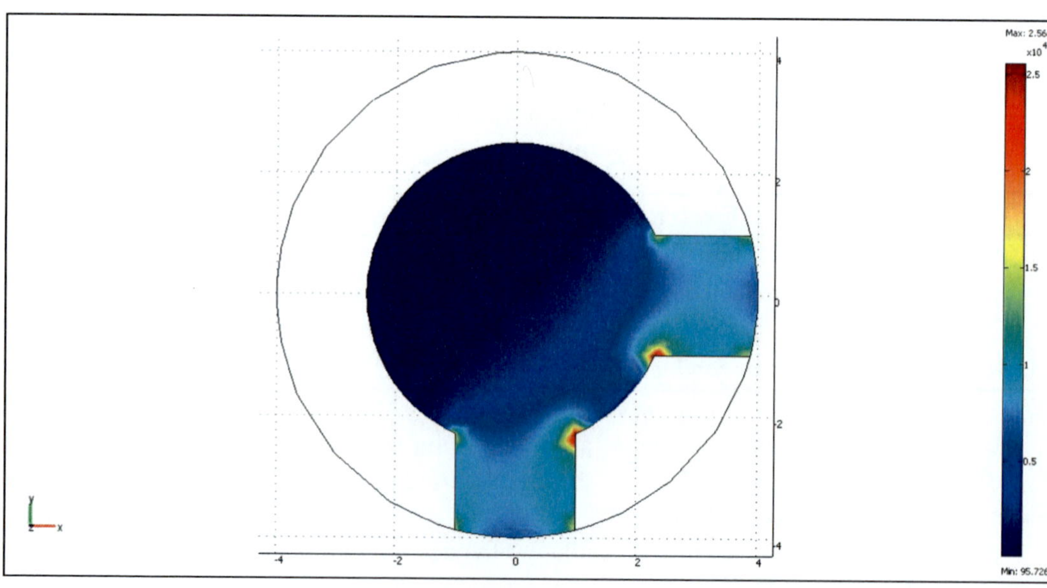

Figure 7. Heat flux [$W \cdot m^{-2}$] , produced by Joule effect, for a constant input voltage of 0.2V.

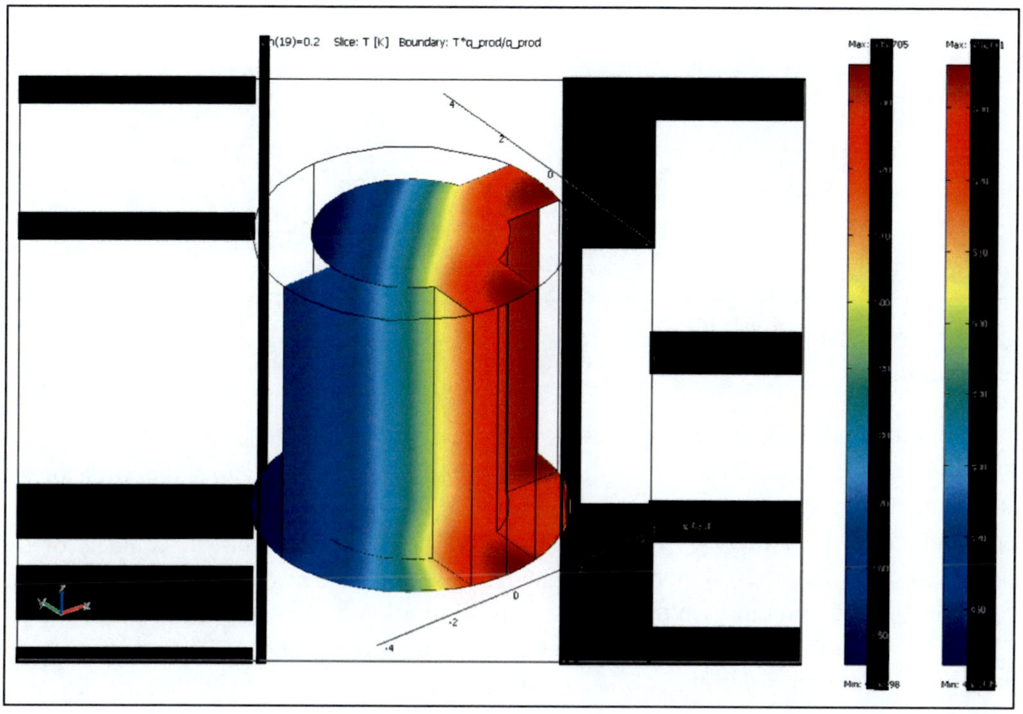

Figure 8. Temperature profile of the quartz [K], at an input voltage of 0.2V

The simulations have been performed looking for the steady state conditions corresponding to a fixed input voltage of 0.2V.

First of all, the distribution of the electric potential, together with the total current density along the electrode, have been considered. The value of the electric resistance of such a structure results to be approximately 0.5**Ω**.

As depicted in Figure 6, the voltage drops from the input value to the ground along a path being symmetric to a -45° tilted axis, while the current lines are much more packed in the region of the electrode close to the two contacts, due to the less resistive path. For this reason, the heat produced in the electrode area is absolutely not uniform, having the main peaks on the electrode geometry edges (Figure 7), where the current is more intense. The temperature profile coming from the produced heat flux is consequently inhomogeneous on the surface of the quartz, while the gradient of the temperature over the thickness is negligible (Figure 8).

This structure has the main advantage of being very simple and practical, nevertheless it presents some disadvantages. The first one is that during the heating, the "ground" electrode is polarized by the voltage (depending on the input) that induces stress in the crystal, not present during the normal QCM working. The second one is linked to the high temperature gradient in the quartz, generally degrading the long term performances of the crystal.

The second device has been fabricated decoupling the heater from the electrode of the QCM (Figure 9). In this way the electrode is secured to the ground during both the normal working of the sensor and the heating process. The geometry of the electrode owns the same characteristics of the previous design, while the heater has an annular, 0.5 mm wide shape surrounding the electrode. It is realized by a gold thin film (1500 Å thick) deposited on the quartz surface.

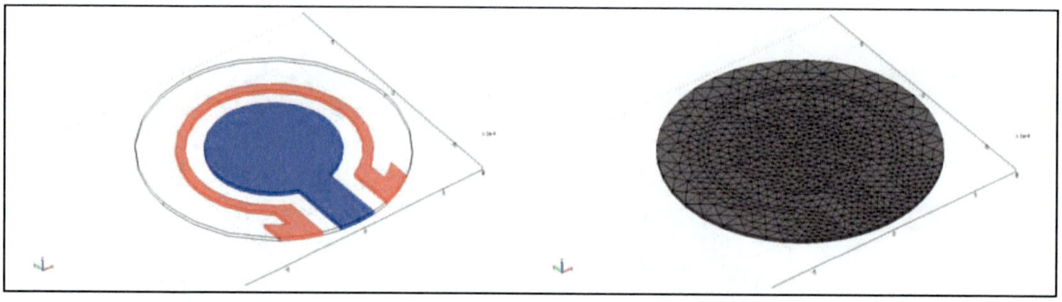

Figure 9. CAD model of the second QCM-Heater system designed, together with the mesh generated for the simulations.

In this situation the voltage drops across the path of the heater obviously not affecting the voltage on the electrode. The current lines are more homogeneous along the wide of the heater compared to the previous design, even if they are more intense in the inner of the ring rather than in its outer region. Moreover, the edges between the electrical contacts of the heater and the ring, are sites of more packed and intense current density (Figure 10) leading to a higher heating flux produced in the same region (Figure 11).

Such a geometry produces an electrical resistance value of about 7.5 **Ω**, fifteen times higher than the previous design, so that the input voltage used in the simulations has been increased up to 1V.

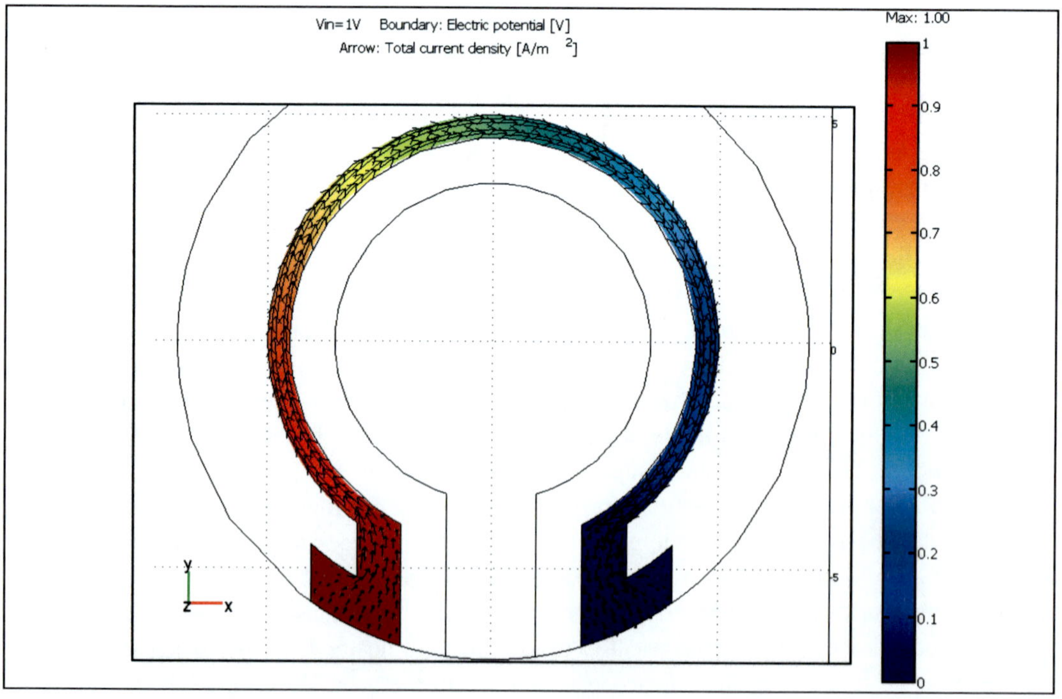

Figure 10. Electric potential distribution [V] and total current density [$A{\cdot}m^{-2}$] along the annular heater, corresponding to an input voltage of 1V.

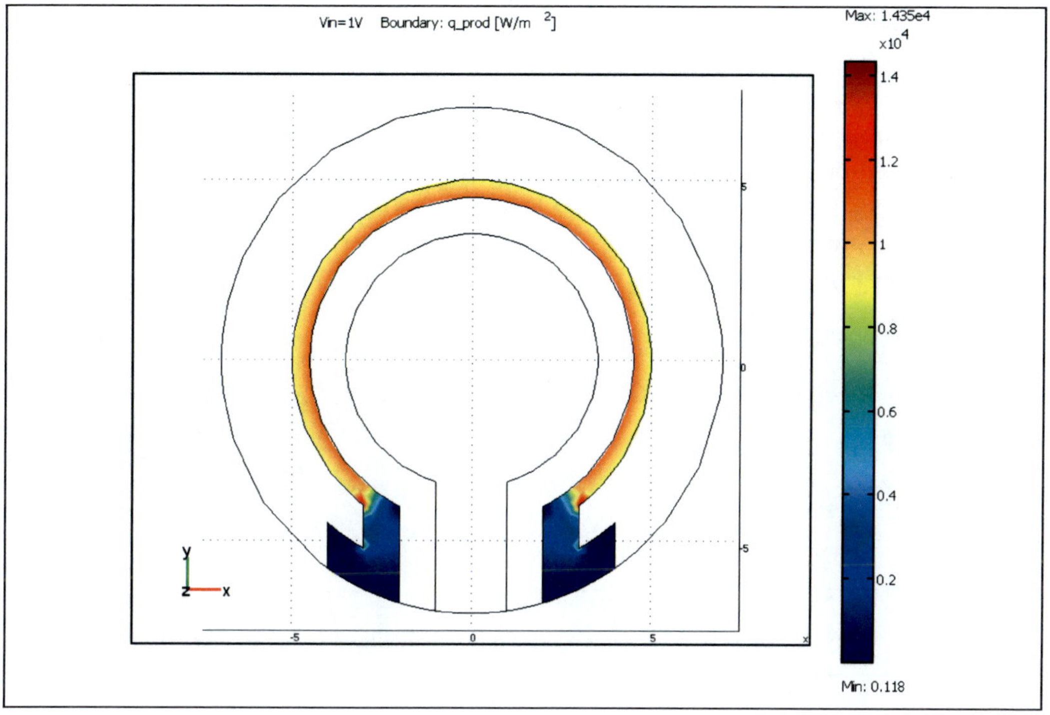

Figure 11. Heat flux [W·m^{-2}] produced by means of Joule effects in the region of the heater, at the simulated input voltage of 1V.

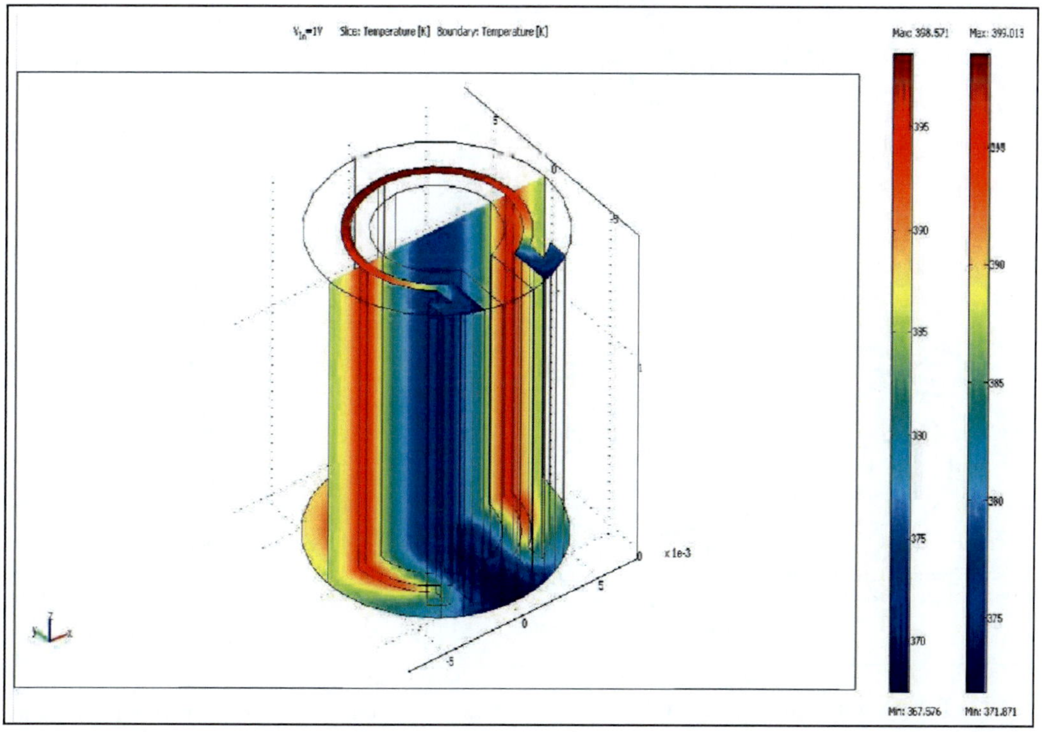

Figure 12. Temperature profile of the quartz [K], for an input voltage of 1V.

The temperature profile in the QCM sensor, shown in Figure 12, is performed by an overall smaller temperature gradient, and an even smaller gradient in the region of the electrode (where the CIM is typically coated).

The advantages of this design are both the reduction of a stress in the crystal during the heating of the sensor and a smoother temperature gradient.

In order to minimize the temperature gradient over the surface of the quartz it is possible to improve the symmetry of the heating device. This result is achievable by both mirroring and reverting the annular heater on the bottom face of the quartz (Figure 13).

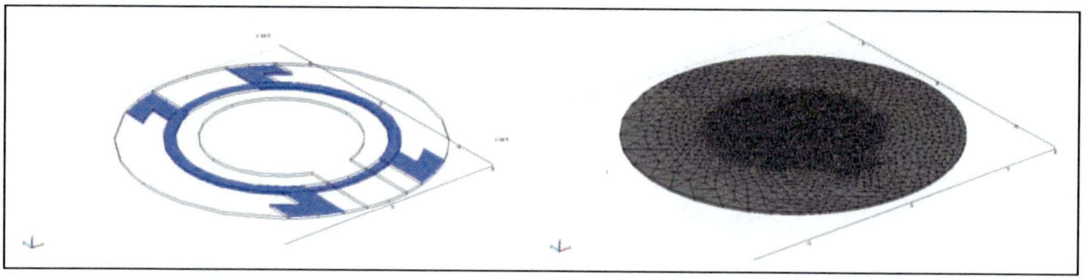

Figure 13. CAD model and the generated mesh of the double annular heater on QCM.

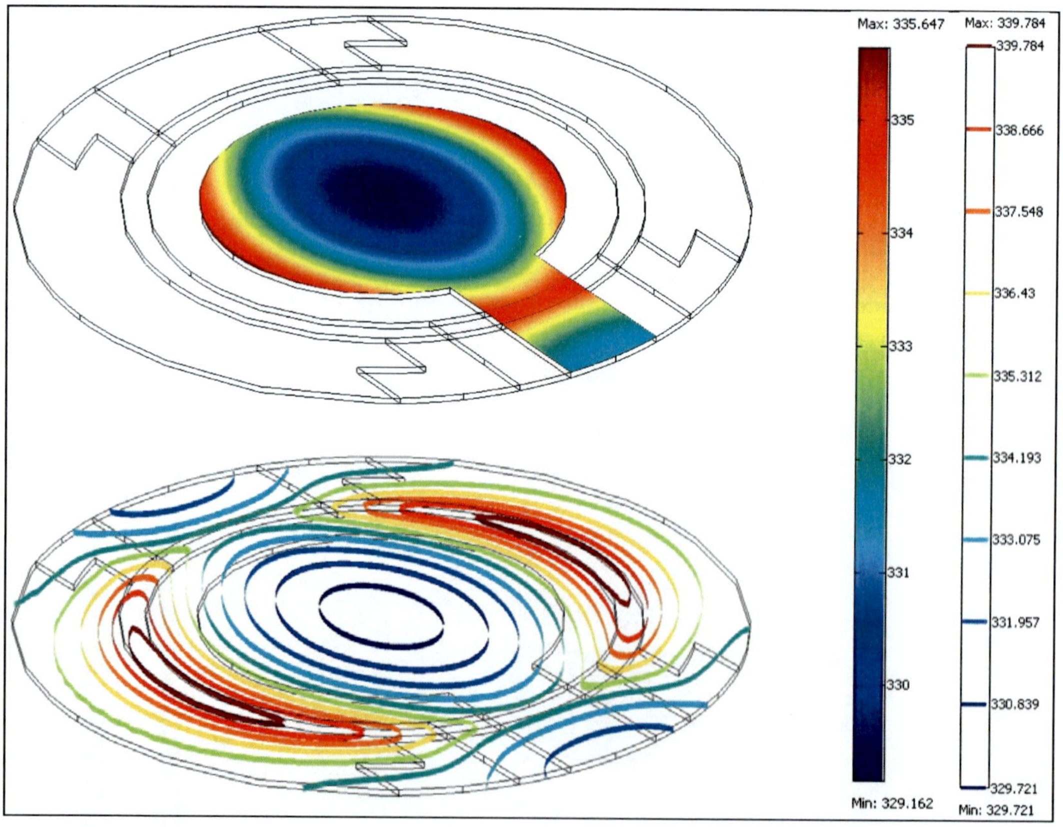

Figure 14. Temperature profile on the QCM sensor, depicted as a continuous colour map in the region of the electrode and as temperature isosurfaces overall the quartz. The input voltage is fixed to 0.5V.

The heating is actuated on both sides of the quartz so that in the regions where the heat flux produced in one face is smaller, there is a compensation originating from the heater on the opposite face. The resulting temperature profile is shown in Figure 14.

The thermal analysis showed a reduced temperature gradient of 10°C in the entire quartz plate, and an even smaller gradient in the region of the QCM electrode, with a temperature profile that is symmetric along the X and Y axes.

QCM-Heater System Applications

After the micro heater calibration, the sensing performances of a QCM sensor to some volatile organic compounds (VOCs) have been valued.

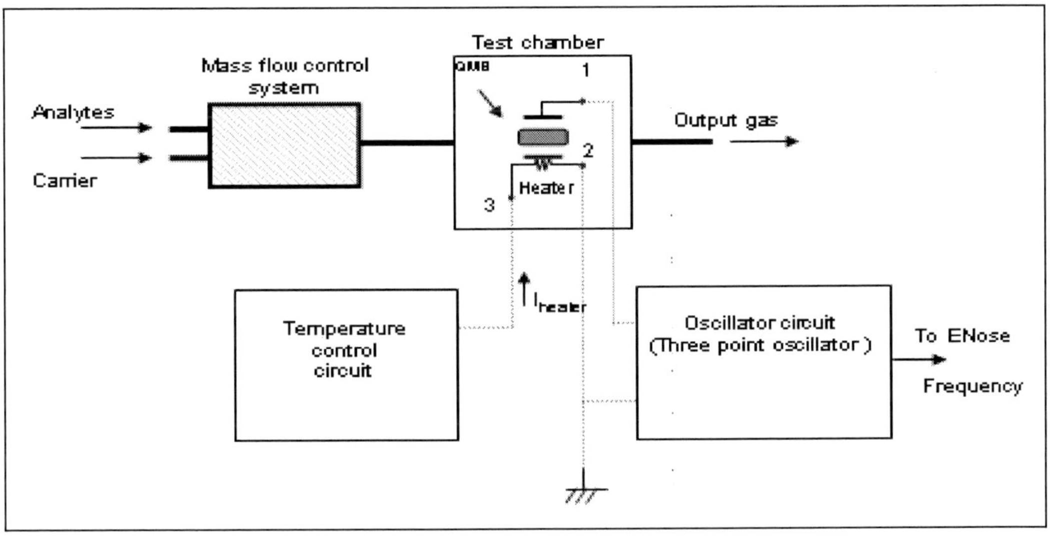

Figure 15. The experimental setup.

The figure 15 shows the experimental set-up, where the temperature control system consists of an electronic circuit regulating the power supplied to the micro- heater. The temperature of the micro-heater and the QCM surface is proportional to this power (Joule's law).

The QCM sensor is a three pin device, where pin 1 is connect to the first electrode of quartz and pin 2 is a common pin for the second quartz electrode and the first micro-heater terminal. The pin 3 is the second terminal of the micro-heater. The output frequency of the oscillator circuit has been measured by a suitable digital electronic interface based on a Field Programmable Gate Array (FPGA) designed to acquire, elaborate and transmit data to elaboration unit (PC). A metallo-porphyrin (Zn-5,10,15,20-tetrakis-(4-eptyloxyphenyl) porphyrin) thin film, has been deposited on a quartz surface, by thermal vacuum deposition, as chemical interacting material. Porphyrin film features as sensing layers have been previously studied by the authors [Macagnano et al., 2007-2008, Di Natale et al. 1997-1998, D'Amico 2000]. A 20 MHz quartz crystal has been used as a transducer (80 μm thick, 7.95 mm and 4.5 mm respectively quartz and electrode diameter). The temperature of the measurement chamber was controlled by a Peltier cell mounted on the top of the stainless steel chamber to maintain the measurement system at a stable temperature. Figure 16 shows a 3D model of the QCM plate with the microheater integrated in the electrode and a picture of a prototype.

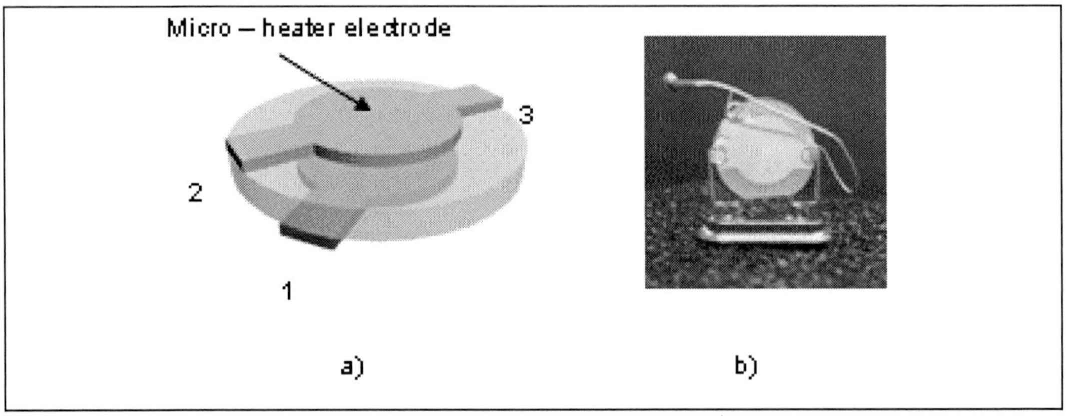

Figure 16. a) A 3D model of QCM sensor with an integrated micro-heater, b) a prototype of a heated QCM

Such an evaporated Au-Cr thin film (40 nm thick) worked both as electrode for the piezoelectric system and as micro-heater. The electrical resistance calculated at room temperature (T = 20 °C) was about 5 Ω.

The effects of this micro-heater on the QCM have been measured. Figure 17 shows the frequency trend with respect to the fundamental frequency of the AT-quartz, and Figure 18 shows the frequency variation occurring at several powers supplied to the micro-heater.

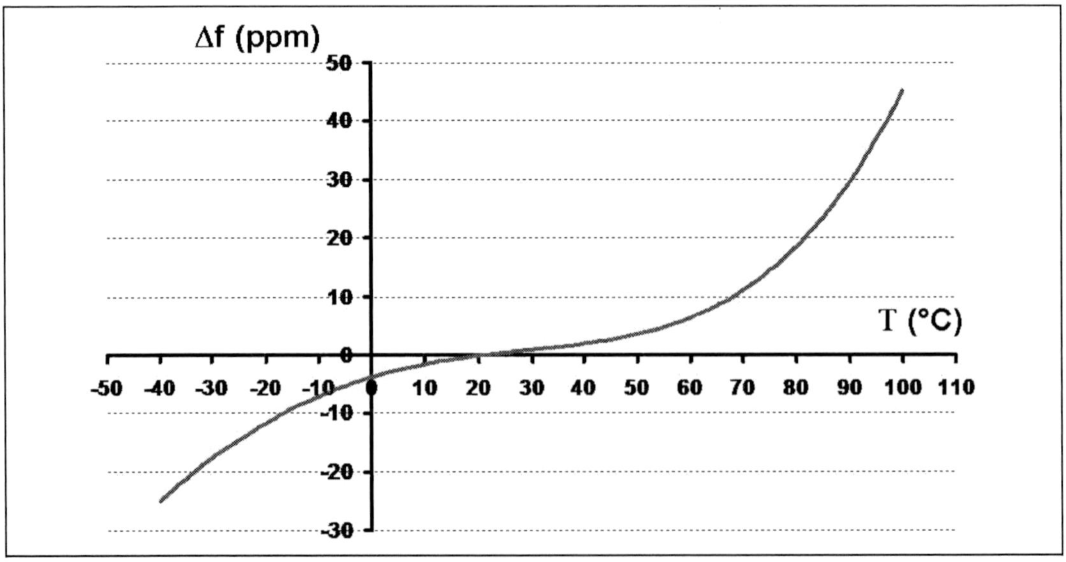

Figure 17. Frequency shift vs. temperature for AT quartz at 20 MHz fundamental frequency. The heating power between 0 and 12 mW corresponds to a temperature range between 20 and 50 °C. The maximum frequency variation in this range is about 4 ppm.

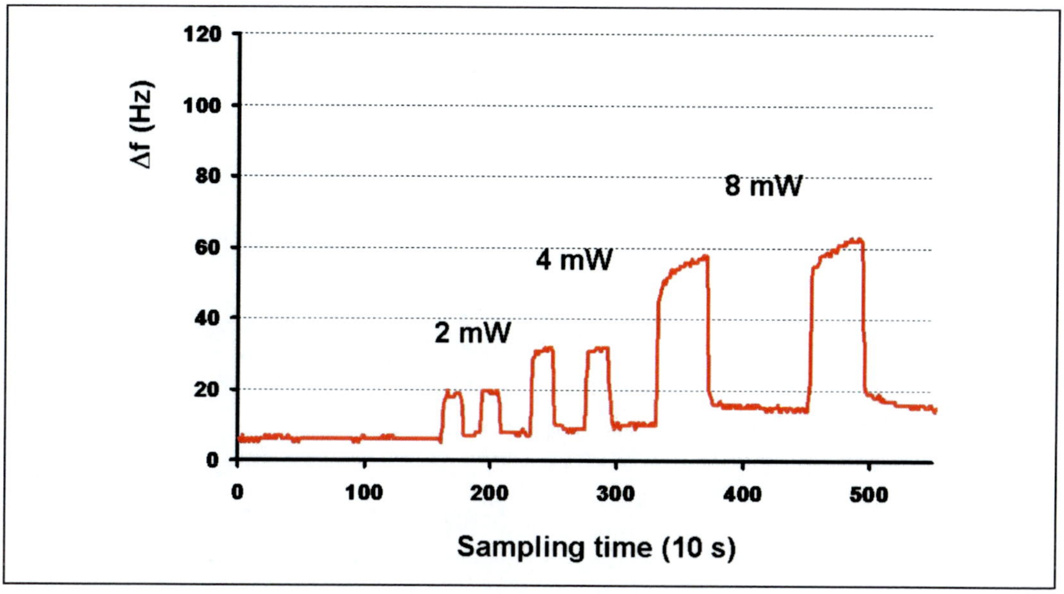

Figure 18. Fundamental frequency variation corresponding to heating powers from 2, 4 to 8 mW.

The calibration of this micro-heater was plotted in terms of both temperature and frequency shift related to the dissipation power (Figs. 19, 20).

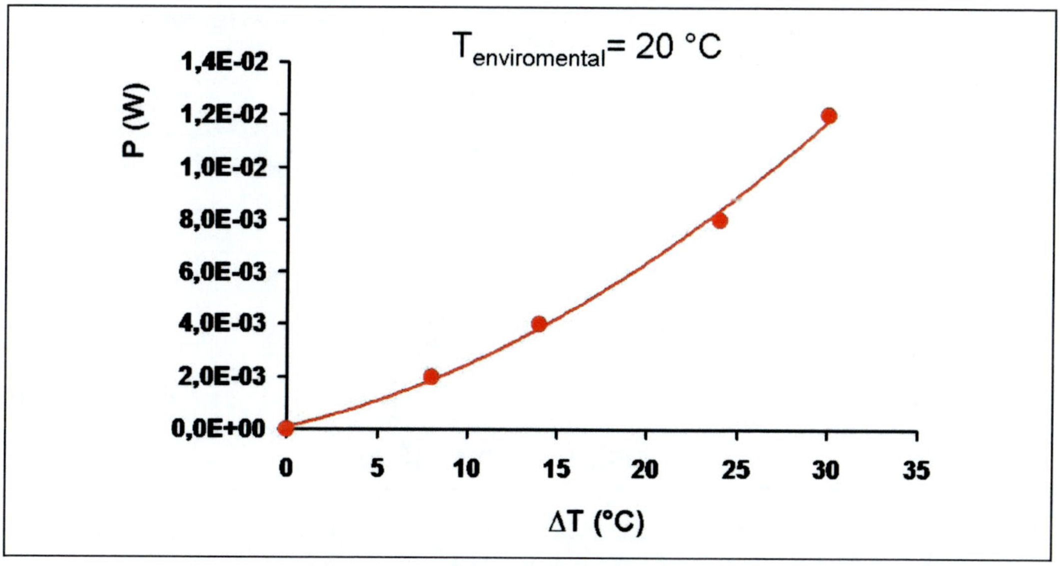

Figure 19. Characterization of the QCM temperature shift versus the heating power.

As shortly described previously, the QCM sensors are typically coated by thin films of chemicals that can selectively interact with the analytes, producing an oscillation frequency reduction. The complete recovery of the fundamental frequency after a measure depends on the strength of the physical-chemical bond between the CIM and the analyte and on their concentration. Furthermore the desorption processes depend on the working temperature of the chemical film.

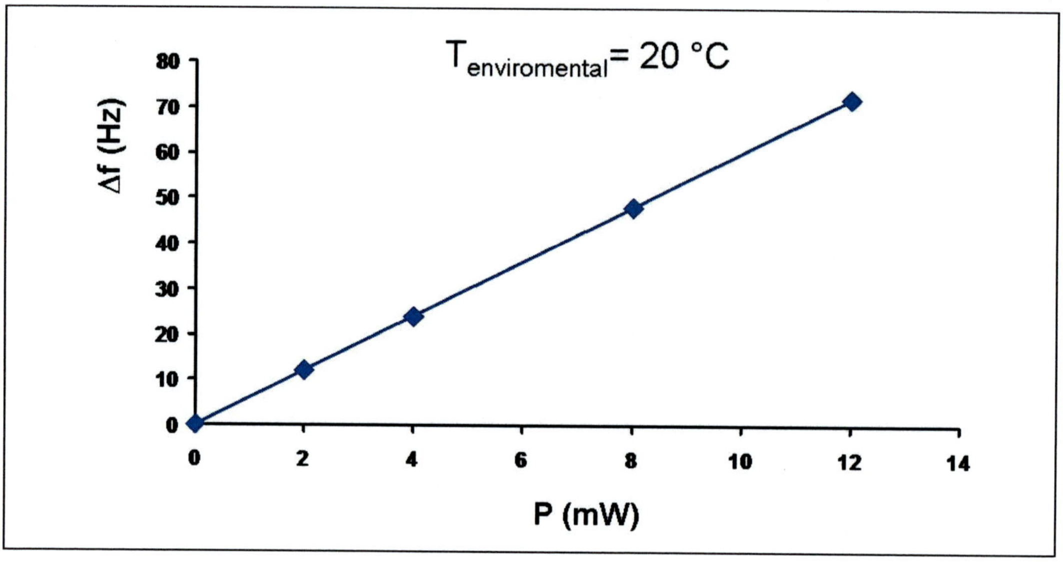

Figure 20. Characterization of the QCM frequency shift versus the heating power.

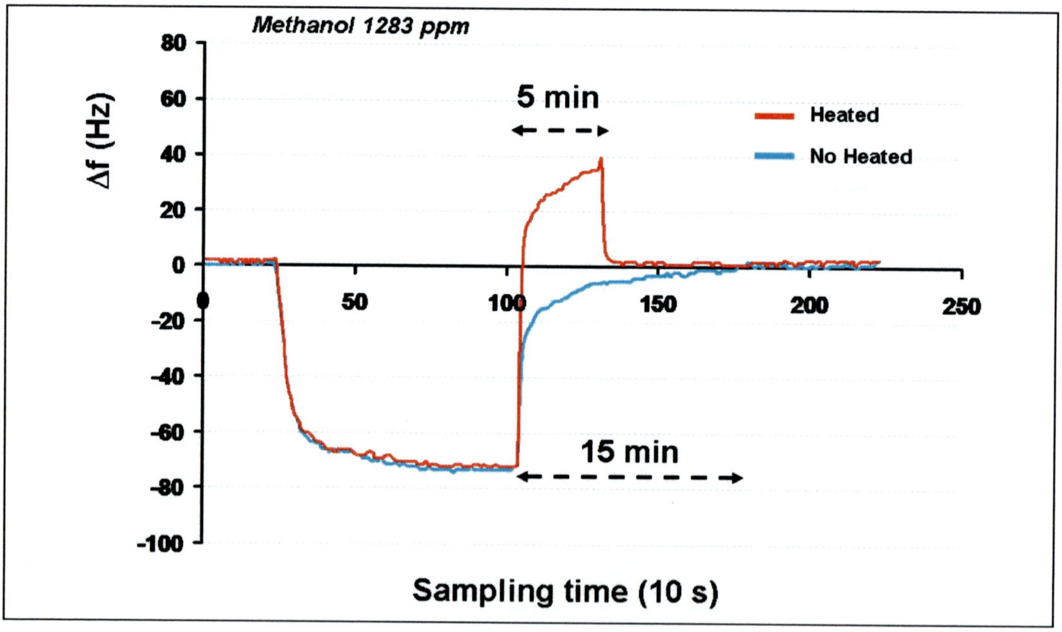

Figure 21. Response curves of the QCM sensor to 1283 ppm of methanol in nitrogen. The measure is performed either without using the heater or pulling it on for about 5 minutes during the desorption phase.

In order to test the effectiveness of the heater used jointly with the QCM sensor, several experiments have been performed. The first set of measurements has been done in order to observe the time reduction of the sensor recovery. The adsorbed species can remain trapped for a long time at room temperature, causing either poor recovery of the sensor or very long recovery times. Actuating the heater during the desorption step, the CIM is speeded up to set free the adsorbed molecules and to quicken the process itself. The experiment has been

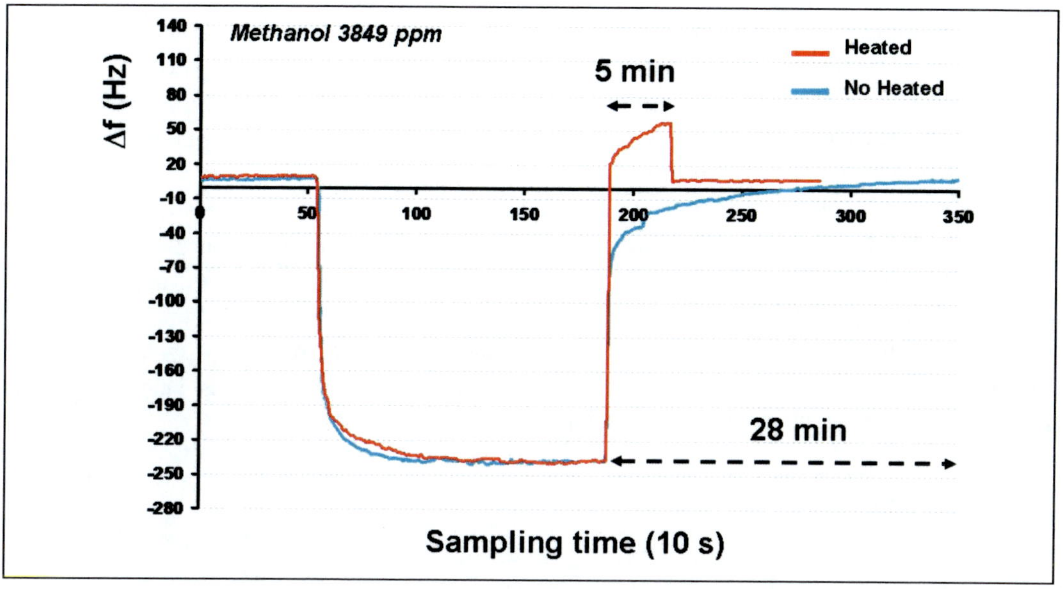

Figure 22. Response curve of the QCM sensor to 3849 ppm of methanol in nitrogen. The measure is performed either without the use of the heater or pulling it on for about 5 minutes during the desorption phase.

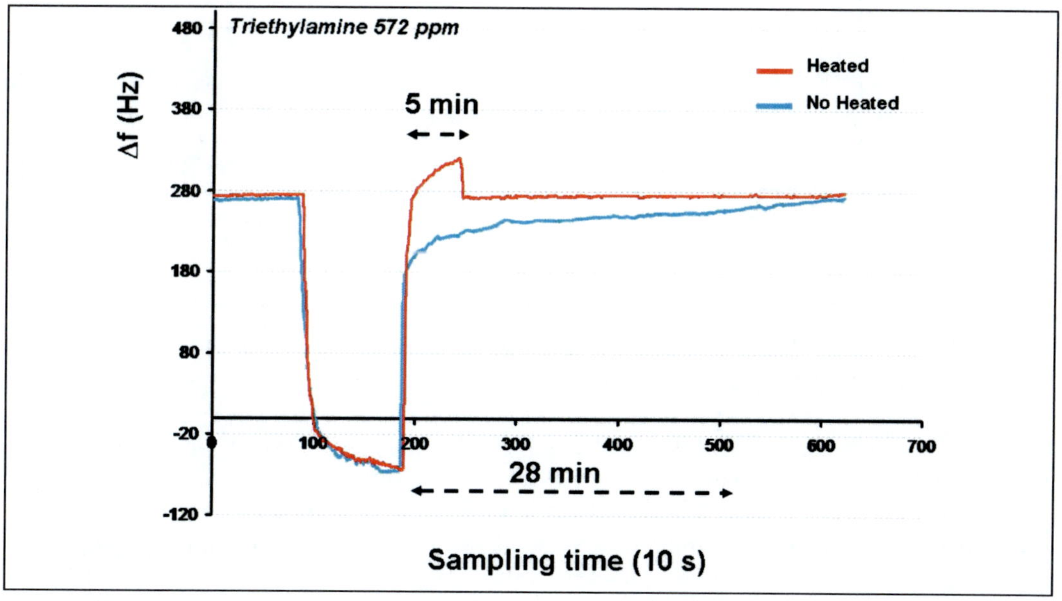

Figure 23. Response curve of the QCM sensor to 572 ppm of triethylamine in nitrogen. The measure is performed either without the use of the heater or pulling it on for about 5 minutes during the desorption step.

carried out, coating the simplest QCM-heater of a thin film of the selected metalloporphyrin, and then flowing in the measurement chamber two different chemicals, methanol and triethylamine. The same measurements have been replicated within and without the microheater usage, in particular switching the heater on during the cleaning phase, for a fixed time of 5 minutes, together with the pure nitrogen flux. After this time, the heater has been switched off while the nitrogen flux has been held.

Figures 21, 22, 23 show the sensor responses to methanol and triethylamine vapors. Here the decreasing recovery time, ranging from 66% up to 93% of the room temperature recorded values, is evidenced.

Then several isotherms of responses have been calculated versus increasing normalized concentrations of methanol and triethylamine, at temperatures varying from 20 °C up to 50 °C. As depicted in Figure 24, according to the theory, as the temperature increases, the response of the sensor decreases varying the sensitivity and the selectivity of the sensor. This result is very important because it is possible to design a temperature control system for such a heater choosing a suitable temperature value on the CIM to vary its selectivity and sensitivity to different chemical species, yielding to a more versatile chemical sensor.

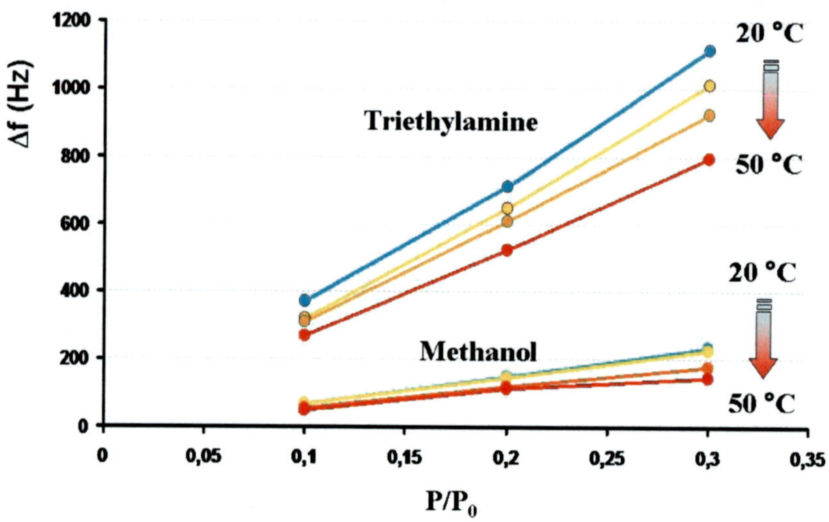

Figure 24. Isotherms calculated with respect to increasing normalized concentrations of methanol and triethylamine at temperatures varying between 20 °C and 50 °C.

The latest experiment, performed with the second designed heater and the QCM coated with the same porphyrin, aims to use the QCM sensor without a reference gas. In many practical applications of chemical sensors (e.g. in long term space missions) enough amounts of reference gases are not available and the standard measurement methodology cannot be applied.

Continuous flows at room temperature of toluene at different concentrations in nitrogen carrier were carried throughout the sensor measurement chamber. Without removing the toluene, the desorption were achieved by heating the sensor surface. The increased temperature allows the desorption of the trapped molecules and inhibits any further adsorption. Three different concentrations of toluene in nitrogen were fluxed over the QCM sensor and several measurement cycles were performed.

The frequency shifts corresponding to the thermally actuated adsorption-desorption cycles could be used to define the response of the sensor to the under test chemical.

Figures 25-26 show, respectively, the sensor signal corresponding to the thermal cycles and the frequency shifts corresponding to the three different toluene concentrations.

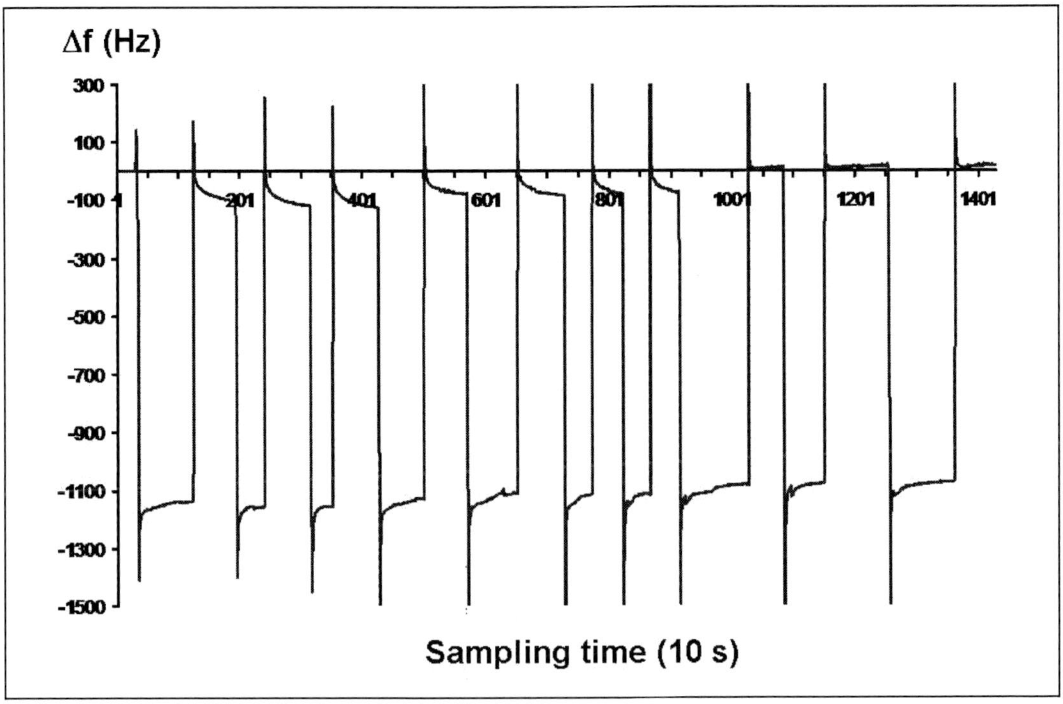

Figure 25. Sensor signal corresponding to 3 different concentrations of toluene in nitrogen. The frequency shift, corresponding to absorption-desorption cycles, are thermally actuated

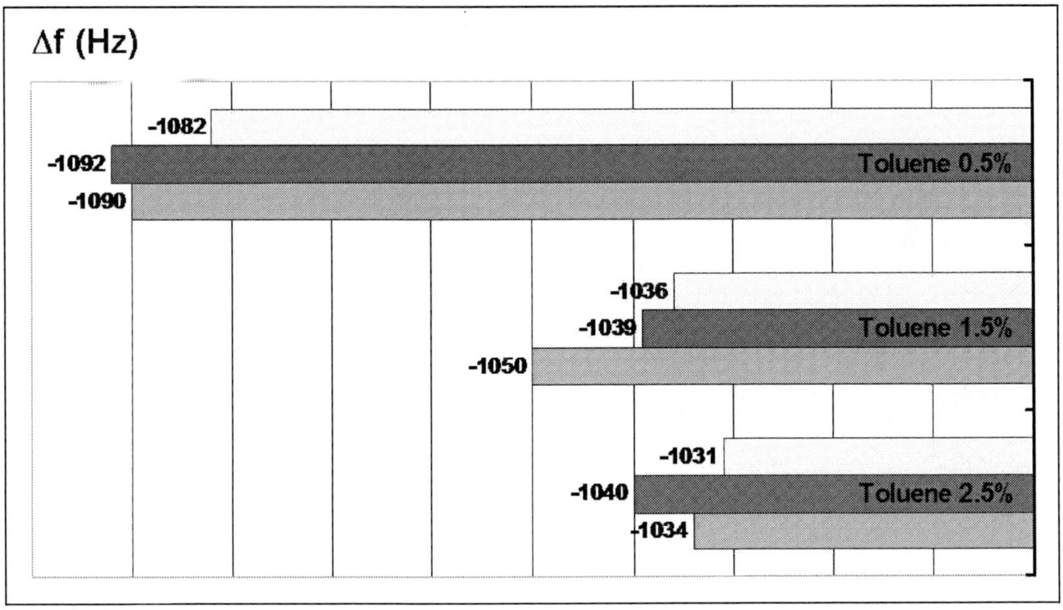

Figure 26. Frequency shifts corresponding to three different toluene concentrations. Each concentration was measured three times.

QCM-Heater System as a Flow Sensor

QCMs also show a significant cross-sensitivity to the flow velocity. Generally, the rate of adsorption is governed by the rate of arrival of molecules at the surface and by the proportion of incident molecules which undergo adsorption. The rate of adsorption (per unit area of surface) can be expressed as a product of the incident molecular flux, *F*, and the sticking probability *S* as follows

$$R_{ads} = S \cdot F \qquad [\text{molecules m}^{-2}\ \text{s}^{-1}] \tag{18}$$

while the flux of incident molecules is given by the Hertz-Knudsen law

$$F = \frac{P}{\sqrt{(2\pi mKT)}} \qquad [\text{molecules m}^{2}\text{s-1}] \tag{19}$$

where P is the gas pressure [N m^{-2}], m is mass of one molecule [Kg] and T is the temperature [K].

Although, in principle, the gas chamber housing QCMs may be preliminarily characterized, the introduction of the quartz disks changes the flow distribution and each microbalance is exposed to a different flow, making both different the adsorption rate for each sensors and more complex the flow velocity measurement and/or control. As an example, Figure 27 shows the flow velocity field distribution inside two different measurement chambers housing 8 QCM sensors.

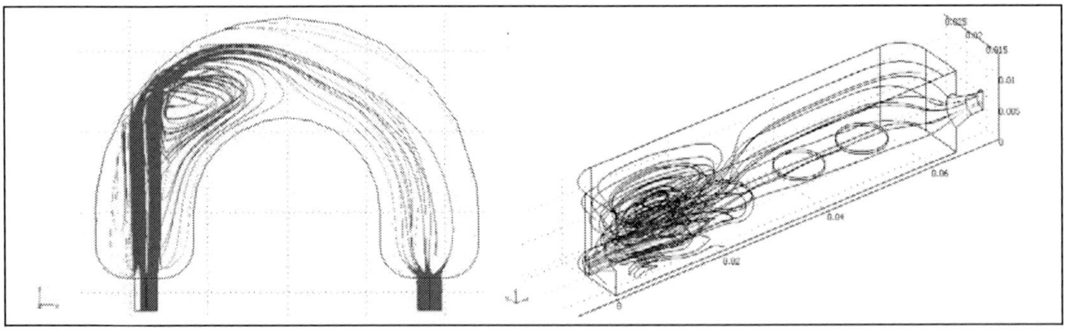

Figure 27. Two adopted different shapes of gas chambers for QCM arrays.

Therefore it is important to know the flow in the test chamber to evaluate correctly a QCM response. Figure 27 shows that the flow is not homogeneous and then the question is: where the flow sensor must be placed in order to perform a systematic measurement? The answer is supposed to be: the flow sensor must be placed close to QCM sensors to have the best information about the local flow. This solution could not be suitable because, although the commercial flow sensors have a small size, the local fluid dynamic can be altered [Kulkarnib et al. 2005, Nie et al. 2003, Falcitelli et al. 2002]. A different solution to perform a right flow measurement is the integration of a flow sensor on QCM plate. A simple integrated flow sensor can be obtained by using a suitable control of the integrated heater, in this case the flow sensor is working as a hot wire type.

Thermal ΣΔ Modulation for Temperature Control and Flow Measure

There are two main operation method to control a heater: the differential constant temperature mode and the constant power mode [Verhoeven 1995, Kulkarnib 2005]. In a differential constant temperature mode, the thermal feedback is used to have a constant difference between the sensor and the environment. In contrast to the constant power mode, the flow response does not depend on both fluid temperature and sensor properties [Dorf et al. 1986]. A simple control technique, working on differential temperature mode, is the thermal sigma delta [Kofi et al 2004] (Figure 28). The comparator device matches up to the output voltages (TS_1 and T_{ref}) coming from temperature sensors. TS_1 is the object temperature and T_{ref} is the desired temperature. The D type flip flop enables (high state) or disables the heater (low state), depending on the output of the comparator. So, this control technique fixes the desired temperature on the device and, contemporary, the flow velocity correlated with the power heating, can be deduced from the digital output of the flip flop. Finally, the thermal filter, constituted by the thermal resistance and the thermal capacitance of the heater (R_{TH}, C_{TH}), acts as the integrator, shaping the thermal noise.

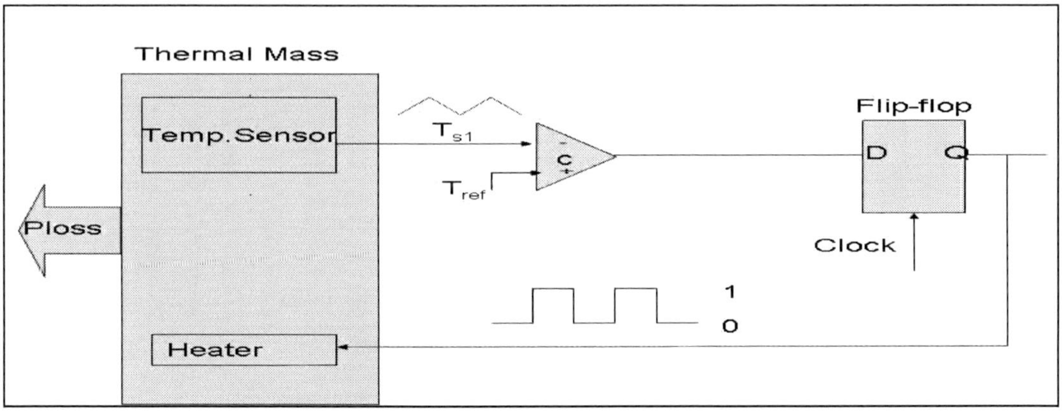

Figure 28. Thermal sigma delta modulation working principle.

In such a way, when the P_{loss} increases, the high flip flop state width increases too, heating the heater device. Observing the high pulse width of the flip flop it is possible to estimate the flow changes. In other words, after an accurate system calibration, it is possible to obtain a flow measure in a simple way.

To evaluate the working of this flow sensor, an integrated heater on QCM, having a simple layout (Figure 29), has been controlled by Thermal ΣΔ [Falconi et al. 2006].

Figure 29 shows the simple integrated heater fabricated on a QCM, where the top and the bottom electrode layouts are different. Here, the top electrode works only as electrode, while the bottom one works as electrode, heater and temperature sensor at once. Figure 30 shows a schematic view of the implemented sigma delta controller. It is possible to see that the heater is driven by the M_1 current only when the voltage Vs_2 is different from V_{S1} voltage, where

Vs_2 is the heater voltage and V_{S1} is the selected voltage corresponding to the desired temperature.

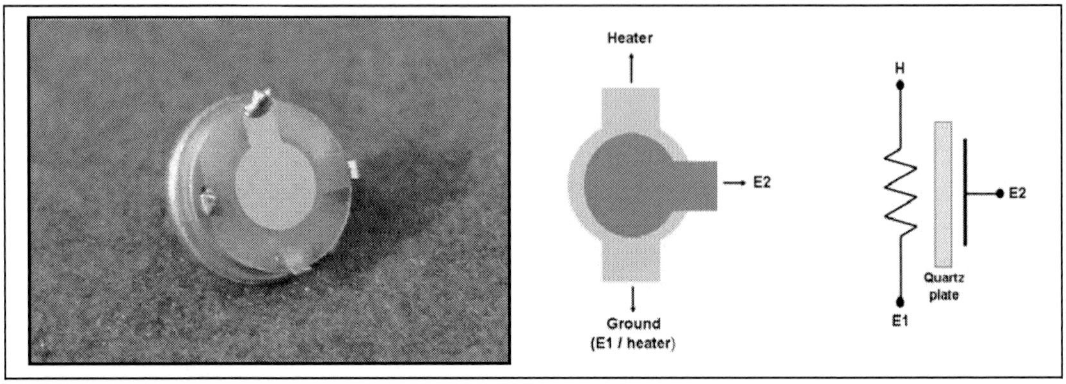

Figure 29. QCM with an integrated heater working as an hot wire.

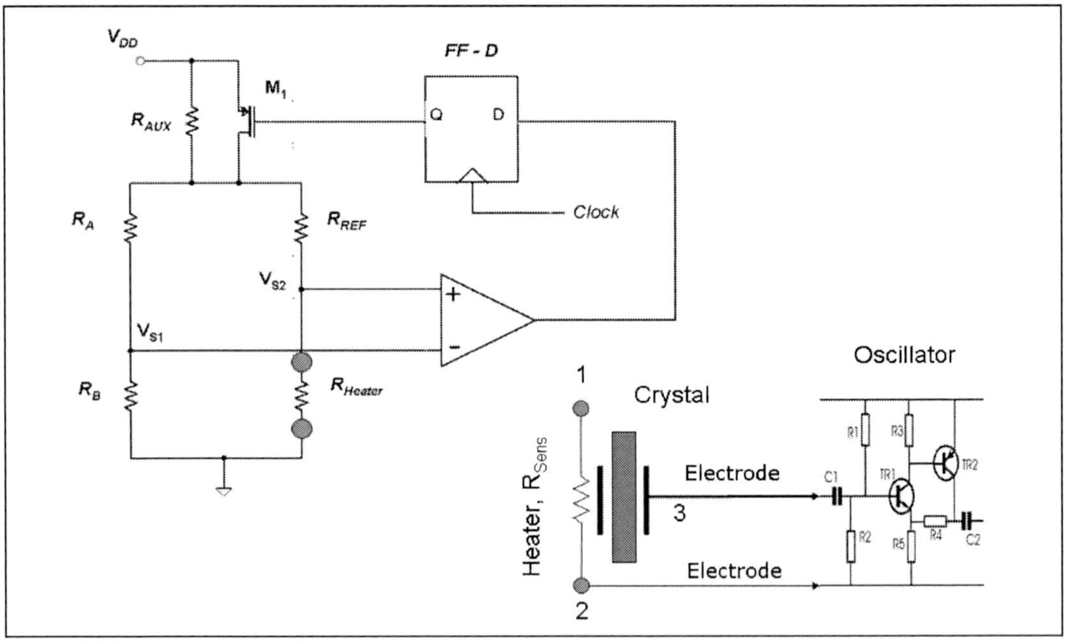

Figure 30. Schematic view of the thermal ΣΔ modulator and a schematic draw of the QCM/Heater connection.

In Figure 31 the temperature values of quartz plate regulated by the thermal ΣΔ circuit, is depicted. The measure is performed by a K type micro thermocouple (40 μV/°C) in contact to the quartz surface. The graph shows that the rise time of the temperature is very fast, depending on the driven circuit power characteristics, while the fall time is slower related to heater thermal parameters (R_{TH}, C_{TH}). In Figure 32 the number of the heating impulses coming from the flip flop output (RHCR), which are supplied to the heater from the sigma delta circuit under flow increasing changes, is shown. In fact when the flow, generated by DC pump (see Figure 33) increases, the circuit compensates the power loss (flow changes) increasing the number of the pulse rate.

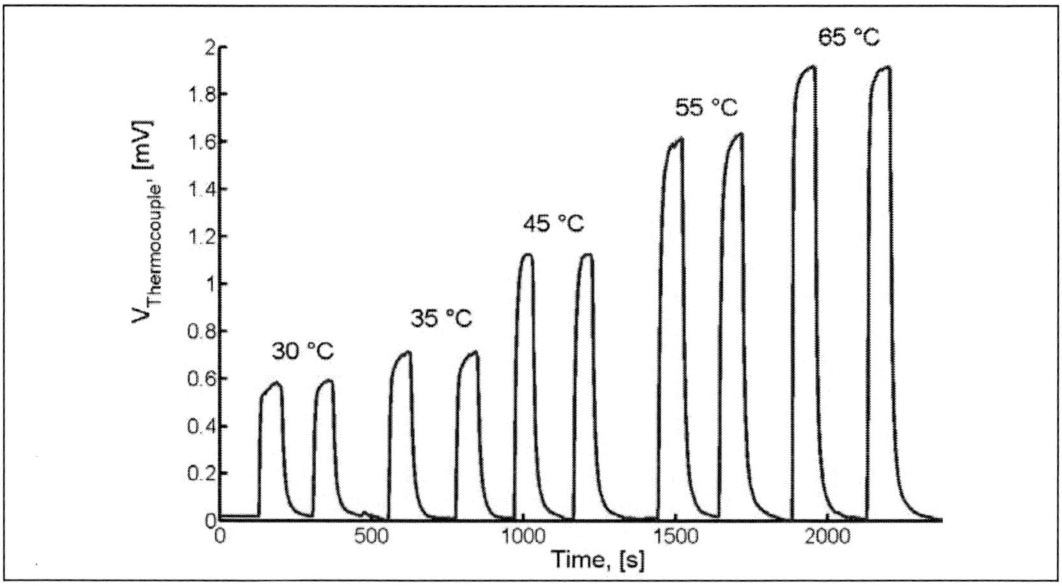

Figure 31. Measured temperature on QCM plate.

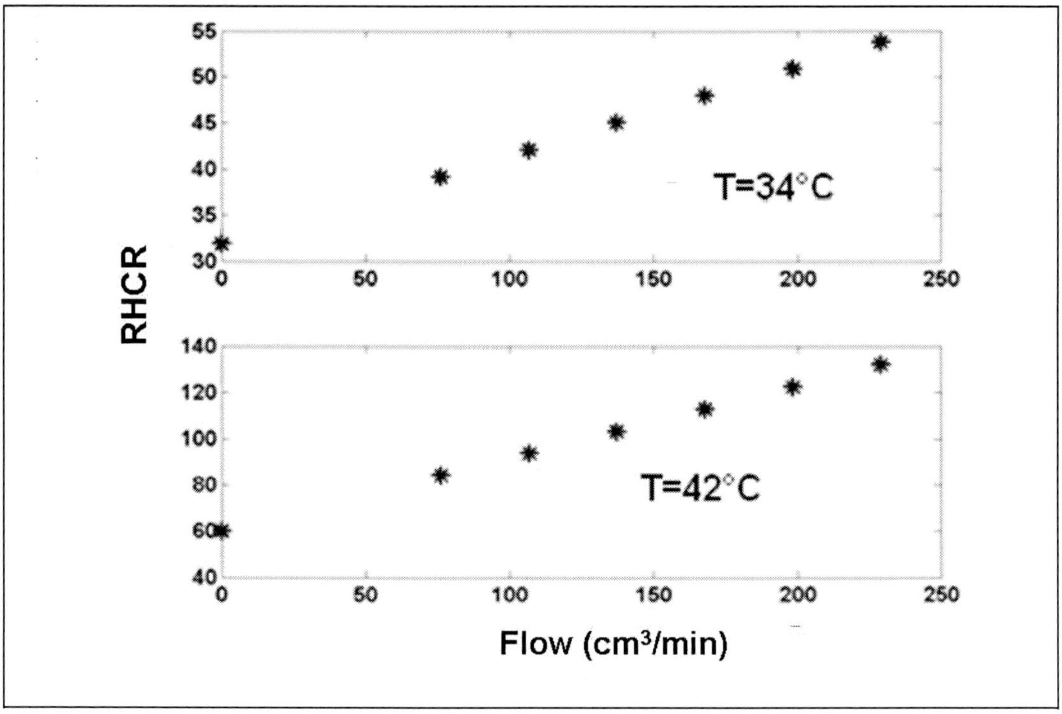

Figure 32. Number of heating impulse per second (RHCR) versus flow in two different bias temperatures.

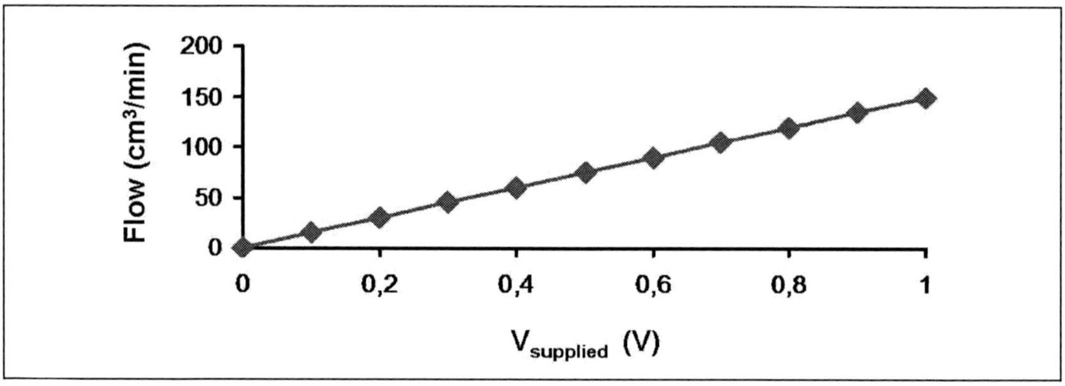

Figure 33. Measured flow by a commercial flow sensor. Flow is generated by a DC pump.

Acknowledgements

The authors are sincerely grateful to Dr. Christian Falconi, Prof. Arnaldo D'Amico and Prof. Corrado Di Natale of "Tor Vergata" University of Rome, for their constant scientific support.

References

Ballantine D.S. et al., "Acoustic Wave Sensors-Theory, Design, and Physico-Chemical Applications", Academic Press, (1996).

Ballato A. and Lukaszek T., "Higher order temperature coefficients of frequency of mass-loaded piezoelectric crystal plates", *Proc. 29th AFCS,* 10-25, (1975)

Bechman R, "Frequency-temperature-angle characteristics of AT- and BT-type quartz oscillators in extended temperature range", *Proc. IRE,* Vol. 48, No. 8, 1494, (1960)

Bechmann R., "Elastic and piezoelectric constants of Alpha-Quartz", *Physical Review* **110**, 5 1060-1061, (1958)

D'Amico A., Di Natale C., Paolesse R, Macagnano A., Mantini A., "Metalloporphyrins as basic material for volatile sensitive sensors£, *Sensors and Actuators B: Chemical* **65** 209-215, (2000)

Di Natale C., Paolesse R., Macagnano A., Mantini A., Goletti C., D'Amico A., "Characterization and design of porphyrins-based broad selectivity chemical sensors for electronic nose applications", *Sensors and Actuators B: Chemical* **52** 162-168, (1998)

Di Natale C., Paolesse R., Macagnano A., Mantini A., Mari P., D'Amico A., "Qualitative structure–sensitivity relationship in porphyrins based QMB chemical sensors", *Sensors and Actuators B: Chemical*, **68** 319-323, (2000)

Falcitelli M., Benassi A., Di Francesco F., Domenici C., Marano L., Pioggia G., "Fluid dynamic simulation of a measurement chamber for electronic noses" , *Sensors and Actuators* **B 85** 166–174, (2002)

Falconi, C.; Zampetti, E.; Pantalei, S.; Martinelli, E.; Di Natale, C.; D'Amico, A.; Stornelli, V.; Ferri, G, "Temperature and flow velocity control for quartz crystal microbalances", *Proceedings of ISCAS 21-24 May 2006, IEEE International Symposium on Circuits and Systems* (10.1109/ISCAS.2006.1693604).

Kofi A.A. Makinwa, "Flow Sensing with Thermal Sigma-Delta Modulators", *OPTIMA Ghrafische Communicatie*, Rotterdam, ISBN:90-77595-61-9, (2004)

Kosinski J.A, Gualtieri J.G. and Ballato A.,"Thermoelastic Coefficients of Alpha Quartz", IEEE transactions on ultrasonic, *ferroelectrics, and frequency control*, **39**, 4 (1992)

Kulkarnib A., Patilb S., Karekara R., Aiyerb R., "Fabrication and characterization of innovative gas flow sensor", *Sensors and Actuators A* **122**, 231–234, (2005)

Lu C. and Czanderna A.W., in "Applications of Piezoelectric Quartz Crystal Microbalances", Czanderna and Lu (Eds.) Elsevier, New York, (1984).

Lucklum R., Hauptmann P., "The DF-DR QCM technique: an approach to an advanced sensor signal interpretation", *Electrochimica Acta* **45**, 3907-3916, (2000)

Macagnano A., Sgreccia E., Paolesse R., De Cesare F., D'Amico A., Di Natale C., "Sorption and condensation phenomena of volatile compounds on solid-state metalloporphyrin films", *Sensors and Actuators B: Chemical* **124** 260-268, (2007)

Macagnano A., Sgreccia E., Zampetti E., Pantalei S., Di Natale C, Paolesse R., D'Amico A., "Potentials and limitations of a porphyrin-based AT-cut resonator for sensing applications*", Sensors and Actuators B: Chemical* **130** 411-417, (2008)

Mason W.P., "Piezoelectric crystals and ultrasonics", New York: Van Nostrand, (1950)

Nie J., Hamacher T., Schulze Lammers P., Weber E., Boeker P.,"A miniaturized thermal desorption unit for chemical sensing below odour threshold", *Sensors and Actuators B* **95**, 1–5, (2003)

O'Sullivan C. K. and Guilbault G.G., "Commercial Quartz Crystal Microbalances-Theory and Applications", *Biosensors and Bioelectronics* **14**, 663 (1999)

R.C Dorf et al., "Modern Control Systems", Addison-Wesley Publishing Company, (1986)

Sekimoto H., Goka S., Ishizaki A. and Watanabe Y.,"Frequency-Temperature Behavior of Spurious Vibrations of Rectangular AT-Cut Quartz Plates", *IEEE Transactions on Ultrasonics, Ferroelectrics and Frequency Control* **45**, (1998)

Verhoeven H. J., "Smart thermal flow sensors", Ph.D. Thesis, Delft University of Technology, Delft, (1995)

INDEX

A

B

D

E

F

G

H

I

J

K

L

M

N

O

P

Q

R

S

T

U

V